PARALLEL
PROCESSING FOR
SCIENTIFIC
COMPUTING

SERIES LIST FOR ALL PROCEEDINGS

Neustadt, L.W., Proceedings of the First International Congress on Programming and Control (1966)

Hull, T.E., Studies in Optimization (1970)

Day, R.H., & Robinson, S.M., Mathematical Topics in Economic Theory and Computation (1972)

Proschan, F., & Serfling, R.J., Reliability and Biometry: Statistical Analysis of Lifelength (1974)

Barlow, R.E., Reliability & Fault Tree Analysis: Theoretical & Applied Aspects of System Reliability & Safety Assessment (1975)

Fussell, J.B., & Burdick, G.R., Nuclear Systems Reliability Engineering and Risk Assessment (1977)

Duff, I.S., & Stewart, G.W., Sparse Matrix Proceedings 1978 (1979)

Holmes, P.J., New Approaches to Nonlinear Problems in Dynamics (1980)

Erisman, A.M., Neves, K.W., & Dwarakanath, M.H., Electric Power Problems: The Mathematical Challenge (1981)

Bednar, J.B., Redner, R., Robinson, E., & Weglein, A., Conference on Inverse Scattering: Theory and Application (1983)

Voigt, R.G., Gottlieb, D., & Hussaini, M. Yousuff, Spectral Methods for Partial Differential Equations (1984)

Chandra, Jagdish, Chaos in Nonlinear Dynamical Systems (1984)

Santosa, F., Pao, Y-H., Symes, W.W., & Holland, C., Inverse Problems of Acoustic and Elastic Waves (1984)

Gross, Kenneth I., Mathematical Methods in Energy Research (1984)

Babuska, I., Chandra, J., & Flaherty, J., Adaptive Computational Methods for Partial Differential Equations (1984)

Boggs, Paul T., Byrd, Richard H., & Schnabel, Robert B., Numerical Optimization 1984 (1985)

Angrand, F., Dervieux, A., Desideri, J.A., & Glowinski, R., Numerical Methods for Euler Equations of Fluid Dynamics (1985)

Wouk, Arthur, New Computing Environments: Parallel, Vector and Systolic (1986)

Fitzgibbon, William E., Mathematical and Computational Methods in Seismic Exploration and Reservoir Modeling (1986)

Drew, Donald A., & Flaherty, Joseph E., Mathematics Applied to Fluid Mechanics and Stability: Proceedings of a Conference Dedicated to R.C. DiPrima (1986)

Heath, Michael T., Hypercube Multiprocessors 1986 (1986)

Papanicolaou, George, Advances in Multiphase Flow and Related Problems (1987)

Wouk, Arthur, New Computing Environments: Microcomputers in Large-Scale Computing (1987)

Chandra, Jagdish, & Srivastav, Ram, Constitutive Models of Deformation (1987)

Heath, Michael T., Hypercube Multiprocessors 1987 (1987)

Glowinski, R., Golub, G.H., Meurant, G.A., & Periaux, J., First International Symposium on Domain Decomposition Method for Partial Differential Equations (1988)

Salam, Fathi M.A., & Levi, Mark L., Dynamical Systems Approaches to Nonlinear Problems in Systems and Circuits (1988)

Datta, B., Johnson, C., Kaashoek, M., Plemmons, R., & Sontag, E., Linear Algebra in Signals, Systems and Control (1988)

Ringeisen, Richard D., & Roberts, Fred S., Applications of Discrete Mathematics (1988)

McKenna, James, & Temam, Roger, ICIAM '87 — Proceedings of the First International Conference on Industrial and Applied Mathematics (1988)

Rodrigue, Garry, Parallel Processing for Scientific Computing (1989)

PARALLEL
PROCESSING FOR
SCIENTIFIC
COMPUTING

Edited by Garry Rodrigue

Philadelphia

PARALLEL PROCESSING
FOR SCIENTIFIC COMPUTING

Proceedings of the Third SIAM Conference on Parallel Processing for Scientific Computing, Los Angeles, California, December 1–4, 1987.

This conference was sponsored by the SIAM Activity Group on Supercomputing and was supported in part by the Department of Energy under grant DE-FG02-87ER25050 and the National Science Foundation under grant ASC-87-17064.

Library of Congress Catalog Card Number 88-62233
ISBN 0-89871-228-9

PREFACE

The Third SIAM Conference on Parallel Processing for Scientific Computing was held in Los Angeles, California on December 1–4, 1987. The First and the Second Conferences were held in Norfolk, Virginia in 1983 and 1985 respectively. The purpose of these conferences is to provide a forum for SIAM members to discuss recent developments in the area of high speed scientific computing. The Organizing Committee for the conference consisted of Garry Rodrigue, Chair, Lawrence Livermore National Laboratory and University of California, Davis; Joseph Oliger, Stanford University; Burton Smith, Supercomputing Research Center; John Van Rosendale, Argonne National Laboratory; and Robert Voigt, ICASE-NASA Langley Research Center. The conference was sponsored by the SIAM Activity Group on Supercomputing.

This proceedings contains the papers or abstracts from the 14 invited talks and the 76 contributed talks that were given at the conference. The contributed talks were selected by the organizing committee from submitted 1000 word abstracts. As might be expected at a SIAM conference, there was emphasis on the development and analysis of numerical algorithms for parallel computing. However, in addition to this, a sizable fraction of presentations dealt with scientific programming languages, scientific programming environments, new computing architectures, and parallel computer performance evaluation. There were also a number of presentations dealing with applications arising from a number of scientific fields.

The Organizing Committee, and especially the Chair, would like to acknowledge the work done by Carolyn Hunt of the Lawrence Livermore National Laboratory. She contributed immensely to smooth running of the conference and to the publication of this proceedings.

Garry Rodrigue
University of California, Davis
Lawrence Livermore National Laboratory

LIST OF CONTRIBUTORS

Tilak Agerwala, IBM T.J. Watson Research Center, Yorktown Heights, New York 10598

Richard C. Allen, Jr., Sandia National Laboratories, Albuquerque, New Mexico 87185

Tom Allen, Alphatech, Inc., Burlington, Massachusetts 01803

Henno Allik, BBN Laboratories Inc., New London, Connecticut 06320

Edward Anderson, Center for Supercomputing Research and Development, University of Illinois at Urbana-Champaign, Urbana, Illinois 61801-2932

Michael A. Arbib, Departments of Computer Science, Neurobiology, Physiology, Biomedical Engineering, Electrical Engineering, and Psychology, University of Southern California, Los Angeles, California 90089-0782

Clifford Arnold, ETA Systems, Inc., St. Paul, Minnesota 55108

Daya Atapattu, Department of Computer Science, Indiana University, Bloomington, Indiana 47405

Timothy S. Axelrod, Computational Physics Division, University of California, Lawrence Livermore National Laboratory, Livermore, California 94550

O. Axelsson, Department of Mathematics, University of Nijmegan, Toernooiveld 6525 ED Nijmegan, The Netherlands

Robert G. Babb II, Department of Computer Science and Engineering, Oregon Graduate Center, Beaverton, Oregon 97006

Lorraine S. Baca, Sandia National Laboratories, Albuquerque, New Mexico 87185

Scott B. Baden, Mathematics Department, Lawrence Berkeley Laboratory, University of California, Berkeley, California 94720

David H. Bailey, NASA Ames Research Center, Moffett Field, California 94035

Muhammad S. Benten, Electrical and Computer Engineering Department, University of Colorado at Boulder, Boulder, Colorado 80309-0425

Marsha J. Berger, Courant Institute of Mathematical Sciences, New York University, New York, New York 10012

Michael Berry, Center for Supercomputing Research and Development, University of Illinois at Urbana-Champaign, Urbana, Illinois 61801-2932

Christian Bischof, Department of Computer Science, Cornell University, Ithaca, New York 14853

J.P. Bixler, Department of Computer Science, Virginia Polytechnic Institute & State University, Blacksburg, Virginia 24061

James L. Blue, Center for Applied Mathematics, National Bureau of Standards, Gaithersburg, Maryland 20899

Ken W. Bosworth, Mathematics Department, Utah State University, Logan, Utah 84322-3900

Grant Bowgen, Active Memory Technology Ltd., Reading RG6 1AZ, United Kingdom

Eugene D. Brooks III, Computational Physics Division, University of California, Lawrence Livermore National Laboratory, Livermore, California 94550

G. Brussino, Data Systems Division, IBM Corporation, Kingston, New York 12401

W. Thomas Cathey, Center for Optoelectronic Computing Systems and Department of Electrical and Computer Engineering, University of Colorado at Boulder, Boulder, Colorado 80309-0425

William Celmaster, Bolt Beranek and Newman, Advanced Computers Inc., Cambridge, Massachusetts 02238

C. David Chan, Division of Science and Mathematics, University of Minnesota-Morris, Minnesota 56267

Sharat Chandran, Center for Automation Research, University of Maryland, College Park, Maryland 20742-3411

Raymond C.Y. Chin, University of California, Lawrence Livermore National Laboratory, Livermore, California 94550

Shui-Nee Chow, Department of Mathematics, Michigan State University, East Lansing, Michigan 48824

Z. Christidis, Data Systems Division, IBM Corporation, Kingston, New York 12401

Eleanor Chu, Department of Computer Science, University of Waterloo, Waterloo, Ontario, Canada

Luis Crivelli, Department of Mechanical Engineering, University of Colorado at Boulder, Boulder, Colorado 80309-0429

Kay Crowley, Department of Computer Science, Yale University, New Haven, Connecticut 06520

George Cybenko, Department of Computer Science, Tufts University, Medford, Massachusetts 02155

Afshin Daghi, Sun Microsystems, Inc., Mountain View, California 94043

Gregory A. Darmohray, Computational Physics Division, University of California, Lawrence Livermore National Laboratory, Livermore, California 94550

Larry S. Davis, Center for Automation Research, University of Maryland, College Park, Maryland 20742-3411

Elise de Doncker, Department of Computer Science, Western Michigan University, Kalamazoo, Michigan 49008-5021

Sudarshan K. Dhall, Parallel Processing Institute, School of Electrical Engineering and Computer Science, University of Oklahoma, Norman, Oklahoma 73019

J. Diaz, Parallel Processing Institute, School of Electrical Engineering and Computer Science, University of Oklahoma, Norman, Oklahoma 73019

David C. DiNucci, Department of Computer Science and Engineering, Oregon Graduate Center, Beaverton, Oregon 97006

Jack Dongarra, Mathematics and Computer Science Division, Argonne National Laboratory, Argonne, Illinois 60439-4844

Gary Doolen, Los Alamos National Laboratory, Los Alamos, New Mexico 87545

Kenneth Dowers, Parallel Processing Institute, School of Electrical Engineering and Computer Science, University of Oklahoma, Norman, Oklahoma 73019

Jeremy Du Croz, Numerical Algorithms Group Ltd., Oxford, OX2 7DE, United Kingdom

Iain Duff, Computer Science and Systems Division, Harwell Laboratory, Oxfordshire OX11 ORA, United Kingdom

Robert J. Dunki-Jacobs, G.E. Corporate Research and Development Center, Schenectady, New York 12301

Timo Eirola, Institute of Mathematics, Helsinki University of Technology, SF-02150, Espoo, Finland

Elizabeth A. Eskow, Department of Computer Science, University of Colorado at Boulder, Boulder, Colorado 80309

Vance Faber, Los Alamos National Laboratory, Los Alamos, New Mexico 87545

Charbel Farhat, Department of Aerospace Engineering Science and Center for Space Structures and Controls, University of Colorado at Boulder, Boulder, Colorado 80309-0427

David C. Fisher, Department of Mathematics, Harvey Mudd College, Claremont, California 91711

Jeffrey W. Frederick, Department of Mechanical Engineering, University of Virginia, Charlottesville, Virginia 22903

Daniel D. Gajski, Department of Information and Computer Science, University of California at Irvine, Irvine, California 92717

E. Gallopoulos, Center for Supercomputing Research and Development, University of Illinois at Urbana-Champaign, Urbana, Illinois 61801-2932

Dennis Gannon, Department of Computer Science, Indiana University, Bloomington, Indiana 47405

George A. Geist, Mathematical Sciences Section, Oak Ridge National Laboratory, Oak Ridge, Tennessee 37831

Alan George, Department of Computer Science, University of Tennessee, Knoxville, Tennessee 37996, and Oak Ridge National Laboratory, Mathematical Sciences Section, Oak Ridge, Tennessee 37831

S. Gerbi, Department of Computer Science, Yale University, New Haven, Connecticut 06520

David Gottlieb, Department of Computer Science, Yale University, New Haven, Connecticut 06520

Leslie Greengard, Department of Computer Science, Yale University, New Haven, Connecticut 06520

William D. Gropp, Department of Computer Science, Yale University, New Haven, Connecticut 06520

Dale B. Haidvogel, The Chesapeake Bay Institute, The Johns Hopkins University, Baltimore, Maryland 21211

Sven Hammarling, Numerical Algorithms Group Ltd., Oxford, OX2 7DE, United Kingdom

Steven W. Hammond, G.E. Corporate Research and Development Center, Schenectady, New York 12301

Robert M. Hardy, G.E. Corporate Research and Development Center, Schenectady, New York 12301

S. Harimoto, Department of Computer Science, Virginia Polytechnic Institute & State University, Blacksburg, Virginia 24061

Bill Harrod, Center for Supercomputing Research and Development, University of Illinois at Urbana-Champaign, Urbana, Illinois 61801-2932

D.J. Hebert, Department of Mathematics and Statistics, University of Pittsburgh, Pittsburgh, Pennsylvania 15260

G.W. Hedstrom, University of California, Lawrence Livermore National Laboratory, Livermore, California 94550

B.J. Helland, Ames Laboratory-USDOE, Iowa State University, Ames, Iowa 50011

R. Herbin, Ecole Polytechnique Fédérale de Lausanne, GASOV/ASTRID, Lausanne, 1015, Switzerland

Hsiao-Ming Hsu, Department of Physical Oceanography, Woods Hole Oceanographic Institution, Woods Hole, Massachusetts 02543

Phillip Q. Hwang, ACSD Laboratory, Naval Surface Weapons Center, Silver Spring, Maryland 20903-5000

Kai Hwang, Computer Research Institute, University of Southern California, Los Angeles, California 90089-0781

Ilse C.F. Ipsen, Department of Computer Science, Yale University, New Haven, Connecticut 06520

E.R. Jessup, Department of Computer Science, Yale University, New Haven, Connecticut 06520

Harry F. Jordan, Electrical and Computer Engineering Department, University of Colorado at Boulder, Boulder, Colorado 80309-0425

John Kapenga, Department of Computer Science, Western Michigan University, Kalamazoo, Michigan 49008-5021

Ken Kennedy, Department of Computer Science, Rice University, Houston, Texas 77251

David E. Keyes, Department of Mechanical Engineering, Yale University, New Haven, Connecticut 06529

Chisato Konno, Central Research Laboratory, Hitachi Ltd., Kokubunji, Tokyo 185, Japan

R.J. Krueger, Ames Laboratory-USDOE, Iowa State University, Ames, Iowa 50011

V.K. Prasanna Kumar, Department of Electrical Engineering Systems, University of Southern California, Los Angeles, California 90089-0781

S. Lakshmivarahan, Parallel Processing Institute, School of Electrical Engineering and Computer Science, University of Oklahoma, Norman, Oklahoma 73019

D. Lee, Center for Supercomputing Research and Development, University of Illinois at Urbana-Champaign, Urbana, Illinois 61801-2932

Mann Ho Lee, Department of Computer Science, Indiana University, Bloomington, Indiana 47405

Charles E. Leiserson, Laboratory for Computer Science, Massachusetts Institute of Technology, Cambridge, Massachusetts 02139

Michael R. Leuze, Department of Computer Science, Vanderbilt University, Nashville, Tennessee 37235

John G. Lewis, Boeing Computer Services Company, Seattle, Washington 98124

H.M. Liddell, Centre for Parallel Computing, Queen Mary College, University of London, London, E1 4NS United Kingdom

Avi Lin, Department of Mathematics, Temple University, Philadelphia, Pennsylvania 19122

Ahmed Louri, Computer Research Institute, University of Southern California, Los Angeles, California 90089-0781

Shing C. Ma, Department of Computer Science, Duke University, Durham, North Carolina 27706

James G. Malone, Engineering Mechanics Department, General Motors Research Laboratories, Warren, Michigan 48090-9057

Dan C. Marinescu, Department of Computer Sciences, Purdue University, West Lafayette, Indiana 47907

Oliver A. McBryan, Department of Computer Science, University of Colorado at Boulder, Boulder, Colorado 80309-0430

Piyush Mehrotra, Department of Computer Sciences, Purdue University, West Lafayette, Indiana 47907

Gérard Meurant, Département de Mathématiques Appliquées, Centre d'Etudes de Limeil-Valenton, B.P. N°27, 94190 Villeneuve St. Georges, France

Gerard G.L. Meyer, Electrical and Computer Engineering Department, The Johns Hopkins University, Baltimore, Maryland 21218

Ravi Mirchandaney, Department of Computer Science, Yale University, New Haven, Connecticut 06520

K.W. Morton, Oxford University Computing Laboratory, Oxford, OX1 3QD, United Kingdom

Vijay K. Naik, Institute for Computer Applications in Science and Engineering, NASA Langley Research Center, Hampton, Virginia 23665

Olavi Nevanlinna, Helsinki University of Technology, Institute of Mathematics, SF-02150 Espoo, Finland

Lionel M. Ni, Department of Computer Science, Michigan State University, East Lansing, Michigan 48824

D.M. Nicol, Department of Computer Science, College of William and Mary, Williamsburg, Virginia 23185

B. Nour-Omid, Mechanics and Materials Engineering Laboratory, Lockheed Palo Alto Research Laboratory, Palo Alto, California 94304

Elizabeth J. O'Neil, Bolt Beranek and Newman, Advanced Computers, Inc., Cambridge, Massachusetts 02238, and Department of Mathematics and Computer Science, University of Massachusetts at Boston, Boston, Massachusetts 02125

M. Ortiz, Division of Engineering, Brown University, Providence, Rhode Island 02912

D. Parkinson, Centre for Parallel Computing, Queen Mary College, University of London, London, E1 4NS, United Kingdom

Merrell L. Patrick, Department of Computer Science, Duke University, Durham, North Carolina 27706

Jih-Kwon Peir, Computer Science Department, IBM Thomas J. Watson Research Center, Yorktown Heights, New York 10598

Rick Pennington, Electrical Engineering Department, Utah State University, Logan, Utah 84322-4120

Louis J. Podrazik, Bendix Environmental Systems Division, Allied-Signal Inc., Baltimore, Maryland 21284-9840

Eugene L. Poole, Awesome Computing Inc., Hampton, Virginia 23665

J.A. Puckett, Institute for Scientific Computation, University of Wyoming, Laramie, Wyoming 82071

Stewart F. Reddaway, ICL, Herts, SG7 5QE, United Kingdom

Daniel A. Reed, Department of Computer Science, University of Illinois at Urbana-Champaign, Urbana, Illinois 61801

M.D. Rees, Oxford University Computing Laboratory, Oxford, OX1 3QD, United Kingdom

Calvin J. Ribbens, Department of Computer Science, Virginia Polytechnic Institute and State University, Blacksburg, Virginia 24061

John R. Rice, Department of Computer Sciences, Purdue University, West Lafayette, Indiana 47907

Charles H. Romine, Mathematical Sciences Section, Engineering Physics and Mathematics Division, Oak Ridge National Laboratory, Oak Ridge, Tennessee 37831

Dirk Roose, Departement Computerscience, Katholieke Universiteit Leuven, B-3030 Leuven, Belgium

Matthew Rosing, Department of Computer Science, University of Colorado at Boulder, Boulder, Colorado 80309

Youcef Saad, Center for Supercomputing Research and Development, University of Illinois at Urbana-Champaign, Urbana, Illinois 61801-2932

Ali Safavi, Information Sciences Institute and Computer Science Department, University of Southern California, Los Angeles, California 90089

Miyuki Saji, Central Research Laboratory, Hitachi Ltd., Kokubunji, Tokyo 185, Japan

Joel Saltz, Department of Computer Science, Yale University, New Haven, Connecticut 06520

Ahmed Sameh, Center for Supercomputing Research and Development, University of Illinois at Urbana-Champaign, Urbana, Illinois 61801-2932

Stephen R. Schach, Department of Computer Science, Vanderbilt University, Nashville, Tennessee 37235

Mark Schaefer, Center for Supercomputing Research and Development, University of Illinois at Urbana-Champaign, Urbana, Illinois 61801-2932

R.J. Schmidt, Institute of Scientific Computation, University of Wyoming, Laramie, Wyoming 82071

Robert B. Schnabel, Department of Computer Science, University of Colorado at Boulder, Boulder, Colorado 80309

Jeffrey S. Scroggs, Center for Supercomputing Research and Development, University of Illinois at Urbana-Champaign, Urbana, Illinois 61801

Bruce Shei, Department of Computer Science, Indiana University, Bloomington, Indiana 47405

Yun-Qiu Shen, Department of Mathematics, Michigan State University, East Lansing, Michigan 48824

William D. Shoaff, Mathematical and Computer Sciences, Florida Institute of Technology, Melbourne, Florida 32901-6988

Richard Sincovec, Computer Science Department, University of Colorado, Colorado Springs, Colorado 80933-7150

Björn Sjögreen, Department of Scientific Computing, Uppsala University, Sturegatan 4BS-752 23 Uppsala, Sweden

R.M. Smith, Department of Computer Science, Yale University, New Haven, Connecticut 06520

Mitchell D. Smooke, Department of Mechanical Engineering, Yale University, New Haven, Connecticut 06520

V. Sonnad, Data Systems Division, IBM Corporation, Kingston, New York 12401

D.C. Sorensen, Mathematics and Computer Science Division, Argonne National Laboratory, Argonne, Illinois 60439-4844

G.S. Stiles, Electrical Engineering Department, Utah State University, Logan, Utah 84322-4120

Francis Sullivan, Center for Applied Mathematics, National Bureau of Standards, Gaithersburg, Maryland 20899

Daniel B. Szyld, Department of Computer Science, Duke University, Durham, North Carolina 27706

Wei Pai Tang, Department of Mathematics, Stanford University, Stanford, California 94305

Terry M. Topka, G.E. Corporate Research and Development Center, Schenectady, New York 12301

Yukio Umetani, Central Research Laboratory, Hitachi Ltd. Kokubunji, Tokyo 185, Japan

Stefan Vandewalle, Departement Computerscience, Katholieke Universiteit Leuven, B-3030 Heverlee, Belgium

Sven Van Den Berghe, Active Memory Technology Ltd., Reading RG6 1AZ, United Kingdom

H.C. Wang, Computer Research Institute, University of Southern California, Los Angeles, California 90089

Layne T. Watson, Department of Computer Science, Virginia Polytechnic Institute & State University, Blacksburg, Virginia 24061

Michael Wolfe, Kuck and Associates, Inc., Champaign, Illinois 61874

David E. Womble, Sandia National Laboratories, Albuquerque, New Mexico 87185

Min-You Wu, Department of Information and Computer Science, University of California at Irvine, Irvine, California 92717

Michiru Yamabe, Central Research Laboratory, Hitachi Ltd., Kokubunji, Tokyo 185, Japan

He Zhang, Department of Mathematics, Temple University, Philadelphia, Pennsylvania 19122

CONTENTS

PART I

Matrix Computations

CHAPTER 1

A Pipelined Block QR Decomposition Algorithm*

Christian Bischof†

Abstract. Block algorithms are effective for obtaining efficient and portable algorithms on computers with special vector processing hardware. We show that these advantages extend to parallel architectures. Combining block oriented algorithms and pipelining, we obtain programs that have good processor utilization and load balancing properties and are easy to port across different architectures. We illustrate this technique with an algorithm for computing the QR factorization.

1. Introduction. A typical block oriented algorithm for computing a matrix factorization proceeds as follows:

> for $i = 1$ to # of blocks
>   (1) Compute factorization of block i
>   **touching the rest of the matrix as little as possible**
>   (2) Apply the resulting update to the rest of the matrix
> end for

For some examples see Bischof and Van Loan [2], Dongarra and Sorensen [6] and Van Loan [14]. Step (2) typically is rich in matrix-matrix operations like those identified in the BLAS 3 proposal (Dongarra, DuCroz, Duff and Hammarling [5]). By using these operations on a block of size k the ratio of data movement to arithmetic operations is $O(1/k)$. This *surface-to-volume effect* (Dongarra and Sorensen [6]) prevents degradation of performance by processors being idle waiting for data. In addition a block algorithm is easily adapted to changing processor characteristics by adjusting the block size and introducing efficient implementations of the BLAS 3 kernels. There is no need to change the algorithm.

---

*This work was supported by the U.S. Army Research Office through the Mathematical Sciences Institute of Cornell University, by the Office of Naval Research under contract N00014-83-K-0640 and by NSF contract CCR 86-02310. Computations were performed at the Advanced Computing Facility at the National Supercomputing Facility at Cornell which is supported by the National Science Foundation and New York State

† Department of Computer Science, Cornell University, Ithaca, NY 14853.

2. Pipelined Block Algorithms. These advantages of block algorithms extend to the parallel setting. For some examples see Bischof [1], Schreiber [12] and Van Loan [15]. In designing an algorithm for a parallel machine one must strive to minimize communication cost and load imbalance. Due to the surface-to-volume effect for a block oriented algorithm the relative cost of communication can be lessened by increasing the block size. Transferring data in larger chunks also decreases the impact of startup costs and message buffer fragmentation.

One of the easiest ways to achieve load balance is to stagger the computation across several processors and this technique has been widely used (Coleman and Li [4], Geist and Heath [8], Ipsen, Saad and Schultz [9] and Moler [10]). By assigning blocks to processors in a round-robin fashion, all processors should be busy after a short startup time. This scheme allows simple static assignment of data to processors and is for the most part synchronized by the flow of data between processors. For many factorization algorithms (including the QR factorization) this assignment is a very natural one since the algorithm steps through the given matrix from left to right.

3. The Compact WY Factorization. We illustrate this technique with a pipelined block algorithm for computing the QR factorization of a matrix A. To arrive at a block algorithm, Bischof and Van Loan [2] expressed the product

$$Q \equiv P_1 \dots P_k$$

of k Householder matrices

$$P \equiv I - 2vv^T, \ \|v\|_2 = 1, v \in \mathbb{R}^m$$

in the form

$$Q = I + WY^T,$$

where W and Y are m-by-k matrices. Schreiber and Van Loan [13] and independently Du Croz [7] refined this representation by expressing $W = YT^T$ where T is a k-by-k upper triangular matrix. Schreiber and Van Loan called the representation

$$Q = I + YTY^T$$

the compact WY representation since it requires only about half as much storage as the original WY representation at only a slight increase in arithmetic complexity. To accumulate W and T, observe that a Householder matrix is a special case of the compact WY representation and that we can write

$$Q^+ \equiv QP = I + Y^+ T^+ Y^{+\mathrm{T}}$$

where

$$Y \leftarrow [\, Y, v \,]$$

$$z \leftarrow -2T \, Y^T v$$

$$T \leftarrow \begin{bmatrix} T & z \\ 0 & -2 \end{bmatrix}$$

Bischof and Van Loan [2] used the WY factorization for computing the QR decomposition on a vector machine and Van Loan [15] used it for computing the QR factorization on a tightly coupled parallel system.

4. A Pipelined Block QR Algorithm. To compute the QR factorization in the distributed setting, we partition A into k block columns, i.e. $A = [\, A_1, \dots, A_k \,]$.

```
/* after distributing columns, the array B contains the block columns dedicated to this processor. */
/* B(i) denotes the ith block column in this processor and for the sake of simplicity we           */
/* assume that every processor houses l block columns. Furthermore "the diagonal" refers            */
/* to the location of the diagonal of the original A in a block column.                             */
wycount ← 0
if ( I am processor no. 1) then
        compute Y and T such that (I + Y T Yᵀ )ᵀ B(1) is zero below the diagonal
        send (Y, T) to the right neighbour
        update B(2: l) ← (I + Y T Yᵀ )ᵀ B( 2: l )
        wycount ← 1
endif
while (wycount < l) do
        receive (Y, T) from left neighbour
        if( ( (Y, T) were not generated by the right neighbour) then
                send (Y, T) to right neighbour
        endif
        if (it is my turn to solve the next subproblem) then
                wycount ← wycount + 1
                compute Y and T such that (I + Y* T* Y*ᵀ )ᵀ BCA(wycount) is zero below the diagonal
                if (this was not the last subproblem to be solved) then
                        send (Y*, T* ) to right neighbour
                endif
                update B(wycount + 1: l) ← (I + Y T Yᵀ )ᵀ B( wycount + 1: l)
                update B(wycount + 1: l) ← (I + Y* T* Y*ᵀ )ᵀ B( wycount + 1: l)
        else
                update B(wycount + 1: l) ← (I + Y T Yᵀ )ᵀ B( wycount + 1: l )
        endif
end while
```

FIG. 1: *Pipelined Block QR Factorization Algorithm*

Assuming a logical ring of p processors, processor i receives block columns A_j where

$$(i - 1) = j \bmod p.$$

Then each processor executes the algorithm given in simplified form in Figure 1. The application and generation of the Y and T factors is staggered so that the computation proceeds in a pipelined fashion. The Y and T factors get computed and sent out as soon as possible to minimize processor idle time. Since the sending of Y and T can be done at the same time as the local updates, all processors should be busy after a short startup period.

5. Experimental Results. We performed experiments with that algorithm on a 16-node Intel iPSC hypercube taking advantage of the hypercube topology only by using a Gray code to embed the ring of processors and by broadcasting the very first Y and T factors to all other processors to cut down on pipeline startup overhead. The code was written entirely in Fortran, compiled with the Ryan-McFarland Compiler Version 2.20a using the huge memory model and executed under the NX node operating system release 3.1.1. in single precision. Taking a matrix of size 750-by-375 as representative example and grouping it into 75 block columns of size five each, we obtained the performance shown in Fig.2. Here t_{max} (t_{min}) is the longest (shortest) time in seconds that any processor took to completion, t_{avg} is the average execution time in seconds,

$$\Delta t := \frac{t_{max} - t_{min}}{t_{avg}} *100$$

is a measure of the load imbalance, $eff(p)$ is the efficiency of the p-processor run compared to the one-processor run, pt_{comm} is the average communication overhead and K_{sust} is the sustained KFlop rate per processor (each addition and multiplication was counted as a separate flop).

p	t_{max}	t_{avg}	Δt	$eff(p)$	pt_{comm}	K_{sust}
0	6999	–	–	100	0	25.3
2	3547	3545	0	98.5	1.1	25.0
4	1824	1820	0.5	96.0	3.6	24.3
8	967	956	2.3	90.5	8.0	23.2
16	548	525	9.3	80.0	15.8	21.1

FIG 2: *Performance of QR Algorithm on a 750×375 matrix*

We observe that the algorithm scales nicely. Communication overhead is low due to the inherent locality of block algorithms and due to the asynchronous sending of Y and T factors. The values of Δt show that the computational load is indeed balanced with only a slight degradation as the work per processor decreases. A more detailed analysis of the algorithm and the experimental results will be reported elsewhere. Again it would be easy to port this algorithm to a different machine. The pipelining can easily be implemented in software even if the underlying architecture is not a ring. In particular it is applicable to shared memory machines. By adjusting the blocking factor, the granularity of the block algorithm can be adapted to machines that exhibit a different ratio of arithmetic to communication speed. Chamberlain and Powell (1986) and Pothen, Jha and Vemulapati (1987) suggested other algorithms for computing the QR factorization on a hypercube architecture. However, their algorithms do not use block tranformations and hence would not be as well suited for a machine with special vector processing hardware.

6. Conclusions. We have illustrated the advantages of block algorithms and pipelining in the parallel setting. While block algorithms ensure low communication overhead and good arithmetic performance in each node, pipelining exhibits good load balancing properties. By combining these techniques we arrive at algorithms that are easily ported across architectures with different system characteristics. An example of a pipelined block algorithm for computing the QR decomposition and numerical results on the Intel iPSC illustrated these points.

REFERENCES

[1] C. H. BISCHOF, *Computing the Singular Value Decomposition on a Distributed System of Vector Processors*, Technical Report TR 87-869, Cornell University, Dept. of Computer Science, submitted to Parallel Computing, 1987.

[2] C. H. BISCHOF and C. F. VAN LOAN, *The WY Representation for Products of Householder Matrices*, SIAM J. Scientific and Statistical Computing, 8 (1987), pp. s2-s13.

[3] R. M. CHAMBERLAIN and M. J. D. POWELL, *QR Factorization for Linear Least Squares Problems on the Hypercube*, Report CCS 86/10, Chr. Michelsen Institute, Dept. of Science and Technology, 1986.

[4] T. F. COLEMAN and G. LI , *A New Method for Solving Triangular Systems on Distributed Memory Message-Passing Multiprocessors*, Report TR 87-812, Cornell University, Dept. of Computer Science, 1987.

[5] J. J. DONGARRA, J. DUCROZ, I. DUFF and S. HAMMARLING, *A Proposal for a Set of Level 3 Basic Linear Algebra Subprograms*, Argonne National Laboratory Report ANL-MCS-TM-41 (Revision 3), 1987.

[6] J. J. DONGARRA and D. C. SORENSEN, *Block Reduction to Tridiagonal Form for the Symmetric Eigenvalue Problem*, Argonne National Laboratory Technical Memorandum, 1987.

[7] J. DUCROZ , *Private Communiation*, 1987.

[8] G. A. GEIST and M. T. HEATH, *Parallel Cholesky Factorization on a Hypercube Multiprocessor*, Oak Ridge Report ORNL-6190, 1985.

[9] I. IPSEN, Y. SAAD and M. SCHULTZ, *Dense Linear Systems on a Ring of Processors*, Lin. Alg. Applic. 77 (1986), pp. 205-239.

[10] C. MOLER, *Matrix Computation on Distributed Memory Multiprocessors*, Intel iPSC report, 1986.

[11] A. POTHEN, S. JHA and U. VEMULAPATI, *Orthogonal Factorization on a Distributed Memory Multiprocessor*, Technical Report CS-86-32, The Pennsylvania State University, Dept. of Computer Science, 1986.

[12] R. SCHREIBER, *Solving Eigenvalue and Singular Value Problems on an Undersized Systolic Array*, SIAM J. Scientific and Statistical Computing, 7 (1986), pp. 441-451.

[13] R. SCHREIBER and C. F. VAN LOAN, *A Storage Efficient WY Representation for Products of Householder Transformations*, Technical Report TR 87-864, Cornell University, Dept. of Computer Science, 1987.

[14] C. F. VAN LOAN, *The Block Jacobi Method for Computine the Singular Value Decomposition*, Technical Report TR 85-680, Cornell University, Dept. of Computer Science, 1985.

[15] C. F. VAN LOAN, *A Block QR Factorization Scheme for Loosely Coupled Systems of Array Processors*, Technical Report TR 86-797, Cornell University, Dept. of Computer Science, 1986.

Parallel QR Factorization on a Hypercube Multiprocessor

Eleanor Chu and Alan George

We describe a new algorithm for computing an orthogonal decomposition of a rectangular $m \times n$ matrix A on a hypercube multiprocessor. The algorithm uses Givens rotations, and requires the embedding of a two–dimensional grid on the hypercube network. We design a global communication scheme which uses redundant computation to maintain the data proximity, and we employ a mapping strategy for data allocation so that the processor idle time remains constant for a fixed number of processors regardless of the size of a square matrix. We describe how the algorithm is easily generalized to include the case when $m > n$. The proposed global communication scheme and the data mapping strategy result in reduced computation and communication cost compared to other known results for the same problem. Complexity results and numerical experiments on an Intel iPSC hypercube will be presented.

CHAPTER 2

Communication Requirements of Sparse Cholesky Factorization with Nested Dissection Ordering

Vijay K. Naik[*]

Merrell L. Patrick[*]

Abstract. Load distribution schemes for minimizing the communication requirements of the Cholesky factorization of dense and sparse, symmetric, positive definite matrices on multiprocessor systems are presented. The total data traffic in factoring an n x n sparse symmetric positive definite matrix representing an n-vertex regular 2-D grid graph using n^α, $\alpha \leq 1$, processors is shown to be $O(n^{1+\frac{\alpha}{2}})$. It is $O(n^{\frac{3}{2}})$, when n^α, $\alpha \geq 1$, processors are used. Under the conditions of uniform load distribution these results are shown to be asymptotically optimal.

1. Introduction

Consider the problem of solving a system of linear equations $Ax = b$ where A is an n x n sparse, symmetric, positive definite matrix, x is an n x 1 vector of variables, and b is an n x 1 vector of constants on a multiprocessor system with both shared and local memory. In this paper the total data traffic in factoring the matrix A using the column version of Cholesky decomposition is analyzed. The analysis is restricted to systems for 2-D regular grid graphs. The schemes developed here, however, can be applied to a wider class of problems [7].

In [2], George et al. have given a parallel sparse factorization scheme for local memory multiprocessor systems that has a total data traffic of $O(n^{1+\alpha} \log_2 n)$ using n^α processors. This result was improved to $O(n^{1+\alpha})$ in [4]. In this paper a scheme that has a total data traffic of $O(n^{1+\frac{\alpha}{2}})$ using n^α, $\alpha \leq 1$, processors is presented; with n^α, $\alpha \geq 1$, processors the resulting data traffic is $O(n^{\frac{3}{2}})$. Although a multiprocessor system with both shared and local memory is assumed as the model of computation, the results developed here are equally applicable to systems with only local memories.

In Section 2, the model of computation and some preliminary requirements are stated. In Section 3, expressions for the total data traffic required in factoring dense matrices are developed. The total data traffic for sparse systems is analyzed in Section 4. Conclusions are given in the final section.

2. Some Preliminaries

The model of computation assumed here is that of a multiprocessor system with a two level memory hierarchy. Each processor has local memory and all processors have access to a shared

* Institute for Computer Applications in Science & Engineering, Hampton, Va 23665 and Computer Science Dept., Duke University, Durham, NC 27706. Research supported by the National Aeronautics and Space Administration under NASA contract No. NAS1-18107.

memory. The access cost per nonzero element in the shared memory is unity for any processor. The total number of shared memory accesses from the beginning to the end of an algorithm is defined as the *communication requirement* or *data traffic* of that algorithm implemented on the multiprocessor system.

The $n \times n$ matrix A considered here represents a system corresponding to a 2-D regular grid graph, the vertices of which are ordered using a nested dissection method [3]. Such an ordering has an optimal sequential operation count and fill-in. The basic algebraic scheme used to factor A is the column version of the Cholesky decomposition method [5]. In the next two sections, work distribution schemes among processors are developed and their communication requirements are analyzed. The analysis assumes some familiarity with graph theory concepts related to matrix representations of systems of equations, in particular, the notion of vertices, edges, separators, subgraphs of a graph, and the correspondences between the vertices and the rows and columns of the matrix, the edges and the nonzero elements, and addition of edges and fill-in during the factorization. For details on these concepts see [3] and [6] and the references therein.

Finally, the standard notations "O", and "Ω" are used to characterize the asymptotic growth rates of computation and communication requirements. Precise definitions of these notations can be found in elementary textbooks on the analysis of algorithms, e.g. [8].

3. Communication Requirements of a SPOCC Factorization Scheme for Dense Matrices

First, a scheme based on the column version of the Cholesky decomposition [5] for factoring an $m \times m$ symmetric, positive definite, dense matrix using p processors is described. Without loss of generality, assume that $p = \frac{1}{2}(r^2 + r)$ where r is an integer. The lower triangular part of the matrix is partitioned into p partitions such that all, except r partitions, are $s \times s$ square blocks, where $s = \frac{m}{r}$. The remaining r partitions which lie on the diagonal of the matrix are $s \times s$ triangular blocks. Each partition is assigned to a unique processor. Initially, each processor reads the data for its partition from shared memory into its local memory. The r processors in charge of the r leftmost $s \times s$ blocks of the matrix commence computations. As soon as an element of the factor is computed, it is written into shared memory. As the necessary data becomes available, the remaining processors initiate computations on their blocks. Each processor accesses from shared memory only those elements that are needed for local computation and each element is read no more than once by any processor. This parallel factorization scheme is termed as the *submatrix-partition oriented column Cholesky* (SPOCC) factorization scheme.

Next the communication requirements of the above scheme are analyzed. It is shown that the total data traffic involved is $O(m^2 \sqrt{p})$ and that this result is asymptotically optimal. Consider the data traffic associated with a generic $s \times s$ square block I. Let block I be bounded by columns $(i-1)s + 1$ and $i \cdot s$, and by rows $(j-1)s + 1$ and $j \cdot s$ where $1 \leq i \leq r$ and $1 \leq j \leq r$.

LEMMA 1 : *A total data traffic of* $(2i - \frac{1}{2}) \cdot s^2 + \frac{1}{2} \cdot s$ *is necessary and sufficient for factoring the square block* I; *it is the same for all square blocks bounded by the columns* $(i-1)s + 1$ *and* $i \cdot s$. *The data traffic associated with a* $s \times s$ *triangular diagonal block bounded by the columns* $(i-1)s + 1$ *and* $i \cdot s$, *is* $(i - \frac{1}{2}) \cdot s^2 + \frac{1}{2} \cdot s$.

PROOF: See [7] for a proof.
□

Using Lemma 1 an upper bound on the total data traffic is obtained.

THEOREM 1 : *The total data traffic involved in the SPOCC factorization of an* $m \times m$ *dense matrix using* p *processors is* $O(m^2 \sqrt{p})$.

PROOF: See [7] for a proof.
□

Now consider the data dependencies in computing an element of the factor. To compute the final value of an element $a_{i,j}$ in the factor, values at all the elements in row j and the values of elements in columns 1 through j of row i are needed. Thus, any off-diagonal element $a_{i,j}$ requires values of $2 \cdot j$ elements for completing the computations and any diagonal element (when $i = j$) requires values of j elements. Two observations follow:

(1)   The values at all the elements in any row i can be used to complete the computations at exactly one element, namely, the diagonal element $a_{i,i}$.

(2)   If i and j are any two rows, such that $i > j$, then the values of the elements in these two rows can be used to complete computations at exactly one off-diagonal element $a_{i,j}$. No other information is needed to complete the computations at that element.

These observations lead to the following lemma on the number of elements at which the computations can be completed by acquiring information from exactly k rows.

LEMMA 2 : *If the required values of all the elements in any k rows are available then the computations of at most $\dfrac{k^2}{2} + \dfrac{k}{2}$ elements can be completed; conversely, the computations corresponding to any $\dfrac{k^2}{2} + \dfrac{k}{2}$ elements depend on at least one element from each row of at least k distinct rows.*

PROOF: See [7] for a proof.
□

Next the total data traffic in the above described scheme is shown to be asymptotically optimal.

THEOREM 2 : *Assuming that the computational load is to be uniformly distributed among p processors, the data traffic involved in computing the Cholesky factor of an $m \times m$ dense matrix is $\Omega(m^2 \sqrt{p})$.*

PROOF: See [7] for a proof.
□

4. Communication Requirements of a SPOCC Factorization Scheme for Sparse Matrices

In this section the total data traffic in factoring a sparse, symmetric, positive definite matrix A corresponding to a regular n-vertex 2-D grid graph is analyzed. A 9-point stencil, unless otherwise stated, is used. In the following discussion L denotes the lower triangular Cholesky factor of the matrix A. First, the worst case communication requirement of $O(n^{\frac{3}{2}})$, which is independent of the number of processors used, is obtained and then a scheme that reduces the communication requirement to $O(n^{1+\frac{\alpha}{2}})$, when n^α, $\alpha \leq 1$, processors are used, is presented. The result is shown to be asymptotically optimal under the condition of uniform load distribution.

Clearly, the communication requirement is the worst when the use of local memory is not allowed. Now consider the computations involved in computing a nonzero element $l_{i,j} \in L$. Recall that the expression $a_{i,j} - \sum_{k=1}^{j-1} l_{i,k} \cdot l_{j,k}$ is computed first, followed by a single division by $l_{j,j}$. Thus, for each multiplication there is one subtraction operation, at most one division and three memory references and a constant overhead such as index computation. Let the values of all the elements of the lower triangular part of matrix A and those of L be stored in the shared memory. Any number of processors are allowed to participate in computing a nonzero element of the factor provided that no single operation is performed by more than one processor. The following theorem gives a bound on the total data traffic under these assumptions. Although the result is obvious, it is useful because it is independent of the number of processors used and gives the worst case bound on the data traffic even for models of computation that are more restrictive.

THEOREM 3 : *The worst case data traffic associated with factoring the matrix A is $O(n^{\frac{3}{2}})$.*

PROOF: Under the assumed model of computation, each multiplication operation requires at most a constant number of memory references. Thus, the total data traffic is
$$\leq const \cdot number\ of\ multiplication\ operations.$$
From [1], the number of multiplication operations associated with factoring matrix A is $O(n^{\frac{3}{2}})$. Hence, the total worst case data traffic is $O(n^{\frac{3}{2}})$.
□

With the model of computation assumed for dense matrices, the bound on the data traffic is improved when n^α, $\alpha < 1$, processors are used. This is possible because now local memory is used to store the data that is needed in more than one computation.

First, bounds are obtained on the data traffic associated with computing the elements of the Cholesky factor along the columns corresponding to a generic "+" shaped separator in the nested dissection ordering of a 2-D regular grid graph. Let the ordering scheme be such that the vertices on the vertical part of a "+" separator are ordered after those on the horizontal part. Let each of the two horizontal parts of the "+" separator be referred to as the *horizontal sub-separator* and the single vertical part be referred to as the *vertical sub-separator*.

Let $\eta_i^j = \{ k \mid k \le j \text{ and } l_{i,k} \ne 0, l_{i,k} \in L \}$; i.e., η_i^j is the set of all columns of the factor L to the left of the column $j+1$ such that the elements in row i of these columns are nonzero. Let $\overline{\eta}_{i,k}^j = \bigcup_{s=i}^{k} \eta_s^j$; i.e., $\overline{\eta}_{i,k}^j$ is the set of all columns to the left of column $j+1$ such that on each of these columns there is a nonzero element in at least one of the i through k rows of the factor. Let Γ represent any m-vertex sub-separator. It is assumed that all the vertices in any sub-separator are ordered consecutively. Let $low(\Gamma)$ and $high(\Gamma)$ be the indices of the lowest and the highest ordered vertices, respectively, on the sub-separator Γ. Note that $high(\Gamma) - low(\Gamma) + 1 = m$. The following lemma establishes some basic sub-separator related properties that are useful in analyzing the communication requirements.

LEMMA 3 : *Let Γ be any m-vertex sub-separator. (i) Corresponding to the vertices of Γ there is a dense $m \times m$ triangular diagonal block in the Cholesky factor. (ii) In the factor L, the columns $low(\Gamma)$ through $high(\Gamma)$ contain at most four off-diagonal rectangular blocks with nonzero elements. Each of these blocks is of size at most $(c_1 \cdot m + c_2) \times m$ where, $c_1 \le 2$ and $c_2 \le 3$ are integer constants. Any nonzero element in these columns is in one of these four blocks or in the diagonal triangular block and no where else.*

PROOF: See [7] for a proof.
□

The next lemma quantifies the data dependencies of the nonzero elements in any of the five blocks specified in Lemma 3. The lemma shows that the number of nonzero elements in any row i of the factor L is less than $c \cdot m$ where c is an integer constant and m is the size of the sub-separator to which the vertex corresponding to row i belongs. It is then shown that for any row i, the computations at all the elements $l_{i,j} \in L$, $low(\Gamma) \le j \le high(\Gamma)$, for some m-vertex sub-separator Γ require a total of less than $c \cdot m$ nonzero elements from that row. Note that this count is independent of the sub-separator to which the vertex corresponding to row i belongs. Thus, the computations at all the elements in a row of any of the five blocks specified in Lemma 3, require only $c \cdot m$ elements from that row irrespective of the relative location of the off-diagonal blocks in the factor.

LEMMA 4 : *Let Γ be any m-vertex sub-separator. The nonzero elements from row i, $i \ge low(\Gamma)$, required in completing the computations of all elements $l_{i,j} \in L$ such that $low(\Gamma) \le j \le high(\Gamma)$, are those elements in row i on the columns in the set given by, $\eta_{low(\Gamma),high(\Gamma)}^{high(\Gamma)} \cap \eta_i^{high(\Gamma)}$. For all $i \ge low(\Gamma)$, the number of such useful nonzero elements in row i is $\le c \cdot m$ for some constant c.*

PROOF: See [7] for a proof.
□

The scheme for completing computations in all five dense blocks associated with an m-vertex sub-separator Γ using p processors is now described. If $p = 1$, that processor accesses necessary values and completes all computations. If $p > 1$, the factorization corresponding to the $m \times m$ triangular diagonal block is first completed using all the processors and then the factorization corresponding to the four off-diagonal blocks. For the first part, as stated in the previous section, the $m \times m$ dense diagonal block is partitioned into $\frac{r^2}{2} - \frac{r}{2}$ square blocks and r diagonal triangular blocks each of size $\frac{m}{r} \times \frac{m}{r}$ where, $p = \frac{r^2}{2} + \frac{r}{2}$, and each of these p partitions is assigned to a unique processor. Each processor completes computations corresponding to its partition by accessing the required data from shared memory. For the purpose of factoring, the off-diagonal blocks are treated as if they were adjacent, and the resultant rectangular block is partitioned into p sub-blocks each of size $\frac{c \cdot m}{\sqrt{p}} \times \frac{m}{\sqrt{p}}$, where $c \le 6$ for a horizontal sub-separator and $c \le 4$ for a vertical sub-separator. Again each partition is assigned to a unique processor.

Using the results of Lemma 4, a bound is obtained on the data traffic associated with the computations performed as outlined in the above scheme.

LEMMA 5 : *The total data traffic associated with completing the computations, using p processors, in all the dense blocks within the columns low(Γ) through high(Γ), for some m-vertex horizontal sub-separator Γ, is bounded by* $(53 + 11\sqrt{2})m^2 \cdot \sqrt{p}$. *It is bounded by* $(28 + 8\sqrt{2})m^2 \cdot \sqrt{p}$ *for an m-vertex vertical sub-separator.*

PROOF: See [7] for a proof.

□

Applying the results from the above lemma, a bound on the total data traffic in factoring the sparse matrix A is obtained. First the overall load distribution scheme is presented. For the sake of simplicity, assume that the n-vertex 2-D grid graph is a $n^{\frac{1}{2}} \times n^{\frac{1}{2}}$ square grid and that there are n^{α} processors, $\alpha \leq 1$. The vertices of the grid are ordered using the nested dissection ordering scheme such that the vertices in each square sub-grid are consecutively ordered. This ordering results in n^{α} sub-grids each of size $n^{\frac{1}{2}-\frac{\alpha}{2}} \times n^{\frac{1}{2}-\frac{\alpha}{2}}$. The computations at all the nonzero elements corresponding to the vertices of each such sub-grid are assigned to a unique processor. After completing the work, two processors from adjacent sub-grids combine to complete the work corresponding to the $n^{\frac{1}{2}-\frac{\alpha}{2}}$-vertex horizontal sub-separator that separates the two sub-grids. This is followed by the work on the $2n^{\frac{1}{2}-\frac{\alpha}{2}}$-vertex vertical sub-separator. Four processors work on each such vertical sub-separator. This process is continued until the factorization of the entire matrix is completed. Whenever more than one processor is working on any sub-separator, the scheme presented earlier for a generic m-vertex sub-separator is applied.

Let $\tau_h (m, p, k)$ represent the data traffic, using p processors, in completing the computations at all the nonzero elements $l_{i,j} \in L$ in the columns corresponding to an m-vertex horizontal sub-separator that is surrounded by higher ordered vertices on k sides. Let $\tau_v (m, p, k)$ represent the same for an m-vertex vertical sub-separator. Let $\tau_g (m, p, k)$ represent the total data traffic, using p processors, in completing the computations corresponding to all sub-separators within an m-vertex sub-grid that is surrounded by higher ordered vertices on k sides.

THEOREM 4 : *The total data traffic in factoring the $n \times n$ sparse matrix A using n^{α} processors is* $O(n^{1+\frac{\alpha}{2}})$; *i.e.,* $\tau_g (n, n^{\alpha}, 0) = O(n^{1+\frac{\alpha}{2}})$.

PROOF: On a $n^{\frac{1}{2}} \times n^{\frac{1}{2}}$ regular grid there is an $n^{\frac{1}{2}}$-vertex vertical sub-separator and two $\frac{1}{2}n^{\frac{1}{2}}$-vertex horizontal sub-separators. The vertical sub-separator is not surrounded by any vertices that are ordered after the vertices on the vertical sub-separator. Each of the two horizontal sub-separators are surrounded on one side. These three sub-separators subdivide the n-vertex grid graph into four sub-grids of size $\frac{1}{2}n^{\frac{1}{2}} \times \frac{1}{2}n^{\frac{1}{2}}$, each surrounded on two sides by higher ordered vertices. Thus, the total data traffic in factoring the matrix A is given by,

$$\tau_g (n, n^{\alpha}, 0) = \tau_v (n^{\frac{1}{2}}, n^{\alpha}, 0) + 2\tau_h (\frac{1}{2}n^{\frac{1}{2}}, \frac{1}{2}n^{\alpha}, 1) + 4\tau_g (\frac{1}{4}n, \frac{1}{4}n^{\alpha}, 2).$$

A recursive expansion of the above expression contains data traffic terms for vertical sub-separators of different sizes that are surrounded on zero sides, two sides, three sides (in two different ways), and on all four sides by higher ordered vertices. It also contains data traffic expressions for horizontal sub-separators of different sizes surrounded in five different ways. To keep the analysis simple, it is assumed that all the four sub-grids of size $\frac{1}{2}n^{\frac{1}{2}} \times \frac{1}{2}n^{\frac{1}{2}}$ are surrounded on all four sides. This simplification results in a conservative expression for the data traffic, but affects only the constant terms in the bound. Thus,

$$\tau_g (n, n^{\alpha}, 0) \leq \tau_v (n^{\frac{1}{2}}, n^{\alpha}, 0) + 2\tau_h (\frac{1}{2}n^{\frac{1}{2}}, \frac{1}{2}n^{\alpha}, 1) + 4\tau_g (\frac{1}{4}n, \frac{1}{4}n^{\alpha}, 4).$$

Expanding the individual terms using Lemma 5 we get,

$$\tau_g (n, n^{\alpha}, 0) \leq \frac{1}{2}(78 + 71\sqrt{2}) \cdot n^{1+\frac{\alpha}{2}}.$$

□

In the following theorem the communication bound of the above presented scheme is shown to be asymptotically optimal.

THEOREM 5 : *Under the condition of uniform load distribution, the data traffic in factoring the n x n sparse matrix A, using n^α processors, $\alpha \le 1$, is $\Omega(n^{1+\frac{\alpha}{2}})$.*

PROOF: For a regular 2-D grid graph with n vertices, the separator size for nested dissection ordering is $n^{\frac{1}{2}}$ [6]. From Lemma 3, it follows that the factor L has an $n^{\frac{1}{2}}$ x $n^{\frac{1}{2}}$ dense triangular diagonal block incorporated in it. From Theorem 2, the data traffic involved in completing the computations associated with the elements of this dense triangular block, under the condition of uniform load distribution using n^α processors, is $\Omega(n^{1+\frac{\alpha}{2}})$. Since the factorization of A cannot be completed without completing the factorization of this dense block, the result follows.

□

5. Conclusions

In this paper load distribution schemes for multiprocessor systems with local and shared memories are presented for factoring dense and sparse symmetric, positive definite matrices. The communication requirements of these schemes are analyzed and are shown to be asymptotically optimal for the systems corresponding to regular 2-D problems where the vertices of the grid graph are ordered using the nested dissection ordering methods. Under these schemes, the total data traffic in factoring an n x n dense, symmetric, positive definite matrix is $O(n^{2+\frac{\alpha}{2}})$ when n^α processors are used. The total data traffic in factoring the n x n sparse, symmetric, positive definite matrix corresponding to a 2-D grid problem, is $O(n^{1+\frac{\alpha}{2}})$ when n^α, $\alpha \le 1$, processors are used. When n^α, $\alpha > 1$, processors are used the total data traffic is $O(n^{\frac{3}{2}})$. Under the condition of uniform load distribution, these results are optimal in an order of magnitude sense.

ACKNOWLEDGEMENTS

The authors would like to thank Bob Voigt for his support and encouragement.

REFERENCES

(1) J. A. George, *Nested dissection of a regular finite element mesh*, SIAM J. Numer. Anal., vol. 10, pp. 345-363, 1973.

(2) J. A. George, M. T. Heath, J. W. H. Liu, and E. Ng, *Sparse Cholesky factorization on a local-memory multiprocessor*, Technical report ORNL/TM-9962, Oak Ridge National Laboratory, Oak Ridge, Tenn, 1986.

(3) J. A. George and J. W. H. Liu, *Computer Solution of Large Sparse Positive Definite Systems*, Prentice-Hall, Inc., Englewood Cliff, NJ, 1981.

(4) J. A. George, J. W. H. Liu, and E. Ng, *Communication reduction in parallel sparse Cholesky factorization on a hypercube*, in Hypercube Multiprocessors 1987, ed. M. T. Heath, SIAM Publication, 1987.

(5) G. H. Golub and C. F. Van Loan, *Matrix Computations*, The Johns Hopkins University Press, Baltimore, MD, 1983.

(6) R. J. Lipton, D. J. Rose, and R. E. Tarjan, *Generalized nested dissection*, SIAM J. Numer. Anal., vol. 16, pp. 346-358, 1979.

(7) V. K. Naik and M. L. Patrick, *Communication requirements of parallel sparse Cholesky factorizations*, ICASE report in preparation, 1988.

(8) E. M. Reingold, J. Nievergeld, and N. Deo, *Combinatorial Algorithms: Theory and Practice*, Prentice-Hall, Inc., Englewood Cliff, NJ, 1977.

CHAPTER 3

LU Factorization on Distributed-Memory Multiprocessors*

George A. Geist[†]
Charles H. Romine[†]

Abstract. We discuss and compare two methods of improving the efficiency of LU factorization with pivoting on distributed-memory multiprocessors. The first method uses a dynamic load balancing scheme to avoid work imbalances caused by pivoting. While this method increases the amount of communication required, it significantly reduces the execution time. The second method uses pipelining to mask the cost of pivoting. The dynamic load balanced and pipelined algorithms have achieved efficiencies of 92% and 97% respectively.

Key words. Gaussian elimination, LU factorization, parallel algorithms, distributed-memory multiprocessors

AMS(MOS) subject classifications. 65F,65W

1. Introduction. Many papers discussing the parallel implementation of *LU* factorization of dense matrices ignore the deleterious effect that pivoting has on the speedup obtained, whether the coefficient matrix is stored by rows or by columns. While the cost of pivoting is relatively small for the serial algorithm, the same is not necessarily true for parallel implementations. In particular, an equitable distribution of the rows of the coefficient matrix to the processors no longer guarantees that the computational load remains balanced throughout the algorithm. For example, Geist and Heath [2] show that on an Intel iPSC, if the rows of the matrix are wrapped onto the processors and pivoting is done implicitly, the penalty for pivoting averages between ten and fifteen percent of the total execution time. Whatever their initial distribution, the distribution of the rows of the matrix to the processors becomes "randomized" by pivoting, thus inducing an imbalance in the computational load.

An alternative, which avoids this problem, is to interchange the rows of the matrix explicitly in order to maintain a wrapped mapping (see Chu and George[1]). Naturally, in a distributed-memory environment, this will entail a (perhaps large) increase in the overall communication cost of the algorithm. Nevertheless, results on an Intel iPSC show that the explicit interchange strategy can outperform the implicit strategy.

* This research was supported by the Applied Mathematical Sciences Research Program, Office of Energy Research, U.S. Department of Energy under contract DE-AC05-84OR21400 with Martin Marietta Energy Systems Inc.

† Mathematical Sciences Section, Oak Ridge National Laboratory, P.O. Box Y, Oak Ridge, Tennessee 37831.

Another method for ensuring that the load remains balanced throughout the computation is initially to wrap the coefficient matrix onto the processors by columns rather than by rows. Any pivoting during the algorithm, whether implicit or explicit, will leave the mapping unaffected. However, the pivot search and formation of the multipliers at each stage must now be done serially by the processor containing the appropriate column. This serial bottleneck will obviously affect the efficiency of the parallel implementation.

We describe two approaches for implementing efficient LU factorization with partial pivoting on a distributed-memory multiprocessor. The first, which maps the matrix to the processors by rows, is an improvement over the explicit interchange strategy of Chu and George. Specifically, we show that the number of interchanges required can be reduced by relaxing the restrictions on the mapping of rows to processors. This causes the computational load to be balanced dynamically and eliminates much of the communication required to preserve wrapping, but the final mapping will no longer be a wrap mapping. The second approach uses the good load balancing properties of the column-wrapped mapping. The effect of the induced serial bottlenecks is then masked through the use of pipelining. The algorithms we describe are valid on any distributed-memory multiprocessor; however, our empirical results were obtained on an Intel iPSC hypercube.

2. Row Storage. We first describe LU factorization of an $n \times n$ dense matrix that is stored by rows on p processors. The initial mapping is unimportant, provided the rows are distributed evenly to the processors, since the wrap mapping will ultimately be produced by the explicit interchange of rows between processors.

At each major stage of the algorithm, the pivot row must first be determined. This requires communication among all the processors, since the pivot column is scattered. The best means of effecting this global communication will depend upon the underlying topology of the processor network. An effective strategy for performing global communication on a hypercube is through the use of a minimal spanning tree embedded in the hypercube network (for an illustration, see Geist and Heath [2]). This allows information either to be disseminated (fanned-out) from one processor to all, or collected (fanned-in) from all processors into one, in $\log_2 p$ steps. In the current context, each processor searches its portion of the pivot column for the element of maximum modulus. The leaf nodes of the spanning tree send these local maxima to their parents. The parents compare these received values to their own local maxima, forwarding the new maxima up the tree. When the fan-in is complete, the pivot row will have been determined by the root processor in the spanning tree, which must then send this information back down the tree. In the Chu and George algorithm, if the next processor in the wrap mapping contains the pivot row, no exchange is necessary; otherwise, this processor exchanges rows with the processor that currently contains the pivot row. Finally, the processor that now contains the pivot row must fan it out to the other processors. Hence, three logarithmic communication stages and (possibly) one exchange are performed before updating of the submatrix can begin.

Note that at the first stage of the above algorithm, if the first processor does not contain the first pivot row then an exchange is performed. However, the choice of the "first" processor in a hypercube is arbitrary. Any of the processors can serve as the first, and thus no exchange is required at the first stage to preserve load balancing. Similarly, an exchange should be required at the second stage only if the second pivot row lies in the same processor as the first.

A generalization of this concept produces what we call *dynamic pivoting*. Dynamic pivoting requires only that rows kp through $(k+1)p - 1$ $(0 \leq k < n/p)$ lie in distinct processors for each k, with the order in which they are assigned unconstrained. That is, a processor that already contains one of these pivot rows cannot contain another, and must exchange rows with a processor that does not already contain one. This scheme produces any one of a family of mappings that have the load balancing properties of wrapping in that the rows assigned to a processor are more or less uniformly distributed in the matrix. This scheme allows considerable leeway in the choice of mapping, and hence reduces the number of exchanges required during pivoting.

The implementation of dynamic pivoting raises a further question. If a processor finds itself with two pivot rows when only one is allowed, with which processor should it exchange rows? Any processor that does not yet contain a pivot row in the current set of p rows is a valid choice. Since dynamic pivoting is designed to reduce the communication cost due to explicit pivoting, the most reasonable procedure is to exchange with the nearest valid neighbor in the topology of the processor array. A breadth-first search of the minimal spanning tree rooted at a particular node yields a list of processors in increasing order of distance from the root node. Exchanging with the first valid processor on this list should help to reduce further the cost of the exchange. This was found to be true in practice, and hence such an exchange strategy was adopted for dynamic pivoting. The results of implementing this algorithm are summarized in Table 1. The table reveals that the cost of pivoting is a small fraction of the total execution time for factorization.

TABLE 1
Execution time (in seconds) for row-stored algorithms.

Matrix of order 1024 on 32 processors	
Without pivoting	816.4
With implicit pivoting	945.8
With dynamic pivoting	852.2

It is important to make certain that deviating from the wrap mapping does not cause undue overhead during the triangular solution stages, since this may negate any savings obtained during the factorization. The most efficient parallel algorithms known for the solution of a triangular system on a hypercube rely heavily on the wrap mapping for their performance (see Heath and Romine [3], and Li and Coleman [4]). However, the performance of the *cube fan-out* algorithm is also reasonably good and is largely unaffected by the choice of mapping. For even moderately large problems ($n \geq 500$ on 64 processors) on the iPSC, the savings from dynamic pivoting is an order of magnitude greater than the potential savings in the triangular solution from using the wrap mapping.

3. Column Storage. Another means of ensuring that the computational load remains equitably distributed to the processors is to store the coefficient matrix by columns. The inital mapping of the columns of the matrix to the processors will be unaffected by pivoting, so we assume that the columns are wrapped onto the processors. Unlike the previous algorithm, no extra communication is required for pivoting, since the entire pivot column resides within a single processor. However, both the search for the pivot element and the computation of the multipliers must now be done serially. Experimental results on an Intel iPSC show that this serial bottleneck can severely degrade the efficiency of the factorization.

Fortunately, most of the serial overhead in the algorithm can be masked through the use of pipelining. We use the term pipelining to mean a reduction in latency obtained when a processor, rather than continuing its current computation, sends already computed values to other processors. For example, a high degree of pipelining is achieved if the processor containing the next pivot column, before updating its portion of the submatrix, first computes and sends each multiplier one at a time. This minimizes the latency that prevents the other processors from beginning their computations, but drastically increases the communication cost. A moderate degree of pipelining, but with lower communication cost, occurs when the processor containing the next pivot column, before updating its portion of the submatrix, first computes and then sends the whole column of multipliers. This was the method we used to implement a pipelined *LU* factorization algorithm. It should be noted that in the row storage algorithm pipelining is infeasible, since the pivoting stage requires the cooperation of all the processors. As the results in Table 2 indicate, the large latency time induced by the serial pivot search and serial computation of the multipliers in the algorithm has been almost entirely eliminated by pipelining. The cost of pivoting is now a negligible percentage of the total factorization time.

TABLE 2
Execution time (in seconds) for column-stored algorithms.

Matrix of order 1024 on 32 processors		
	basic algorithm	pipelined algorithm
Without pivoting	843.3	802.7
With pivoting	929.7	804.2

4. Conclusions. We have presented two parallel algorithms for the LU factorization with partial pivoting of a dense matrix. The first assumes that the coefficient matrix is stored by rows and implements dynamic pivoting. This pivoting scheme is designed to ensure that the computational load remains balanced throughout the algorithm, while attempting to minimize the extra communication required to effect this balance. The second assumes that the coefficient matrix is wrapped by columns (thus ensuring a good load balance), and uses pipelining to mask the high serial cost induced by pivoting.

Whether row storage with dynamic pivoting or column storage with pipelining is used, results on an Intel iPSC indicate that the cost of partial pivoting in parallel is an insignificant fraction of the total cost of the factorization. Moreover, these algorithms attain remarkably high efficiencies on the iPSC. The execution times presented in Tables 1 and 2 above give efficiencies of 92% and 97% for the dynamic pivoting and pipelined algorithms, respectively.

REFERENCES

[1] CHU, E. AND A. GEORGE, *Gaussian elimination with partial pivoting and load balancing on a multiprocessor*, Parallel Computing, 5 (1987), pp. 65–74.

[2] GEIST, G.A. AND M.T. HEATH, *Matrix factorization on a hypercube multiprocessor*, in Hypercube Multiprocessors 1986, M. Heath, ed., Society for Industrial and Applied Mathematics, Philadelphia, 1986, pp. 161–180.

[3] HEATH, M.T. AND C.H. ROMINE, *Parallel solution of triangular systems on distributed-memory multiprocessors*, SIAM J. Sci. Stat. Comput. (to appear), (1988). Also Tech. Rept. ORNL/TM-10384, Mathematical Sciences Section, Oak Ridge National Laboratory, Oak Ridge, Tennessee 37831, March 1987.

[4] LI, G. AND T.F. COLEMAN, *A parallel triangular solver for a hypercube multiprocessor*, in Hypercube Multiprocessors 1987, M. Heath, ed., Society for Industrial and Applied Mathematics, Philadelphia, 1987, pp. 539–551. (Also Tech. Rept. TR 86-787, Dept. of Computer Science, Cornell University, October, 1986).

An Efficient Fixed Size Array for Solving Large Scale Toeplitz Systems

Afshin Daghi, V.K. Prasanna Kumar and Ali Safavi

In this paper systolic implementation and partitioning of the Bareiss algorithm for solving an $(n + 1)$ by $(n + 1)$ Toeplitz system of equations is studied. While systolic solutions to Toeplitz systems have been well studied, the main problem with the known results is that they cannot solve any arbitrary Toeplitz system of equations on a fixed size array of processors. Our systolic implementation solves the Bareiss algorithm on a fixed size linear array. Our algorithm achieves $O\left(\frac{n^2 \log_2(p)}{p}\right)$ computing time and $O(n^2)$ storage on p (p may be fixed) processors. Even without the partitioning, our implementation has superior time performance compared to known results within a constant factor. Furthermore, our implementation achieves a processor utilization of almost 100%. Our solution allows the real time solution of large Toeplitz matrices which occur in most Toeplitz applications.

CHAPTER 4

Gaussian Techniques on Shared Memory Multiprocessor Computers*

Gregory A. Darmohray†
Eugene D. Brooks III‡

Abstract. We present performance results for parallel Gauss and Gauss-Jordan elimination algorithms on a shared memory multiprocessor. The Cerberus multiprocessor simulator, a simulator for a scalable shared memory multiprocessor with fully pipelined functional units, is used to evaluate algorithm performance. Our parallel implementations of these linear system solvers make extensive use of barrier synchronization. We show the need for barrier synchronization supported directly in hardware for tightly coupled algorithms. For a fixed problem size, the performance of Gauss-Jordan elimination crosses that of Gauss elimination as we increase the number of processors, even though the latter algorithm has a lower operation count. Sometimes, one can profit by trading operations for a better load balance and lower relative synchronization cost in a parallel algorithm.

1. Introduction. The performance of processor synchronization, relative to the speed at which computational work can be done, is a critical issue for tightly coupled parallel programs. On a shared memory machine built with microprocessor technology, where a floating point operation can take hundreds of machine cycles, relatively fast processor synchronization is easy to achieve. The situation is very different for a shared memory multiprocessor based on supercomputer technology. With pipelined functional units, vectorized floating point operations can be streamed at one (or more) per machine cycle. Synchronization operations, which are inherently scalar in nature, can take hundreds or even thousands of machine cycles. The relative cost of processor synchronization in a multiprocessor which efficiently pipelines its computational work can be very high.

We consider parallel implementations of non-pivoting Gauss and Gauss-Jordan elimination using the barrier as the basic processor synchronization technique. The performance of the algorithms is evaluated using the Cerberus multiprocessor simulator [1], a simulator for a scalable shared memory multiprocessor with fully pipelined functional units. We compare the execution times of the algorithms, using both barrier synchronization implemented in software and in hardware, for problems sizes ranging from 32 to 128 unknowns. The algorithms are executed on machines with from 1 to 32 processors. We find that high performance synchronization is critical if one wants to obtain good speedups on such tightly coupled problems. We also find that the Gauss-Jordan algorithm, which has an operation count 50% higher than Gauss elimination, can outperform Gauss elimination as the number of processors is increased for a fixed problem size. This is due to the better load balancing characteristics and lower synchronization

---

* Work performed, in part, under the auspices of the U. S. Department of Energy by the Lawrence Livermore National Laboratory under contract No. W-7405-ENG-48.

† Graduate Group in Computing Science, University of California at Davis in Livermore, Livermore, California 94550
‡ Parallel Processing Project, Lawrence Livermore National Laboratory, Livermore, California 94550

costs in the Gauss-Jordan algorithm.

2. Processor Synchronization. In our parallel implementations of the linear system solvers, barrier synchronization was used exclusively. In barrier synchronization, all of the processors are required to *meet at the barrier* before any are allowed to proceed. In general, the use of barrier synchronization is a rather generic synchronization strategy where barriers are used to separate two sets of computations which have data dependencies between them. It might pay to more carefully exploit the data dependencies themselves, by using a synchronization strategy that is more custom tuned to the algorithm.

In our performance evaluations two barrier implementations are used. The first is a software technique, the butterfly barrier [2], which is a fast bottleneck free algorithm. It is the method of choice for our production work on several shared memory multiprocessors. On a microprocessor based machine, the butterfly barrier requires the time equivalent of 5 to 10 floating operations to execute, making it suitable for even the most tightly coupled problems. On the Cerberus machine, using 2×2 switch nodes in the shared memory server, the algorithm takes from 82 clocks for 2 processors to 294 clocks for 32 processors. Computational work is efficiently pipelined by the Cerberus processor achieving an issue rate of one instruction per clock. The result is a requirement of a hundred or more floating point operations, per processor, to break even with software barrier cost. Given the high memory latency of some recently introduced supercomputers, and their capability to pipeline 2 or more floating point operations per clock, the cost of the butterfly barrier synchronization algorithm can be even greater.

The second barrier implementation used on the Cerberus machine is a hardware barrier in which all of the processors are released one clock after the last processor has arrived. One would call this an *ideal barrier*, which is in fact quite possible to construct. Essentially all that is required is an N input AND gate, one input for each participating processor. Each processor would assert its input when it arrives at the barrier then wait until the output was true. Using a hardware assist for the barrier operation, a latency of just a few machine clocks is easy to obtain. Considering the number of AND gates contained in a computer, adding a few more for barrier support seems like a cost effective thing to do. With barrier synchronization implemented in hardware, the number of floating point operations required between barriers is greatly reduced. This improves the performance of tightly coupled algorithms as we will see in the following sections.

3. Parallel Gauss Elimination. Gauss elimination consists of a reduction (reducing a square matrix into an upper triangular matrix) followed by a back substitution (solving for the unknowns). Consider the system of equations

$$a_{11}x_1 + a_{12}x_2 + a_{13}x_3 + a_{14}x_4 + a_{15}x_5 = b_1 \tag{1}$$

$$a_{21}x_1 + a_{22}x_2 + a_{23}x_3 + a_{24}x_4 + a_{25}x_5 = b_2 \tag{2}$$

$$a_{31}x_1 + a_{32}x_2 + a_{33}x_3 + a_{34}x_4 + a_{35}x_5 = b_3 \tag{3}$$

$$a_{41}x_1 + a_{42}x_2 + a_{43}x_3 + a_{44}x_4 + a_{45}x_5 = b_4 \tag{4}$$

$$a_{51}x_1 + a_{52}x_2 + a_{53}x_3 + a_{54}x_5 + a_{55}x_5 = b_5 \tag{5}$$

in five unknowns. In the reduction we want to first zero out the coefficients of x_1 below a_{11} using (1), then zero out the coefficients of x_2 below a_{22} using the modified (2), ..., continuing until the matrix operator is reduced to an upper triangular form.

The *SAXPY* operations required to zero out a coefficient of x_1 below a_{11} are independent and can be done in parallel. These operations are interleaved across the processors using a *forall* loop. Once all the coefficients of x_1 below a_{11} have been zeroed out, the work to zero out the coefficients of x_2 below a_{22} can start. These operations are again done in parallel using a *forall* loop, but we use a barrier between the two *forall* loops to prevent the second step of the reduction from starting before the first step is complete.

The Cerberus processor architecture which does have fully pipelined functional units does not have vector instructions nor multiple channels to memory, which could be used to stream *SAXPY* operations at 2 floating point operations per clock. Using loop unrolling, the *SAXPY* operations on rows of the matrix are efficiently pipelined at a peak rate of 0.4 floating operations per clock cycle, because each multiply and add must be accompanied by two fetches and a store. In our *SAXPY* implementation, the loop is unrolled 32 ways, and we come very close to the peak speed possible for the processor architecture.

In the reduction, there are $N-1$ steps each of which is followed by a barrier synchronization. The k'th reduction step requires $N-k$ *SAXPY* operations on independent rows of the matrix, and $N-k$ scalar multiply-add operations on the b_i. The vector length of the *SAXPY* operations in the k'th reduction step is $N-k$. This gives a total number of multiply-add operations for the k'th reduction step of $(N-k)^2 + N-k$. These operations must be followed by a barrier synchronization as the next reduction step will need the new values. As the reduction proceeds, the work done in each step drops making the cost of the barrier synchronization more significant. Load balance degrades as the reduction proceeds as well; in the last reduction step only one processor has any work to do at all.

After the reduction, (1) through (5) become

$$a'_{11}x_1 + a'_{12}x_2 + a'_{13}x_3 + a'_{14}x_4 + a'_{15}x_5 = b'_1 \qquad (6)$$

$$a'_{22}x_2 + a'_{23}x_3 + a'_{24}x_4 + a'_{25}x_5 = b'_2 \qquad (7)$$

$$a'_{33}x_3 + a'_{34}x_4 + a'_{35}x_5 = b'_3 \qquad (8)$$

$$a'_{44}x_4 + a'_{45}x_5 = b'_4 \qquad (9)$$

$$a'_{55}x_5 = b'_5 \ , \qquad (10)$$

respectively. In the back substitution, a single processor solves for x_5 using (10). Once this is done, (6) through (9) may be simplified by substituting for x_5 doing the simplifications on the four equations in parallel. While one processor is solving for x_i, the other processors which need this value for their share of the equation simplifications must wait at a barrier. This is repeated for each x_i until x_1 is finally solved. We make sure that the same processor that solves for x_i handles the simplification for equation $(i-1)$. This reduces the barrier requirement to one barrier per back substitution step, which would otherwise be two.

The total work done in a back substitution step, in which we solve for x_i and then simplify $i-1$ equations, is a divide to solve for x_i and then $i-1$ multiply-adds for the equation simplifications. A barrier must precede the equation simplifications to safeguard the data dependency on x_i. The number of operations between barriers scales linearly in the problem size; a situation which is far worse than the quadratic scaling of work between barriers for the reduction. Because of this, the back substitution is more tightly coupled than the reduction and limits the speedup which we can obtain as the number of processors is increased for a given problem size.

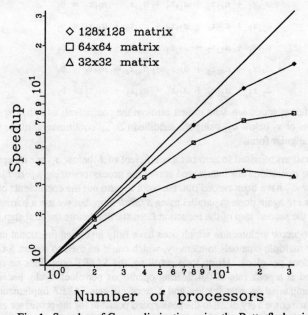

Fig. 1. Speedup of Gauss elimination using the Butterfly barrier.

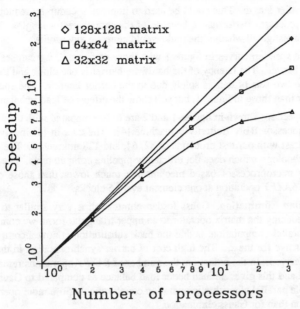

Fig. 2. Speedup of Gauss elimination using the hardware barrier.

We show in figures 1 and 2 the speedups for Gauss elimination using the software and hardware barrier implementations, respectively. The speedup curves for the software barrier are dismal; in fact, it takes longer to solve the 32 unknown problem with 32 processors than with 16 processors. This is directly attributed to the high cost of the software barrier. The shape of these speedup curves matches the predictions of an analytic model [3].

The cause of such poor speedups can be attributed to a number of factors. An important one, for the software barrier, is the amount of useful work which is performed per barrier. An amount of work between barriers which takes the same time as the barrier latency will at best result in a speedup of 50% of the number of processors. In Gauss elimination, the real culprit in this regard is the back substitution where the amount of work done between barriers scales only linearly in the number of unknowns. The back substitution, which is a small part of the execution time for Gauss elimination on a single processor, can dominate the execution time when the number of processors matches the problem size.

A second factor influencing speedups is load imbalance. If the computational work between two barriers is not equally divided among the processors, the processors with less work arrive at the barrier too soon and spin, wasting cpu cycles that could have been applied to the workload. Because the number of processors may not divide evenly into the number of *SAXPY* operations in a reduction step, and the number of *SAXPY* operations in a reduction step shrinks as the reduction proceeds, the Gauss elimination algorithm has some serious load imbalance problems when the number of processors approaches the number of unknowns.

A third factor affecting speedups is the growth of shared memory latency as the number of processors is increased. Fortunately, on the Cerberus machine the latency of shared memory increases only logarithmically with the number of processors. For the machine configuration we used, the base of this logarithm is 2 due to the 2×2 switch nodes used to construct the memory subsystem. Memory subsystems, that are practical to construct and cost effective, could be constructed using 4×4 and 8×8 switch nodes. These would reduce the growth rate of shared memory to a base 4 or base 8 logarithm and substantially reduce the memory latency for larger machine sizes. Because of the rather extreme 32 way loop unrolling, and a minimum memory latency ranging from 7 clocks for two processors to 15 clocks for 32 processors, the growth of memory latency as the number of processors is increased has little effect on our speedup results. The only exception to this would be in the final stages of the reduction, where vector lengths are rather short and in the back substitution which is not vectorized.

One could try and restructure the algorithm, say by dividing the *SAXPY* operations up into multiple vector operations of smaller length. This could be used to improve speedup by reducing load imbalance, but would also reduce processor performance because of the shorter vector lengths. There is a trade off here, and we have not investigated whether the trade would generate any profits.

By comparing the speedup curves in figure 1 with those of figure 2, we can see the effects of high latency in the software barrier. The latency of the hardware barrier is one clock and the differences in the speedups between these two sets of data is solely due to the barrier latency. The speedups presented in figure 2 are much better than those in figure 1, but still show the effects of load imbalance.

The rather poor speedups shown in figures 1 and 2 are to be compared with the near linear speedups obtained with a 128 processor BBN Butterfly™ machine [4]. The data in [4] involved a problem with 1200 unknowns in contrast with our test problems of 32, 64, and 128 unknowns. The BBN machine utilizes microprocessor technology which does not efficiently pipeline computations. The relative synchronization costs on such a microprocessor based machine are much lower than those on Cerberus which efficiently pipelines the *SAXPY* operation at one element every 5 clocks.

4. Parallel Gauss-Jordan Elimination. Gauss-Jordan elimination is very similar to Gauss elimination except that instead of reducing the matrix operator to an upper triangular form, we completely diagonalize it. The big gain, for parallel computation, is that the back substitution is now a completely trivial set of independent divides to solve for the x_i. The high cost of barrier synchronization in the back substitution has been completely removed. In the reduction, the number of *SAXPY* operations remains constant as the reduction steps proceed and this gives a much better load balance as compared to Gauss elimination. We pay fcr this more suitable parallel algorithm in operation count; roughly 50% more operations are done for Gauss-Jordan elimination than for Gauss elimination.

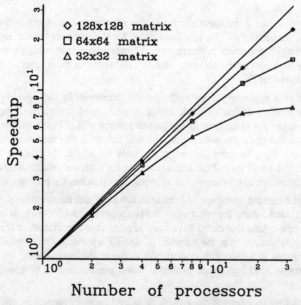

Fig. 3. Speedup of Gauss-Jordan elimination using the Butterfly barrier.

5. Comparison of Gauss and Gauss-Jordan Elimination. The speedup curves for Gauss-Jordan elimination using the software and hardware barrier are shown in figures 3 and 4, respectively. Comparing these curves, we see that the speedups of the Gauss-Jordan algorithm are much better than those of the Gauss algorithm. The differences in speedup are quite startling for the measurements using the software barrier. Speedup curves alone are misleading; we also need to compare the absolute running times of the algorithms. The Gauss-Jordan algorithm requires 50% more operations, and a better speedup on more work is not always a good thing. We compare, in figures 5 and 6, the absolute running times of the two algorithms versus the number of processors, for software and hardware barrier support, respectively. The ordinate in the graphs is the running time multiplied by the number of processors, so that an *ideal linear*

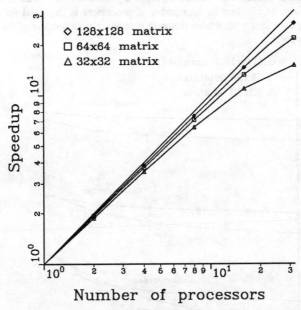

Fig. 4. Speedup of Gauss-Jordan elimination using the hardware barrier.

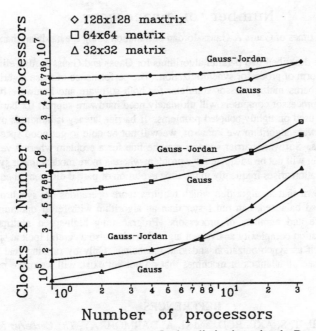

Fig. 5. Clock times of Gauss & Gauss-Jordan elimination using the Butterfly barrier.

speedup would appear as a horizontal line in the plot.

As can be seen, neither algorithm provides a perfect linear speedup; the upward growth in the curves indicates the inefficiency as the number of processors is increased. The Gauss-Jordan curves grow at a slower rate indicating a slower increase in inefficiency as the number of processors is increased. This is because the Gauss-Jordan algorithm is more load balanced and fewer barriers are required. With one processor the Gauss-Jordan algorithm requires 50% more execution time, as predicted from examining operation counts, but the better speedups in the Gauss-Jordan algorithm cause it to catch up to and exceed the performance of the Gauss elimination as the number of processors is increased for a fixed problem size. In figure 6, the cross-over points have been pushed towards the right because of the lower overhead of the hardware barrier support.

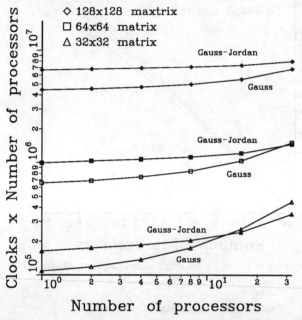

Number of processors

Fig. 6. Clock times of Gauss & Gauss-Jordan elimination using the hardware barrier.

6. Discussion. We have described two parallel algorithms for Gauss and Gauss-Jordan elimination using the barrier as the basic form of processor synchronization. The performance of these algorithms has been examined using the Cerberus multiprocessor simulator for both software and hardware barrier support. Future high speed multiprocessor computers will ultimately need hardware support for barrier synchronization, if they are to be used on tightly coupled problems. If barrier latency is hundreds of clock cycles, as it is for the best software algorithm we know of, we will not be able to get good speedups on small tightly coupled problems. Software barrier support may be fine for a problem where a very large linear system is solved once, but will not be adequate for a problem where a more modest linear system is solved many times. Such a situation arises frequently in the time evolution of partial differential equations.

We have also shown that an algorithm which requires more operations but synchronizes less frequently, and is more load balanced, can run faster than an algorithm with fewer operations for certain choices of problem size and numbers of processors. Programming high-speed multiprocessors will require not only an operation complexity analysis of algorithms, but also a careful look at specific parallel programming issues such as synchronization and load balancing. Only by running real codes on real machines, or very accurate simulations of machines that do not yet exist, will we make progress in this area.

REFERENCES

[1]   E. D. BROOKS III, T. S. AXELROD, and G. A. DARMOHRAY, *The Cerberus Multiprocessor Simulator*, Proc. Third SIAM Conference on Parallel Processing, SIAM, 1987.

[2]   E. D. BROOKS III, *The Butterfly Barrier*, International Journal of Parallel Programming, **15** (4) (1986), pp. 295-397.

[3]   T. S. AXELROD, *Effects of synchronization barriers on multiprocessor performance*, Parallel Computing, **3** (1986), pp. 129-140.

[4]   W. CROWTHER, et al., *Performance measurements on a 128—node Butterfly™ parallel processor*, Proc. 1985 International Conference on Parallel Processing, IEEE, 1985.

CHAPTER 5

Orderings for Parallel Sparse Symmetric Factorization

Charles E. Leiserson*
John G. Lewis†

Abstract. The primary requirement for computing the Cholesky factor of a sparse symmetric positive definite matrix in parallel is to find a reordering that produces a shallow elimination tree with modest storage requirements. To date most approaches to this problem have been adaptations of sequential ordering heuristics. We have attacked the parallel ordering problem directly by applying graph bisection heuristics to general sparse matrices to create nested dissection orderings that have short elimination trees and small space requirements. We report on the empirical effectiveness of our heuristics.

1. Overview. The parallel computation of the Cholesky factorization of a sparse matrix must satisfy the data dependencies among the columns of the Cholesky factor. The *elimination tree* (cf [5,7]) of the factored matrix exhibits in a minimal form these dependencies. In a parallel model where the elimination of a column is a unit time operation, the depth of the elimination tree is essentially the parallel completion time for column or submatrix Cholesky algorithms. In several other important parallel contexts (e.g. [5]), a shallow elimination tree is also critical.

We have developed a heuristic algorithm for finding orderings that result in shallow elimination trees. We generalize George's nested dissection algorithm [2] by finding nodal separator sets through a novel extension of standard graph bisection heuristics. The orderings yield shallower elimination trees than those previously published.

Graph bisection is a partitioning of the nodes of a graph into two sets $\hat{P}_1$ and $\hat{P}_2$, each containing half the nodes, with a minimum number of *cut edges* spanning $\hat{P}_1$ and $\hat{P}_2$. This problem is known to be NP-complete. It is, however, an important problem in VLSI-layout design, where a number of different heuristic algorithms have been proposed. In this work we use the heuristic by Fiduccia and Mattheyses [1], with orthogonal starting vectors [4] to control convergence to local minima.

There are four features incorporated in our heuristics. We contribute new techniques for transforming edge separators into node separators. We introduce the use in

* Laboratory for Computer Science; Massachusetts Institute of Technology, Cambridge MA 02139 and Thinking Machines Corp., 245 First Street, Cambridge MA 02142. This research was supported by Thinking Machines Corp.

† Boeing Computer Services Company, Mail Stop 7L-21, P.O. Box 24346, Seattle WA 98124. This research was supported in part by the Air Force Office of Scientific Research under contract F49620-85-C-0057. This work was begun while this author was a Boeing Fellow in the Center for Advanced Engineering Study at M.I.T. and a visiting researcher at Thinking Machines Corp.

bisection of an *adjacency model* for the sparse matrix rather than the usual incidence model. Our heuristics use the standard technique of allowing bounded imbalance between the two partitions. We also consider means of incorporating the boundaries of subgraphs in the dissection, but we do not report on these.

2. Obtaining Nodal Separators from Edge Separators. Graph bisection creates an *edge* separator: a set of edges whose removal from the graph splits the graph into two disconnected pieces. Dissection requires a *node* separator: a set of nodes whose removal splits the graph into at least two disconnected components. In this section we consider several different ways to transform an edge separator into a node separator.

Suppose that the bisection of the graph has resulted in the partitioning $\{\hat{P}_1, \hat{P}_2\}$. Let S_1 be the set of nodes $s \in \hat{P}_1$ such that at least one edge incident on s is cut by the bisection. Similarly, let S_2 be the subset of $\hat{P}_2$ that is adjacent to $\hat{P}_1$. The removal of $S = S_1 \cup S_2$ removes all the cut edges and exactly the nodes that are the endpoints of the cut edges. Thus, S is a node separator, separating $P_1 = \hat{P}_1 - S_1$ and $P_2 = \hat{P}_2 - S_2$.

Technically S is a *wide separator*, since a path from any node in $\hat{P}_1 - S_1$ to $\hat{P}_2 - S_2$ must be of length at least three, passing through S_1, across a cut edge, and then through S_2. Although such wide separators are of use in solving certain sparse problems (cf [3]), a *narrow separator* with crossing paths of length two is all that is necessary for the Cholesky factorization. A smaller separator is better because the size of the separator is the increment made at this step to the elimination tree height. We consider three different heuristics for choosing a narrow separator from the wide separator S:

- choosing an obvious narrow separator,
- creating a narrow separator by removing nodes of minimum cut degree,
- creating a narrow separator by including nodes of maximum cut degree.

Either S_1 or S_2 alone is a separator since the removal of either would also remove all of the cut edges connecting P_1 and P_2. We shall refer to the choice of the smaller of S_1 and S_2 as the *obvious narrow separator*. The use of the obvious narrow separator and the wide separator are discussed in [3].

The wide separator S may include a narrow separator N that is smaller than either S_1 or S_2, or one that is the same size and results in better balance. We have constructed two heuristics for finding such a narrow separator N by *covering* all of the cut edges — at least one of the endpoints of each cut edge must appear in the separator set. We begin with the node separator S, separating $P_1 = \hat{P}_1 - S_1$ and $P_2 = \hat{P}_2 - S_2$. Consider the wide separator S as a bipartite graph consisting of S_1 and S_2 and the cut edges connecting them.

The *minimal removal* heuristic is based on *removing* nodes from S without destroying the separator property. Remove the node x that has the smallest cut degree (the smallest number of cut edges), and add $x \in S_i$ to the corresponding partition P_i. The set of nodes adjacent to x in S are included in N and removed from S, since all edges incident on x must be covered. The number of nodes added to N is minimized locally. In case of ties, attempt to preserve balance by removing a node x so that it will be added to the partition P_i that currently has the fewest nodes.

The *maximal inclusion* heuristic tries to cover as many cut edges as possible at each step. In the same bipartite graph, add to the separator N the node y with the largest cut degree. That is, cover as many edges as possible at this step. Any neighbor of y that has no other neighbors is removed from further consideration, and added to the appropriate partition P_i. Again, break ties to preserve balance by choosing y on the side of S whose partition P_i currently is largest.

Neither of these heuristics is guaranteed to be optimal or even to produce a narrow separator that is as small or smaller than the obvious narrow separator. (A variation in the minimal removal heuristic is always at least as good as the obvious narrow separator, but empirically produces poorer orderings than the heuristic as stated.) In practice the heuristics produce separators that are often better and almost always at least as good as the obvious narrow separator. It is easy to determine if the obvious narrow separator is smaller, so our standard minimal removal scheme heuristic uses the obvious narrow separator in the very few cases where it is smaller. The maximal inclusion heuristic would make this substitution more often, but performs less well by doing so.

3. Graph Models for Graph Partitioning. Graph bisection algorithms commonly used in VLSI layout minimize cut 'signals' or 'nets' rather than edges. A net is essentially a hyperedge — more than two 'cells' can be connected by the same net, which corresponds to a signal passing from one of these cells to the rest. The obvious model for the bisection algorithm is the usual undirected *incidence graph* $G = \{X, E\}$ of the sparse matrix. The nodes X of the graph are identified with cells, and edges E with nets.

We have chosen instead to use an *adjacency graph* model for the sparse matrix. The cells are the nodes, but we define each net to be a hyperedge connecting each member of $\{x\} \cup \text{Adj}(x)$, with x as the representative of this net. A net is cut if any two of the nodes connected by the net lie in different partitions. Clearly, if an edge is cut in the ordinary graph model, two nets, corresponding to the endpoints of the edge, are cut in this model. Conversely, if a net is cut, then the representative node x of the net must lie in a different partition than one of the other nodes. Hence, there is an edge connecting these two nodes in the original graph, and so there exists at least one cut edge corresponding to the cut net.

The adjacency model has the important characteristic that the representatives of the cut nets are exactly the nodes in the wide separator. Thus, minimizing the number of cut nets minimizes the size of the wide separator. In contrast, the incidence model allows wide separators of differing sizes for the same number of cut edges. The adjacency model does not directly produce mimimal narrow separators, but it performs better in practice. Further, the number of nets is lower, $|X|$ instead of $|E|$, which produces faster running times.

4. Balancing Partitions during Generalized Nested Dissection. To create a nested dissection ordering, we recursively partition the graph of the sparse matrix, obtaining a dissection tree in which the internal nodes correspond to the separators. The height of the dissection tree, weighted by the size of each dissection tree node, is the height of the corresponding elimination tree. Strict bisection yields a balanced dissection tree, but its separators often are larger than necessary, giving a deeper elimination tree. However, badly balanced partitions typically lead to deep dissection trees, also giving deep elimination trees even if the separators are small. Striking the proper balance by allowing some flexibility in the partitioning leads to a reasonably balanced dissection tree with relatively small separators and, hence, a good elimination ordering.

In our graph bisection heuristic, we use a lower bound α on the relative size of the smaller partition P_1, and require that each intermediate partitioning satisfy $\alpha \leq \frac{|P_1|}{(|P_1|+|P_2|)} \leq \frac{1}{2}$. The lower bound α is a parameter to our ordering procedure; by decreasing α we allow the partitioning to become less balanced. There is no reason to think that a single choice of α is optimal for all problems. Further, the heuristic is at best a local optimization, so it is possible for the result of a more flexible partitioning to be worse on both balance and separator size. This occurs relatively infrequently, particularly as the bisection heuristic is normally applied to several different starting partitions, but this behavior does not make the proper choice of α any more obvious. In our empirical study we resorted to testing a small number of different ratios to determine the extent of variability.

5. Empirical Effectiveness. The reason for implementing a generalized nested dissection ordering procedure based on graph bisection is to determine if such an ordering could produce better orderings for parallel computation of the Cholesky factorization. Our evaluation is against the current best sequential ordering heuristic: Liu's multiple minimum degree (MMD) ordering, followed by an application of the Jess and Kees algorithm to minimize elimination tree height [6]. Liu [6] presents statistics for a collection of eleven sparse matrices from the Harwell-Boeing sparse matrix collection, which we use as our benchmark.

The paramount question is whether a single nested dissection heuristic can produce better orderings. There are a number of unspecified parameters in our procedure. We have options for:

- the graph model (adjacency, incidence)
- narrow separator selection (minimal removal, maximal inclusion, obvious narrow separator)
- minimum partition size ratio ($\alpha = .49, .45, .41, .37$ and $.33$)

A systematic comparison of these and several other alternatives resulted in 90 different orderings for each of the eleven test problems. No single combination of parameters was universally best, but an analysis of these orderings led to a single procedure that performs well in general and never performed badly in these tests. The adjacency model was selected over the incidence model because the latter leads occasionally to very poor orderings. The minimal removal heuristic was favored narrowly over the maximal inclusion heuristic, while the obvious narrow separator heuristic was less good generally and was particularly bad in some cases. The combination of the adjacency model and the minimal removal heuristic was always better than MMD for all partition ratios tested. A partition ratio of .45 was chosen narrowly over .41 as a general choice; the best choice is strongly problem dependent.

Table 1 is a brief description of the test problems. In Table 2 we compare the

TABLE 1
Test Matrices from the Harwell-Boeing Sparse Matrix Collection

key	description	order
BCSPWR09	Western US power grid	1723
BCSPWR10	Eastern US power grid	5300
BCSSTK08	TV Studio structure	1074
BCSSTK13	Fluid Flow Stiffness Matrix	2003
BCSSTM13	Fluid Flow Mass Matrix	2003
BLCKHOLE	Geodesic Dome	2132
CAN 1072	Aircraft structure	1072
DWT 2680	Naval destroyer structure	2680
LSHP3466	L-shaped regular grid	3466
GR 40x40	40 by 40 square grid	1600
GR 80x80	80 by 80 square grid	6400

TABLE 2
Ratio of Factorization Costs (Nested Dissection versus MMD)

key	tree height	nonzeros	operations
BCSPWR09	0.76	1.22	1.17
BCSPWR10	0.87	1.24	1.44
BCSSTK08	0.83	1.16	1.31
BCSSTK13	0.79	1.02	0.98
BCSSTM13	0.90	0.99	0.90
BLCKHOLE	0.75	0.93	0.79
CAN 1072	0.87	1.17	1.37
DWT 2680	0.61	1.27	1.73
LSHP3466	0.67	0.94	0.84
GR 40x40	0.66	0.95	0.84
GR 80x80	0.65	0.93	0.77
geometric mean	0.75	1.07	1.06

nested dissection ordering from our designated heuristic — adjacency model, minimal removal separator, $\alpha = .45$ — with the MMD ordering. The data in the table are the elimination tree height, as a parallel complexity measure, and the usual sequential measures of storage and multiplications. The values reported are the values for our heuristic divided by those from MMD. The results are unambiguous. Even without an follow-up application to our orderings of the Jess and Kees algorithm, the tree heights are 25% smaller on average than the MMD tree heights with the Jess and Kees algorithm. Moreover, the sequential measures are very close to those from MMD. Clearly, a nested dissection heuristic can provide good orderings for parallel factorization.

REFERENCES

[1] C. FIDUCCIA AND R. MATTHEYSES, *A linear-time heuristic for improving network partitions*, in Design Automation Conference, Jan. 1982.

[2] J. A. GEORGE, *Nested dissection of a regular finite element mesh*, SIAM J. Numer. Anal., 10 (1973), pp. 345–363.

[3] J. R. GILBERT AND E. ZMIJEWSKI, *A Parallel Graph Partitioning Algorithm for a Message-Passing Multiprocessor*, Tech. Rep. 87-803, Cornell University, Computer Science Department, Jan. 1987.

[4] M. K. GOLDBERG, S. LATH, AND J. W. ROBERTS, *Heuristics for the graph bisection problem*, 1987, Rensselaer Polytechnic Institute.

[5] J. W. LIU, *Computational models and task scheduling for parallel sparse Cholesky factorization*, Parallel Computing, 3 (1982), pp. 327–342.

[6] ———, *Reordering Sparse Matrices for Parallel Elimination*, Tech. Rep. CS-87-01, York University, 1987.

[7] ———, *The Role of Elimination Trees in Sparse Factorization*, Tech. Rep. CS-87-12, York University, 1987.

CHAPTER 6

Performance of Blocked Gaussian Elimination on Multiprocessors

Elizabeth J. O'Neil[*]
Henno Allik[†]

Abstract. A performance analysis of blocked Gaussian elimination of a symmetric positive-definite system on a simple MIMD model is presented. The adopted MIMD model assumes a high bandwidth switch or network with bounded bandwidth ports to global and local memories. The blocking is shown to reduce global references to a non-interfering level, under certain mild restriction, so that synchronization and block-overhead are the important performance considerations. The analysis explains our experimental results on the BBN Butterfly multiprocessor: both the strongly limited speedup of the row-based algorithm and the far more extensive speedup of the block-based algorithm.

Introduction. Blocked methods in linear systems provide algorithms rich in matrix-matrix multiplication and similar operations, concentrating the computation into tasks dependent on a minimal amount of the given data. Historically this property was exploited in small-memory uniprocessors, but it is equally relevant to current multiprocessors, where otherwise parallel algorithms suffer because of excessive internode communication. Our work deals with a blocked version of Gaussian elimination (GE), which is used in the finite element code we have implemented on the Butterfly Parallel Processor. Actually our first implementation involved a row-based GE [1], and the switch to the blocked algorithm was an attemt to improve prformance at small bandwidths as reported in [2]. Here we present a detailed analysis of blocked GE for positive definite symmetric matrices, possibly banded, and show correspondence with actual Butterfly results.

The MIMD model. The multiprocessor has P processors, each with enough local memory for code and a moderate amount of data, and M globally accessible memories. The processors and memories are interconnected with a high bandwidth switch or network providing service bounded by data rate r matrix elements per unit time to/from each globally accessible memory, and similarly rate r to/from each processor-local memory unit. Thus the switch bandwidth is required to be at least max(rP,rM), as provided by, for example,

---
[*] Bolt, Baranek, and Newman, Inc., and Department of Mathematics and Computer Science, University of Massachusetts, Boston
[†] BBN Laboratories, Inc.

the BBN Butterfly switch. The transfers are assumed to have negligable startup/shutdown costs.

The Blocked Algorithm. We divide the $n \times n$ matrix into $s \times s$ blocks, thus obtaining an $(n/s) \times (n/s)$ block matrix. If it is banded with half-band size b, we have b/s blocks across the band: if not, let $b = n$ in the following. The blocked algorithm does a Gaussian elimination (GE) on the pivot block and equivalent operations on the other pivot-row blocks (b/s tasks, asymptotically), and then uses the updated pivot row blocks in the appropriate matrix operation "MOD" on the triangle of blocks below them ($.5(b/s)^2$ tasks). In terms of data-dependence and task precedence, the block operations follow the element-wise GE operations exactly (although element-wise GE is not practical because the tasks are too small). The longest path through this precedence graph is $2(n/s)$, and it is straighforward to devise a schedule, since there are many more MODs than GEs, and only some of the MODs need to precede the next set of GEs:

```
-----------------------------------------
| GE1 | MOD1 | GE2 | MOD2 | GE3 | ...
------                -------
        | MOD1  MOD1 | MOD2  MOD2 | ...
-----------------------------------------
```

Parallel schedule for GEs and MODs for pivot blocks 1, 2, 3,...

The computational complexity of the GEs is $.5ws^3$, and the MODs ws^3, where w is the uniprocessor time for an element-element computation using local memory. Thus 2 GEs can fit across one MOD, approximately, and we can apply more and more processors until the above picture has two MOD tasks in the time-width for each pivot block. This provides the basic synchonization constraint for this method: the number of MOD-sized tasks associated with a pivot block must be at least $2P$. Asymptotically, there are $.5(b/s)^2$ such tasks, so the synchronization condition is $4P <= (b/s)^2$. this constraint argues for small s, i.e. small blocks, to allow more processors to be applied to the work, but both block overhead and communications costs argue for large s, as considered below.

Communications characteristics: a square-cube law. Consider an individual task: it takes ws^3 time, for which it needs 3 blocks transferred from global memory, $3s^2$ data. Thus the required data rate from switch to local memory is $3/(ws)$, which can be quite small for well chosen s – this is the square-cube result underlying the good properties of the block method. Even for small bounds r, we can choose s large enough to maintain $3/(ws) <= r$, as required by the stated switch model.

Now consider switch to global memory transfers. Some blocks are "hard-hit" by the algorithm, in particular, the pivot row blocks. Each of these is referenced by up to $2(b/s)$ tasks. Assuming otherwise perfect speedup, we can show that the data rate out of such a memory (with one or more such blocks) reaches $O(P/(bw))$ if whole blocks are stored in individual memories. Applying the bound in the switch model we see that $P <= O(rwb/2)$, a further, unwanted, restriction on the solution domain, linear in b. The problem here is that, for the usual case of $M > b/s$,some memories are underutilized in this bottleneck situation. Our solution was to scatter the rows of the blocks across memories: the resulting requirement is $P <= (swr)min(M,b)/4$, a much weaker condition. In the Butterfly case that $M = P$, it simplifies to $P <= O((rw)^{2/3}b^{4/3})$, a bound allowing superlinear behavior in b.

A More Realistic Computation Model. Having derived encouraging communications bounds, we return to the computation, later to check the bounds consistency. The block structure itself entails considerable calling overhead, even when run on a uniprocessor. In multiprocessor runs, the task overhead comes in on a block basis. Accordingly, we may approximate the work of a task as $ws^3 + w_1$, for some constant w_1, compared to the ideal ws^3, so the speedup efficiency is bounded by the ratio. The synchronization restricts the speedup to $b^2/(4s^2)$. Thus the speedup is less than the minimum of these, but we are free to choose s to maximize this minimum. The optimal s chosen this way is a function purely of the quantity P/b^2, and over a wide range, it is approximately $.5b/\sqrt{P}$, as expected from the synchronization bound: we simply choose the largest blocksize which provides 2P tasks in the active triangle for each pivot block.

Checking back to the pivot row access constraint, we see that as long as the pivot row blocks are themselves scattered by rows, this bound remains true over a wide range, including that relevant to the Butterfly runs.

Experimental Results. The blocked algorithm was run as the solver for a static finite element analysis program, as reported in [2], on a BBN Butterfly multiprocessor configured with 64 processors. Results for three problem sizes are shown in Fig.1, together with row-

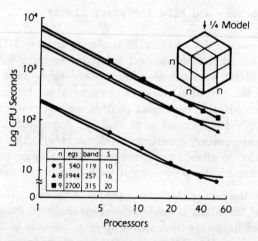

Figure 1: Performance of Blocked GE vs. Row-based GE.

based GE results taken from [1]. Note that the blocked results are for the particular block size size shown in the figure, namely the optimum block size for 55 processors. Speedup results for the 5x5x5 element FEM problem ($b = 119$) is presented in Figure 2 for various block sizes. In contrast, the results for the row-based algorithm shows poor speedups for $P >= \sqrt{b}$, far below the synchronization bound of $P <= b/2$. In particular, speedup is approximately 60when $P/(\sqrt{b}) = 4$, by Figure 6 of [2]. This can be explained in terms of the simple MIMD model as follows. Consider the memory containing the pivot row, itself containing b elements. In doing the calculations associated with one pivot, the tasks on all P processors obtain a copy of this pivot row, reading Pb elements, and taking at least Pb/r time. The total work for the pivot is $wb^2/2$, and if shared equally among P processors, must take time at least $wb^2/(2P)$ time. If the former time is comparable to the latter, the speedup will suffer significantly, and this happens when $P = O(\sqrt{b})$ in terms of b, as observed.

Finally we show in Fig.3 that the combination of the synchronization bound and the block overhead effect explains all of the data we have obtained so far, on runs with band b up to 315.

Speedup vs. Block Size: 5x5x5 Model

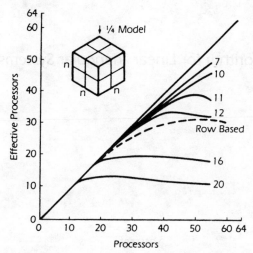

Figure 2: Speedup Curves for various block sizes in a single problem

Blocked GE Speedup Analysis

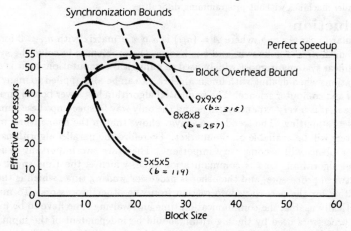

Figure 3: Comparison of Theoretical Bounds and Experimental Results

REFERENCES

(1) H. Allik, S. Moore, E. O'Neil, and E. Tenenbaum, *Finite Element Analysis on the BBN Butterfly Multiprocessor*, Computers and Structures, 27, No.1, (1987), pp. 13-21.

(2) E. O'Neil, H. Allik, S. Moore, E. Tenenbaum, *Finite Element Analysis on the BBN Butterfly Multiprocessor*, Proceedings of the Second International Conference on Supercomputing, 1987.

CHAPTER 7

A New Parallel Algorithm for Linear Triangular Systems

Avi Lin[*]
He Zhang[*]

Abstract

This paper presents a near optimal parallel algorithm for solving linear triangular systems. Some of the more important features of this algorithm are: (1) It works well for any number of processors. (2) It is especially good for relatively small number of processors for which it achieves optimal speed up, optimal efficiency and low communication complexity. (3) It can be used in both distributed parallel computing environments and tightly coupled parallel computing systems. (4) Generally, it can be mapped onto most parallel machines without programming difficulties.

1   Introduction

Given a triangular linear $Ax = b$, where $A = (a_{ij})$ is a $n \times n$ matrix with $a_{ij} = 0$ for $j > i$, x and b are vectors of length n, x is unknown, and b is known, the problem is to solve the system for x.

Triangular linear systems are most widely used in numerical computations as it is one of the basic numerical operations when dealing with linear algebra. It can be also applied to many other branches in mathematics and computer science. The sequential algorithm has lower bound time complexity of $O(n^2)$. For large n, it is a very expensive time, and the only way to speed up these computations is by means of parallel computing. The trend nowadays shows that in the near future, more degenerate parallel computers will be available on the market. Therefore the parallel algorithms for many basic operations or problems will become very important. There are two important issues in design an efficient parallel algorithm. One is communication time, which is the time spent on transferring information between processors, and the other is processor waiting time, which is the processor idle time while waiting for the necessary information from the other processors. To make the parallel algorithm practically usable, the communication time and waiting time have to be minimized, while the number of processors used by the algorithm should be independent of the input size n. Several parallel algorithms for the triangular linear system problem have been derived in the past (see [2,3,4]) but, most of them have the following main disadvantages:

(1) Large amount of processors are needed. (2) Communication complexity is very large.

It is already well recognized that these deficiencies make the parallel algorithm useless. Like many other problems, the very simple sequential algorithm makes the efficient parallel algorithm very difficult to design. In this study, PE denotes the processor element, also P denotes the number of PEs available in the parallel computing system, which are indicated by $PE1$, $PE2$, $\cdots$, PEP. The present parallel computation model consists of these P PEs, and PEi is locally connected to PE_{i-1} and PE_{i+1}, for $2 \leq i < P$, where the last two are the neighbors of PEi, i.e., they form a multiprocessor ring, see [1] for details.

[*]Department of Mathematics, Temple University, Philadelphia, PA 19122

2   The Parallel Algorithm for Triangular System

Let us write first the sequential algorithm.
$x_1 = \frac{b_1}{a_{11}}$; $x_k = \frac{b_k}{a_{kk}} - \sum_{i=1}^{i=k-1} \frac{a_{ki}}{a_{kk}} x_i$; for $k = 2, 3, \cdots, n$. As a general goal in parallel computations, the idea is to let each PE perform the same amount of computation between two successive communication operations, and to make the redundant computations as small as possible. The present algorithm is executed in $m_p + 1$ phases, and there are 2 steps in each phase.

Parallel Algorithm: Assume $k = \{k_i\}_{i=1}^{m_P}$ is a sequence of positive integers, such that $\sum_{i=1}^{m_P} k_i < n$.
Let $h_i = \frac{n - \sum_{i=1}^{j=i} k_j}{P-1}$, and $c_i = \sum_{j=1}^{j=i} k_j$.

The $1'st$ phase

Step 1: $PE1$ calculates $x_1, x_2, \cdots, x_{k_1}$, using the above sequential algorithm.

For $j = 2, 3, \cdots, P$ do (in parallel): PEj evaluates $b_i^{(1)} = b_i - a_{i1}x_1$, for index i in the range of $k_1 + 1 + \lceil (j-2)h_1 \rceil \leq i \leq k_1 + \lceil (j-1)h_1 \rceil$, where k_1 is determined from the equation:

$$\frac{k_1(k_1 + 1)}{2} = \frac{(n - k_1)}{P - 1}. \tag{1}$$

Step 2: $PE1$ transmits $x_1, x_2, \cdots, x_{k_1}$ to all other $P - 1$ PEs, and the value of $b_{k_1+1}^{(1)}$, $b_{k_1+2}^{(1)}$, $\cdots$, $b_{k_1+k_2}^{(1)}$ are transferred to the $PE1$, where k_2 is determined from the following recursive equation for k_i with $i \geq 2$.

$$k_i k_{i-1} + \frac{k_i(k_i + 1)}{2} = \frac{k_{i-1}}{P - 1}(n - k_1 - k_2 - \cdots - k_{i-1}) \tag{2}$$

Also the rows of A matrix that correspond to those $b_i^{(1)}$'s are transferred to the $PE1$.
Let $s(i,j) = c_i + \lfloor (j-2)h_i \rfloor$ and $e(i,j) = s(i,j) + \lfloor h_i \rfloor$, then the values of $b_{s(2,j)+1}, b_{s(2,j)+2}, \cdots,$ $b_{e(2,j)}$ are passed to the PEj, and also the corresponding rows of A matrix are passed to the PEj, for $j = 2, 3, \cdots, P$.

The ith phase, for $2 \leq i \leq m_P$

Step1: $PE1$ calculates $x_{c_{i-1}+1}, x_{c_{i-1}+2}, \cdots, x_{c_i}$ using the newly updated $b^{(i-1)}$s. Parallel to $PE1$, PEj (for all j, $2 \leq j \leq P$) computes $b_m^{(i)} = b_m^{(i-1)} - \sum_{t=c_{i-2}+1}^{c_{i-1}} a_{mt}x_t$, where index m is in the range of $c_i + \lfloor (j-2)h_i \rfloor + 1 \leq m \leq c_i + \lfloor (j-1)h_i \rfloor$.

Step 2: $PE1$ transmits $x_{c_{i-1}+1}, x_{c_{i-1}+2}, \cdots, x_{c_i}$ to all other PEs, and the values of $b_{c_i+1}^{(i)}$, $b_{c_i+2}^{(i)}$, $\cdots$, $b_{c_{i+1}}^{(i)}$ are transferred to the $PE1$ with their corresponding rows of A matrix. Similar to phase 1, the values of $b_{s(ij)+1}, b_{s(ij)+2}, \cdots, b_{e(ij)}$ are passed to the PEj with their corresponding rows of A matrix, for $j = 2, 3, \cdots, P$.

The $(m_P + 1)'th$ phase

$PE1$ finishes all the remaining eliminations.

The equations (1) and (2) above are obtained based on the concept that the processor waiting time is minimized. The left hand side of the equation (1) is the computation time of $PE1$, and the right hand side of it is the computation time of PEj for $j = 1, 2, \cdots, P$.

To implement the algorithm, the integer sequence k has to be calculated by an appropriate outside procedure. The running time for the procedure is very small, since $m_P = O(P log(n))$, as will be proved in next section. The following table gives some numerical results for $P = 2$ case.

n	i	1	2	3	4	5	6
10^4	k_i	139	1400	2814	2337	1434	820
	i	7	8	9	10	11	
	k_i	462	260	146	82	46	
n	i	1	2	3	4	5	6
10^3	k_i	43	213	279	197	116	66
	i	7	8	9			
	k_i	37	21	12			

3 Analysis of The Recursive Sequence k

In the step 2 of each phase of the parallel algorithm, the informations are transmitted and spread among the processors. So it is easy to see that the number m_P determines the computation complexity. To obtain the order of m_P, let's study a slightly different recursive system for u, which is defined by the following equations:

$$\frac{u_1^2}{2} = \frac{n - u_1}{P - 1} \tag{3}$$

$$u_{i-1}u_i + \frac{u_i^2}{2} = \frac{u_{i-1}(n - \sum_{j=1}^{i} u_i)}{P - 1}, \ (for \ i \geq 2) \tag{4}$$

Define: $N_i = \frac{n - \sum_{i=1}^{i-1} u_j}{P-1}$, then we have $N_{i+1} = u_{i+1}(1 + \frac{1}{P-1} + \frac{u_{i+1}^2}{2U_i}$, and using $N_{i+1} = N_i - \frac{u_i}{P-1}$,
we get finally $u_{i+1}(1 + \frac{1}{P-1}) + \frac{u_{i+1}^2}{2u_i} = u_i + \frac{u_i^2}{2u_{i-1}}$.
Define: $t_i = \frac{u_{i+1}}{u_i}$, then the following equation which is equivalent to equation (4) is obtained:

$$\frac{t_i^2}{2} + t_i(1 + \frac{1}{P-1}) - (1 + \frac{t_{i-1}}{2}) = 0 \tag{5}$$

Solving the equation (5) for t_i, we get $t_i = \sqrt{(1 + \frac{1}{P-1})^2 + 2 + t_{i-1}} - (1 + \frac{1}{P-1})$.

Define: $\beta_i = t_i + 1 + \frac{1}{P-1}$, we have $\beta_i = \sqrt{(\frac{1}{P-1})(1 + \frac{1}{P-1}) + 2 + \beta_{i-1}}$. Now the following lemma is obvious.

<u>Lemma 1:</u> $L_\beta = \lim_{n \to \infty} \beta_i = \frac{1}{2} + \sqrt{\frac{9}{4} + \frac{1}{P-1}(1 + \frac{1}{P-1})}$

Now let's define $\mu = \sqrt{[\frac{1}{P-1}(1 + \frac{1}{P-1}) + 2]\frac{1}{L_\beta} + 1}$, and $R = log(L_\beta + \epsilon)$. The following lemma gives an estimation of the rate of convergence of the sequence $\{\beta_i\}$.

<u>Lemma 2:</u> Given $\epsilon > 0$, let $N_\epsilon = log(log(\beta_2)) - log(R)$, then $\beta_i < L_\beta + \epsilon$ for $i > N_\epsilon$.
Proof: Simply by realizing first that $log(L_\beta + \epsilon) - 2log(\mu) > 0$, and secondly, since $\beta_i \geq L_\beta$, then $\beta_i \leq \mu\sqrt{\beta_{i-1}}$.

$\square$

From the definition of β_i, it can be easily seen that $t_i < L_\beta - (1 + \frac{1}{P-1}) + \epsilon$ for $i > N_\epsilon$.

<u>Lemma 3:</u> $u_i \leq u_1$ for $i > O(Plog(n))$.
Proof: Let's choose ϵ such that $L_\beta - (1 + \frac{1}{P-1}) + \epsilon < 1 - \gamma$, where γ is a fixed number, and $0 < \gamma < 1$.
Consider the equation $L_\beta - (1 + \frac{1}{P-1} + \epsilon^* = 1$, then $\epsilon^* = 1.5 + \frac{1}{P-1} - \sqrt{\frac{9}{4} + \frac{1}{P-1}(1 + \frac{1}{P-1})} = O(\frac{1}{(P-1)})$. Letting $\gamma = \frac{\epsilon^*}{2}$ and $\epsilon = \gamma$, then it is obvious that $t_i \leq 1 - \gamma$ when $i > N_\epsilon$. That is $u_{i+1} < (1 - \gamma)u_i$ when $i > N_\epsilon$. The lemma is finally proved by observing that $log(R) = O(log(P))$ and $u^{i+k} \leq (1 - \gamma)^k u_i$ for $i > N_\epsilon$

$\square$

Theorem 1: $m_P = O(\,P log(n)\,)$.

Proof: By induction it can be shown that $\frac{k_i}{u_i} \leq 1 - n^{-\frac{3}{4}}$. Using this result, a lemma for k_i which is similar to lemma 3 for u_i can be formulated, and the result follows.

□

4   Analysis of The Algorithm

Let's define the time needed to broadcast a packet of information of size N linear relation $t_N = \alpha + \Gamma_P N$, where α is start up time, and Γ_P is a function of P. $\Gamma_P = 1$ if the information is sent to the immediate neighbors, and $\Gamma_P = P$ if the information is sent to all other $P - 1$ PEs. $T_c(n)$ is defined as the communication time of the parallel algorithm.

In the parallel algorithm, $PE1$ has to send its result to all other $P - 1$ processors at step 2 of each phase, and all the other processors send only the result to the immediate neighbors. An upper bound for $T_c(n)$ is given by the following theorem.

Theorem 2: $T_c(n) \leq O(\,Pn + \frac{n^2}{2(P-1)}\,)$.

Let's define by $T_1(n)$ the computational sequential time and by $T_P(n)$ the computational time using P PEs. Let's denote by $S_P(n) = \frac{T_1(n)}{T_P(n)}$ the speedup and by $\eta_P = \frac{S_P(n)}{P}$ the efficiency.

In our parallel algorithm, each PE has the same computation load in each of the m_P phases. So the speedup is optimal. The maximum speedup is obtained when $P = O(\sqrt{n})$, this can be easily see from theorem 2.

References

[1] Ipsen, I., Saad, Y., Schultz, M., "Complexity of Dense-linear-System Solution on a Multiprocessor Ring", *LINEAR ALGEBRA AND ITS APPLICATIONS* 77:205-239, 1986.

[2] Lin, A., "A Parallel Algorithm for Linear Triangular Systems", to appear in the Journal of the ACM, 1988.

[3] Lin, A., "On a Parallel Algorithm for Inherent Serial Techniques", Linear Algebra and Applications, 79, *pp.* 229 − 236, 1986.

[4] James M. Ortega, "Introduction to Parallel and Vector Solution of Linear Systems", Plenum Press, 1988.

CHAPTER 8

A Proposal for a Set of Level 3 Basic Linear Algebra Subprograms

Jack Dongarra*
Jeremy Du Croz†
Iain Duff‡
Sven Hammarling†

Abstract. This paper describes a proposal for Level 3 Basic Linear Algebra Subprograms (BLAS). The Level 3 BLAS are targeted at matrix-matrix operations with the aim of providing more efficient, but portable, implementations of algorithms on high-performance computers, especially those with hierarchical memory and parallel processing capability.

1. Introduction. The original basic linear algebra subprograms, now commonly referred to as the BLAS [4] and fully described in [10], have been used in a wide range of software including LINPACK [5] and many of the algorithms published by the *ACM Transactions on Mathematical Software*. In particular, they aid clarity, portability, modularity, and software maintenance, and they have become a *de facto* standard for the elementary vector operations.

An extended set of Fortran BLAS aimed at matrix-vector operations (Level 2 BLAS) was subsequently proposed by Dongarra, Du Croz, Hammarling, and Hanson [7]. The Level 2 BLAS achieve efficiency on vector-processor machines by keeping the vector lengths as long as possible, and in most algorithms the results are computed one vector (row or column) at a time. In addition, on vector register machines, performance is increased by reusing the results of a vector register, and not storing the vector back into memory.

Unfortunately, this approach to software construction is often not well suited to computers with a hierarchy of memory and to true parallel-processing computers (see [8]). For those architectures it is often preferable to partition the matrix or matrices into blocks and to perform the computation by matrix-matrix operations on the blocks. This approach allows full reuse of data while the block is held in the cache or local memory. It also avoids excessive movement to and from memory and gives a surface-to-volume effect for the ratio of operations to data movement. In addition, operations on distinct blocks may be performed in parallel; and within the operations on each block, scalar or vector operations may be performed in parallel.

---

*Mathematics and Computer Science Division, Argonne National Laboratory, Argonne, Illinois 60439-4844. Work supported in part by the Applied Mathematical Sciences subprogram of the Office of Energy Research, U.S. Department of Energy, under Contract W-31-109-Eng-38.

†Numerical Algorithms Group Ltd., NAG Central Office, Mayfield House, 256 Banbury Road, Oxford OX2 7DE.

‡ Computer Science and Systems Division, Harwell Laboratory, Oxfordshire OX11 ORA.

The proposed Level 3 BLAS are targeted at the matrix-matrix operations required for these purposes. Algorithms implemented by calls to the proposed routines can, we believe, be both portable and efficient across a wide variety of vector and parallel computers. Certainly, the efficiency of such algorithms is well documented (see [6] for references). The question of portability has been much less studied, but we hope to encourage research into this aspect.

This proposal does not include any routines for matrix factorization; these are covered by LINPACK [5], and we continue to advocate the use of the LINPACK calling sequences as a standard interface, even if the details of the underlying algorithm are modified to suit particular architectures. Nor does the proposal provide a comprehensive set of routines for elementary matrix algebra. It is intended primarily for software developers and, to a lesser extent, for experienced applications programmers.

2. Scope of the Level 3 BLAS. The routines proposed here have been derived from the Level 2 BLAS by replacing the vectors x and y with matrices B and C. Our motivation is to make it easy for users to remember the calling sequences and parameter conventions, since we believe the resulting operations to be useful on both state-of-the-art and conventional computers.

In real arithmetic the operations proposed for the Level 3 BLAS have the following forms:

a) *Matrix-matrix products*:
$$C \leftarrow \alpha AB + \beta C$$
$$C \leftarrow \alpha A^T B + \beta C$$
$$C \leftarrow \alpha AB^T + \beta C$$
$$C \leftarrow \alpha A^T B^T + \beta C$$

b) *Rank-k updates of a symmetric matrix*:
$$C \leftarrow \alpha AA^T + \beta C$$
$$C \leftarrow \alpha A^T A + \beta C$$

c) *Multiplying a matrix by a triangular matrix*:
$$B \leftarrow TB$$
$$B \leftarrow T^T B$$
$$B \leftarrow BT$$
$$B \leftarrow BT^T$$

d) *Solving triangular systems of equations with multiple right-hand sides*:
$$B \leftarrow T^{-1} B$$
$$B \leftarrow T^{-T} B$$
$$B \leftarrow BT^{-1}$$
$$B \leftarrow BT^{-T}$$

Here α and β are scalars; A, B and C are rectangular matrices (in some cases square and symmetric); and T is an upper or lower triangular matrix (and nonsingular in form d).

Analogous operations are proposed in complex arithmetic: conjugate transposition is specified instead of simple transposition and in form b, C is Hermitian and α and β are real.

3. Naming Conventions. The name of a Level 3 BLAS routine follows the conventions of the Level 2 BLAS. The fourth and fifth characters in the name denote the type of operation: MM - matrix-matrix product; RK - rank-k update of a symmetric or Hermitian matrix; and SM - solve a system of linear equations for a matrix of right-hand sides. Characters two and three in the name denote the kind of matrix involved: GE - general matrix; HE - Hermitian matrix; SY - symmetric matrix; and TR - triangular matrix. The first character in the name denotes the Fortran data type of the matrix: S - REAL; D - DOUBLE PRECISION; C - COMPLEX; and Z - COMPLEX*16 or DOUBLE COMPLEX (if available). The available combinations are indicated in following table:

Complex	Real	MM	RK	SM
CGE	SGE	*		
CHE	SSY	*	*	
CTR	STR	*		*

In the first column, the initial C may be replaced by Z. In the second column, the initial S may be replaced by D. (See Appendix C of [6] for the full subroutine calling sequences.)

The collection of routines can be categorized as *real*, *double precision*, *complex*, and *complex*16*. The routines can be written in ANSI Standard Fortran 77, with the exception of the routines that use COMPLEX*16 variables. These routines are included for completeness and for their usefulness on those systems that support this data type; but because they do not conform to the Fortran standard, they may not be available on all machines.

4. Argument Conventions. We follow a convention for the argument lists similar to that for the Level 2 BLAS, with the necessary adaptations. The order of arguments is as follows: arguments specifying options, arguments defining the sizes of the matrices, input scalar, description of input matrices, input scalar (associated with input-output matrix), and description of the input-output matrix. Note that not each category is present in each of the routines.

The arguments that specify options are character arguments with the names SIDE, TRANS, TRANSA, TRANSB, UPLO, and DIAG. SIDE is used by the routines as follows: 'L' - multiply general matrix by symmetric or triangular matrix on the left; 'R' - multiply general matrix by symmetric or triangular matrix on the right. TRANS, TRANSA, and TRANSB are used by the routines as follows: 'N' - operate with the matrix; 'T' - operate with the transpose of the matrix; 'C' - operate with the conjugate transpose of the matrix. In the real case the values 'T' and 'C' have the same meaning and in the complex case the value 'T' is not allowed. UPLO is used by the Hermitian, symmetric, and triangular matrix routines to specify whether the upper or lower triangle is being referenced as follows: 'U' - upper triangle; 'L' - lower triangle. DIAG is used by the triangular matrix routines to specify whether the matrix is unit triangular as follows: 'U' - unit triangular; 'N' - non-unit triangular. When DIAG is supplied as 'U', the diagonal elements are not referenced. Thus, UPLO and DIAG have the same values and meanings as for the Level 2 BLAS; TRANS, TRANSA, and TRANSB have the same values and meanings as TRANS, where TRANSA and TRANSB apply to the matrices A or B, respectively, except that in the complex routines transposition is not allowed.

We recommend that the equivalent lowercase characters be accepted with the same meaning, although, because they are not included in the standard Fortran character set, their use may not be supported on all systems. See [7, Sec. 7] for further discussion.

The sizes of the matrices are determined by the arguments M, N, and K. It is permissible to call the routines with M, N, or K = 0, in which case the routines exit immediately without referencing their matrix arguments. For rectangular matrices the input-output matrix is always $m \times n$, and for square matrices it is always $n \times n$.

The description of the matrix consists of the array name (A, B, or C) followed by the leading dimension of the array as declared in the calling (sub)program (LDA, LDB, or LDC).

The scalars always have the dummy argument names ALPHA and BETA. The following values of arguments are invalid: any value of the character arguments SIDE, TRANS, TRANSA, TRANSB, UPLO, or DIAG, whose meaning is not specified; M < 0; N < 0; K <

0; LDA < the number of rows in the matrix A; LDB < the number of rows in the matrix B; and LDC < the number of rows in the matrix C.

If a routine is called with an invalid value for any of its arguments, it must report the fact and terminate execution of the program. In our model implementation, each routine, on detecting an error, calls a common error-handling routine XERBLA, passing to it the name of the routine and the number of the first argument in error. Specialized implementations may call system-specific exception-handling and diagnostic facilities, either via an auxiliary routine XERBLA or directly from the routines.

5. Storage Conventions. Unless otherwise stated, matrices are stored in a two-dimensional array with matrix-element $a_{i,j}$ stored in array-element A(I,J). For symmetric and Hermitian matrices, only the upper triangle or the lower triangle is stored. For triangular matrices, the argument UPLO defines whether the matrix is upper (UPLO='U') or lower (UPLO='L') triangular. For a Hermitian matrix, the imaginary parts of the diagonal elements are of course zero, and thus the imaginary parts of the corresponding Fortran array elements need not be set, but are assumed to be zero. In the RK routines these imaginary parts will be set to zero on return, except when $\alpha = 0$, $k = 0$, or $n = 0$, in which case the routines exit immediately.

6. Rationale. Three basic matrix-matrix operations were chosen because they occur in a wide range of linear algebra applications. We have aimed at a reasonable compromise between a much larger number of routines each performing one type of operation (e.g., $B \leftarrow L^T B$), and a smaller number of routines with a more complicated set of options. In each data type there are, in fact, five routines performing altogether 48 different operations.

We have not proposed specialized routines to take advantage of packed storage schemes for symmetric, Hermitian, or triangular matrices, nor of compact storage schemes for banded matrices, because such schemes do not lend themselves to partitioning into blocks. Similarly, we are not aware of a need for simple transposition in the complex case. We also have not proposed a set of extended-precision routines analogous to the ES and EC routines in the Level 2 BLAS, since this would require a two-dimensional array in extended precision. As with the Level 2 BLAS no check has been included for singularity, or near singularity, in the triangular equation-solving routines. Since the requirements for such a test depend on the application, we felt that this should be performed outside the triangular solver.

In a few cases we have deliberately departed from the conventions of the Level 2 BLAS. For example, although the input-output matrix C in the matrix multiply routines is the analog of the vector y in the matrix-vector product routines, here C always has the same dimensions. Also, in the rank-k update routines we have included a parameter β, which we feel is useful in applications; and since the matrix multiply routines can also be viewed as rank update routines, we have consistency between the MM and RK routines. Finally, we have not allowed transposition in the complex case for the Level 3 BLAS.

7. Applications. The primary application of the Level 3 BLAS is in implementing algorithms of numerical linear algebra in terms of operations on submatrices (or blocks). There is a long history of block algorithms (see, for instance, [11]). Recently, several workers have demonstrated the effectiveness of block algorithms on a variety of architectures with vector-processing or parallel-processing capabilities, on which potentially high performance can easily be degraded by excessive transfer of data between different levels of memory [e.g., 1-3].

In [6], we show how the Level 3 BLAS routines can be used to implement three fundamental algorithms of numerical linear algebra—Cholesky factorization, LU factorization, and QR factorization. We also illustrate their use in a high-level matrix-multiply routine.

For the factorization routines the strategy in each case is to compute at each stage a block of consecutive columns of the result. The size of the block is a parameter, nb, which may be varied to suit the size of the problem and the architecture of the machine. There are, of course, other ways to organize the computation. For example, one can compute a block of consecutive rows at each stage. The analysis of [9] can easily be extended to algorithms that work by blocks. We have chosen an organization that works by columns rather than rows, and that involves fewest memory references.

Also, we have implemented the algorithms in such a way that submatrices passed to the Level 3 BLAS routines are kept as large as possible (once the parameter nb has been fixed). This approach gives greatest scope for achieving efficiency. Alternatively, one might partition the matrix into, say, square blocks of size $nb \times nb$: this would require many more calls to the Level 3 BLAS routines, but might allow a more precise control of memory or of parallelism.

A similar strategy is used for the matrix multiplication.

8. Implementation via netlib. A model implementation, in Fortran 77, of the Level 3 BLAS routines is available via netlib by sending an electronic mail message to netlib@anl-mcs.arpa or research!netlib. The message should be of the form *send sblas3 from blas3* (which would return the single-precision real version of the Level 3 BLAS). Alternatively, one may obtain a copy of the routines by sending a request, along with a magnetic tape, to any of the authors.

References

(1) M. Berry et al., *Parallel Algorithms on the CEDAR System*, CSRD Report No. 581, 1986.

(2) C. Bischof and C. Van Loan, *The WY representation for products of Householder matrices*, SIAM SISSC, 8 (1987)

(3) I. Bucher and T. Jordan, *Linear algebra programs for use on a vector computer with a secondary solid state storage device*, in *Advances in Computer Methods for Partial Differential Equations*, R. Vichnevetsky and R. Stepleman, eds., IMACS, 1984, pp. 546-550.

(4) D. Dodson and J. Lewis, *Issues relating to extension of the Basic Linear Algebra Subprograms*, ACM SIGNUM Newsletter, 20 (1985), pp. 2-18.

(5) J. J. Dongarra et al., *LINPACK Users' Guide*, SIAM Pub., Philadelphia, 1979.

(6) J. J. Dongarra et al., *A Proposal for a Set of Level 3 Basic Linear Algebra Subprograms*, Argonne National Laboratory report ANL-MCS-TM-88 (April 1987).

(7) J. J. Dongarra et al., *An Extended Set of Basic Linear Algebra Subprograms*, Argonne National Laboratory report ANL-MCS-TM-41 (Rev. 3), November 1986.

(8) J. J. Dongarra and I. S. Duff, *Advanced Architecture Computers*, Argonne National Laboratory report ANL-MCS-TM-57 (Rev. 1), January 1987.

(9) J. J. Dongarra, F. Gustavson, and A. Karp, *Implementing linear algebra algorithms for dense matrices on a vector pipeline machine*, SIAM Review, 26 (1984), pp. 91-112.

(10) C. Lawson et al., *Basic Linear Algebra Subprograms for Fortran usage*, ACM TOMS, 5 (1979), pp. 153-165.

(11) A. C. McKellar and E. G. Coffman, Jr., *Organizing matrices and matrix operations for paged memory systems*, CACM, 12 (1969), pp. 153-165.

CHAPTER 9

High Performance Linear Algebra on the AMT DAP 510

Stewart F. Reddaway[*]
Grant Bowgen[†]
Sven Van Den Berghe[†]

Abstract. The performance on linear algebra of bit-organised SIMD arrays such as the AMT DAP 510 can be greatly increased by defining at a high level new subroutines and implementing them efficiently. The subroutines are essentially rank k updates to matrices, and they exploit the fact that one vector is multiplied by a succession of numbers from another vector. The subroutines have been used in implementations of matrix multiplication and equation solving. Performance improves more than sixfold (to 50 MFLOPS) in favourable cases.

1. **Introduction.** SIMD arrays of simple bit-organised Processing Elements (PEs) perform floating point arithmetic by low level system software. The AMT DAP 510 produces an array of 1024 32-bit floating point results in about 1000 cycles of 100 nsec each to give a performance of about 10 MFLOPS.

This paper describes the definition of a family of subroutines at a higher level than single array operations, and their efficient implementation. They can be used to greatly increase the performance of many linear algebra problems, and example implementations of matrix multiplication and equation solving are described.

The DAP (Distributed Array Processor) has been described in many papers as it has evolved over the years; for example references 2, 3, 4 and 5. This paper refers to the current product, the AMT DAP 510 [1]. This has an SIMD array of 1024 PEs, and with arithmetic performed by low level software the time an operation takes is related to its complexity. Thus Boolean array operations are very fast (about 10,000 MOPS) but floating point arithmetic is much slower, with the most relevant 32-bit operations performing as:

	Cycles	MFLOPS
vector - vector:		
Add	860	12
Multiply	1400	7
Average Add, Multiply	1130	9
scalar - vector:		
Multiply	800	13

---

[*] 3 Woodforde Close, Ashwell, Baldock, Herts. SG7 5QE, England
[†] Active Memory Technology Ltd., 65 Suttons Park Avenue, Reading RG6 1AZ, England

The average is given because linear algebra uses an equal mix of add and multiply. Many other operations are much faster, and we will exploit integer add being much faster than floating point add. Every floating point precision from 24-bits to 64 bits in 8-bit intervals is available as standard from the high level language (FORTRAN-PLUS), with a smooth trade-off against speed and storage space.

2. Performance of Machines on Linear Algebra. It is well known that on linear algebra problems, such as equation solving with pivoting, the measured MFLOPS of a machine typically falls a long way short of the rated peak MFLOPS. Without the use of the new subroutines of this paper, the DAP programmed (using a 2D mapping) in FORTRAN-PLUS on problems of order 100 - 1000 has system efficiencies in the ranges:

Matrix multiplication	60 - 90%
Equation solving	35 - 75%

This is good by industry standards, but can we do better still?

3. New Subroutines. Most work in linear algebra can be expressed in terms of matrix multiplication (i.e. as an accumulation of outer products of pairs of vectors). This occurs, for example, in what is referred to as rank k updates of a matrix. The family of basic subroutines is:

MMk	"Matrix Multiply"
MMAk	"Matrix Multiply and Add"
MMSk	"Matrix Multiply and Subtract"

k gives the common dimension of the matrices being multiplied, e.g. 1 (outer product), 8, 16.

An example of calling a subroutine is:

CALL MMS8 (AVS, SCVS, VS, N, LM)

The Nx8 matrix SCVS is multiplied by the 8x1024 matrix VS and the result subtracted from an Nx1024 matrix AVS masked by LM, a 1024 element logical vector. SCVS and VS are sets of 8 DAP "long vectors" and AVS is a set of N long vectors; each long vector contains up to 1024 elements. N is up to 1024. The 1024 dimension in the above matrices can be converted into a dimension L (up to 1024) by LM containing L consecutive TRUE bits. (Other uses may be found for LM having arbitrary patterns.) The structure of MMS8 is shown in Fig. 1.

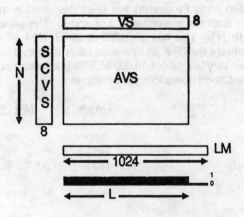

FIG. 1 Structure of MMS8

4. Ideas used inside the new subroutines. Multiplication can be speeded up by noting that in an outer product one vector is multiplied by a succession of scalars from the other vector, and doing appropriate preparation work on both vectors. Many (specifically 32) multiples of the vectors in VS are generated. The numbers in SCVS are analysed into "digits" that will be used serially to select, by global addressing, the appropriate multiple of the VS vector. In this way one vector add can advance the scalar multiplier one digit at a time rather than one bit. By permitting add or subtract and by allowing for gaps between digits, an average digit can deal with n + 3 multiplier bits for 2**n pre-computed multiples. (The code is being implemented in phases, and in the current code digits average n + 2 bits, or 7 bits with n = 5.) The 24-bit mantissa of a 32-bit floating point number requires at most 4 digits. The multiplication is shown schematically in Fig. 2.

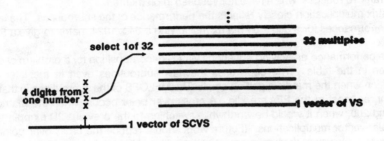

FIG. 2. Schematic for Multiplication in Outer Products

Normal element by element multiplication of 2 vectors with 24-bit mantissas requires 23 adds. On DAP the normal multiplication of a vector by a scalar requires about half as many adds because zero multiplier bits are skipped. The new outer product code requires at most 3 adds, because of the preparation work on both vectors. The technique can barely be used on single scalar-vector multiplies because the overhead of the preparation work cannot be spread over many multiplies.

An outer product contains no adds, but the function MMA1 contains as many adds as multiplies, and these adds are now the bottleneck. When k>1 (e.g. MMS8) the adds in the accumulation are speeded-up by using a form of local block floating point, and only the final subtract uses the standard floating point; in a later phase it is intended to move to a more sophisticated code that adapts to the needs of the data.

In MMS8 8 products are being accumulated, with each product consisting of up to 4 digit multiply and adds. There is advantage in sorting the (up to) 32 digits according to their significance as part of the preparation work, so that the accumulation is more regular. We are thus using sorting to speed up matrix multiplication!

The accumulation now consists of (at most) 4 adds per product. Ignoring the preparation work, all except one of the floating point adds have become one extra (pre-aligned) add; this is nearly an order of magnitude faster. There is a significant requirement for working memory. MMS8 needs 8K bits in every PE for the pre-computed multiples.

5. Examples of Use. The new subroutines have wide potential in linear algebra. We have implemented demonstration examples in matrix multiplication and equation solving via LU decomposition.

Suitable sized matrix multiplication is very easy to code. The FORTRAN-PLUS code to multiply a NxM matrix B by a Mx1024 matrix A to give an Nx1024 matrix C is in essence:

```
        DO 100 J = 1,M,8
        CALL MMA8 (C,A(,,J),B(,,J),N,MASK)
100     CONTINUE
```

The code (with column pivoting) for LU decomposition and equation solving is more complex and will not be described in detail. The main points are:
* 8 pivot columns are extracted and updated to find the next 8 pivot rows ("pivot look-ahead")
* these 8 pivot rows are updated
* the 8 pivot rows and 8 pivot columns are used to perform a rank 8 update of the main data with MMS8.

The above procedure means that if the problem order is large compared with 8, the bulk of the work is done by MMS8.

6. Performance.

The rank 8 routines include 1 normal floating point add/subtract for every 8 multiplies and 7 fast adds. This normal add plus the normalisation at the end of the accumulation account for about 40% of the subroutine time. This means there is significant gain in going to rank 16 routines, which have not yet been implemented.

Matrix multiplication closely reflects the performance of the subroutines. The table below gives the performances for a 500 x 64 matrix multiplying a 64 x 1024 matrix to give a 500 x 1024 result matrix.

The performance on solving equations via LU decomposition for a problem of order 1024 is also given in the table. The code uses the new subroutines even in the later stages of decomposition when the matrix is getting small. The MFLOPS of the subroutines in that regime is getting poor, as most of the PEs are idle. A cross-over point occurs when the matrices are of order around 200, when it would be worthwhile switching to the previous 2D mapping that does not use scalar-vector multiplications. (Future work should reduce the cross-over point to below 100, see below.) However, the improvement achieved by switching for a problem of order 1024 is small.

	actual MFLOPS	% normal performance
Matrix multiplication		
Using MMA1	8	90%
Using MMA8	15.5	170%
(Using MMA16	37.4	420%
	50	560%)
Equation solving		
Old code 2D	6.5	72%
Old code 1D	4.5	50%
Using MMS1	6.6	73%
Using MMS8	19.8	220%
(Using MMS16	26	290%)

TABLE 1. Performance on DAP 510

The subroutines MMA16 and MMS16 have not yet been implemented. The figures given assume that as well as halving the contributions from the normal add and the normalisation, the accumulation advances 8 multiplier bits at a time (instead of 7) and some other minor improvements are made. The latter effects would improve MMA8 and MMS8 performance about 10%. For equation solving with MMS16, the pivot look-ahead work is about 10% of the total; this work could be approximately halved by use of MMS8 and MMS4.

The "% normal performance" column in the table is based on 9 MFLOPS being 100%; this is an equal mix of normal vector adds and multiplies in an ideal situation with no overheads. It is interesting that nearly 6 times that figure can be achieved with the new subroutines.

The accuracy of the equation solving was compared with a normal floating point implementation and found to be very similar on average.

7. **Further Work.** There are obvious extensions to rank 16, 4 and 2. Above rank 16 there are diminishing returns and temporary storage demands are large; in due course rank 32 will be worthwhile. More flexibility in the number of pre-computed multiples would also be helpful, for example for smaller N. It is intended that future codes advance n + 3 multiplier bits at a time instead of n + 2. Extending the floating point precisions to the full range offered by FORTRAN-PLUS is important. 64-bit will be approximately 3 times slower and have more severe storage demands; 24-bit will be nearly 2 times faster.

It is intended to investigate more fully making the subroutines adaptive to the data encountered. The aim is to improve accuracy for awkward data so that it is never worse than normal floating point (and is usually better), and to do so in such a way that extra time is used only when difficult data is encountered. It is thought that the effect on average speed will be nil.

The MFLOPS performance of the new subroutines drops proportionately as the current matrix order falls below 1024. It is hoped to produce (more complex) variants of the subroutines in which 2, 3 or 4 PEs are used for each accumulation; this will give higher performance when the matrix order is less than 512, 340 or 256 respectively. The cross-over point for switching to a 2D mapping could be as low as order 64. Performance on solving equations of order 1024 should improve about 15%, and equations of order 300 should more than double to nearly 20 MFLOPS.

The subroutines can, of course, be used for solving systems larger than 1024, although backing memory would need to be used for much bigger problems.

Acknowledgement. We should like to thank Peter Flanders of AMT for help with the data movement using Parallel Data Transforms [3] and with the sorting.

REFERENCES

[1]   Active Memory Technology, Inc. <u>DAP 510 Attached Processor System</u>, 16802 Aston St, Suite 103, Irvine, Ca.92714, USA.

[2]   P.M. FLANDERS, D.J. HUNT, S.F. REDDAWAY and D. PARKINSON, <u>Efficient High Speed Computing with the Distributed Array Processor</u>, in <u>High Speed Computer and Algorithm Organisation</u>, D.J. Kuck, D.H. Lawrie and A.H. Sameh (eds), Academic Press, New York, 1977, pp. 113-128.

[3]   P.M. FLANDERS and D. PARKINSON, <u>Data Mapping and Routing for Highly Parallel Processor Arrays</u>, to be published in Future Computing Systems (Oxford University Press), volume 1, number 1 (1988).

[4]   S.F. REDDAWAY, <u>DAP - a Distributed Array Processor</u>, Proc. 1st Annual Symposium on Computer Architecture, G.J. Lipovski and S.A. Szygenda (eds), Computer Architecture News, <u>2</u>, 4 (IEE Cat. No. 73CH0824-3C), 1973, pp. 61-65.

[5]   S.F. REDDAWAY, <u>Signal Processing on a Processor Array</u>, in <u>Les Houches, Session XLV, 1985, Traitement du Signal/Signal Processing</u>, J.L. Lacoume, T.S. Durrani and R. Stora (eds), North-Holland, Amsterdam, 1987, pp. 831-858.

Communication Complexity of Matrix Multiplication
David C. Fisher

Suppose a parallel processing machine multiplies two $n \times n$ matrices so only one copy of each input exists at the start of computations. We show that no matter what algorithm is used, it is possible to halve the machine so that the amount of data exchanged between the halves is at least $(4n^4 - L^2)/16n^2$ where L is the maximum number of input entering one port. This supercedes a similar bound of Savage (Journal of Computer and System Science, 22, 230–242 (1981)). This bound is used to find lower bounds on the time needed to do matrix multiplication on various parallel architectures.

CHAPTER 10

Parallel Rapid Elliptic Solvers*

E. Gallopoulos[†]
Y. Saad[†]

Abstract. We present a rapid elliptic solver suitable for vector and parallel machines which is based on a modification of the Block Cyclic Reduction algorithm (BCR). The main bottleneck of BCR lies in the solution of linear systems whose coefficient matrix is the product of tridiagonal matrices. This bottleneck is handled by expressing the rational function corresponding to the inverse of this product as a sum of elementary fractions. This leads to an algorithm which requires the parallel solution of tridiagonal systems. We discuss various implementations of these algorithms on shared memory as well as distributed memory machines.

1 Introduction

This paper is concerned with the solution of separable elliptic equations in two dimensions. As is well-known there are some circumstances, where these problems can be solved by fast direct methods having an operation complexity of $O(N \log_2 N)$ where N is the size of the discretized problem. There techniques are based on either FFT transforms or block cyclic reduction algorithms. When it comes to vector/ parallel implementations, the FFT based methods have received most of the attention in the recent years, due to highly efficient FFT algorithms and tridiagonal system solvers that can best exploit supercomputer environments. The block cyclic reduction algorithm has been neglected because of the highly sequential nature of its traditional implementation. The purpose of this paper is to rehabilitate BCR by showing that a simple transformation can provide sufficient amount of parallelism and vectorization into the BCR algorithm. Similar ideas have also been presented in [11]. The emphasis will be on implementations on shared-memory multivector processors and distributed memory architectures. As will be seen there are many advantages of this approach over the FFT approach. Certainly the most important of these is that the BCR algorithm is applicable to a wider variety of problems than are the FFT based methods. Moreover, its performance is not as sensitive to the number of blocks of the block tridiagonal matrix representation. One can use this approach

*Research supported by the National Science Foundation under Grants No. US NSF DCR84-10110 and US NSF DCR85-09970, the US Department of Energy under Grant No. DOE DE-FG02-85ER25001, by the US Air Force under Contract AFSOR-85-0211, and an IBM donation.

†Center for Supercomputing Research and Development, University of Illinois at Urbana Champaign, Urbana, Illinois 61801.

to write a parallel version of FISHPACK [9], which, in its original form, was entirely based on Block Cyclic Reduction. Perhaps the most promising approach, which is not considered in this paper, is to combine the parallel block cyclic reduction technique with the Approximate Cyclic Reduction algorithm recently proposed by P. Swarztrauber [8].

2   Parallel Block cyclic Reduction

The BCR algorithm for the numerical solution of the discretized PDE - in its stabilized Buneman form - is described in [1], [10]. For the simplest case, of a $n \times m$ grid with $n = 2^\mu - 1$ and m arbitrary, the algorithm proceeds by eliminating half the unknowns at every step. The bulk of the computational work during the reduction step consists in solving the block system

$$\prod_{i=1}^{i=k}(A - \lambda_i I)X = Y \tag{1}$$

in which A is a tridiagonal matrix of dimension $m \times m$, and X and Y are matrices of dimension $m \times \nu$, with ν and k related by $k \times (\nu + 1) = (n + 1)/2$. From this, it appears that the source of parallelism at this level is in the solution of the ν independent systems. Moreover, the factor ν decreases very rapidly as k increases. Once this decomposition is realized however, we still need to solve

$$\prod_{i=1}^{i=k}(A - \lambda_i I)x = y \tag{2}$$

for vectors x and y. As k increases and ν decreases, the method loses its source of parallelism rapidly unless counter measure steps are taken. Fortunately, this is possible by exploiting the following well-known *partial fraction decomposition* on the rational polynomial $r(t) = [(t - \lambda_1)(t - \lambda_2)...(t - \lambda_k)]^{-1}$

$$\left[\prod_{i=1}^{i=k}(t - \lambda_i)\right]^{-1} = \sum_{i=1}^{i=k}\frac{\alpha_i}{t - \lambda_i} \tag{3}$$

where the coefficients α_i are scalars that are known in terms of the roots λ_i. The application of this decomposition to the above problem transforms the product into a sum and therefore, the k independent tridiagonal systems $(A - \lambda_i I)x_i = y$ can be solved in parallel and then linearly combined to yield the solution x of (2), see [3]. For boundary conditions of the Dirichlet, Neumann, and periodic type, the poles λ_i are known analytically, making it possible to derive easily computable formulas for the partial fraction expansion coefficients. As shown in [10], when the number of blocks is not equal to $2^\mu - 1$, additional systems of the form

$$\prod_{i=1}^{i=k}(A - \lambda_i I)x = \prod_{j=1}^{j=l}(A - \mu_j I)y \tag{4}$$

have to be solved at each step. Our technique applies equally well in this case, reducing the rational matrix polynomial

$$\prod_{j=1}^{j=l}(A - \mu_j I) \cdot [\prod_{i=1}^{i=k}(A - \lambda_i I)]^{-1} \tag{5}$$

into a sum of partial fractions.

Swarztrauber has extended the BCR algorithm to handle more general separable elliptic equations [7]. A careful examination of the method shows that it is also based on the solution of equations involving a matrix polynomial or rational form in A, where A is a tridiagonal matrix. In contrast with the previous case however, the roots of the polynomial are not known beforehand in an analytic form. Standard methods can be used for the numerical computation of these roots, followed by the use of the same technique described in section 3 to transform the seemingly sequential problem into a parallel one.

3   Implementation on shared memory mutiprocessors.

The approach described in the previous section has been implemented and tested on an Alliant FX/8 running Concentrix 3.0. We show in Figure 1 a comparison of the times obtained with the original FISHPACK package and the implementation of FISHPACK based on the parallel approach of the previous section. Figure 2 gives an idea of the speed-ups obtained for the parallel algorithm as the number of processors varies. The results are for the solution of Poisson's equation on a $n \times n$ grid with Dirichlet boundary conditions. The timings given compare the current best results with the timings for the original FISHPACK code. A more detailed discussion of the numerical examples as well as the extensions to the other cases that the method can handle can be found in [3].

4   Implementation on distributed memory architectures.

One should note that in the algorithm described in Section 2, the total number linear systems to be solved can be kept high because of the equality $k(\nu + 1) = (n + 1)/2$. By duplicating each tridiagonal matrix for each right hand side, we end up with $k\nu = (n+1)/2 - k \geq (n+1)/4$ linear systems which may be sufficiently large for vectorization to be efficient. Therefore, the only concern is on achieving load balancing during the process. In [4] we describe a mapping of the algorithm onto various loosely coupled architectures.

In what follows we outline the algorithm for hypercube architectures. For hypercubes load balancing can be achieved by initially using a Gray-code mapping of the block tridiagonal systems, emulating scalar cyclic reduction. For the purpose of illustration let us consider the simple case where $n = 15$ and the number of processors $N = 2^p$ is 8. We start by mapping two right hand sides per processor except for the last processor which will only hold one right hand side. The Gray-code mapping here consists of assigning f_1, f_2 to node 000, f_3, f_4 to node 001 , ..., f_{2i+1}, f_{2i+2} to node g_i where $g_0, g_1, ..., g_{2^p-1}$ is the standard binary reflected Gray code sequence. As is easily seen in this example, when combining three successive vectors $f_{2i-1}, f_{2i}, f_{2i+1}$ as in the first step of forward reduction, we only need nearest neighbor communication. After these linear combinations are done, each processor solves a tridiagonal system involving A. In the later reduction steps, subvectors with subscripts differing by a power of two are combined. Because of a well-known property of the binary reflected Gray code, it turns out that communication is kept at distance of exactly two. It can also be seen that the same property holds when $n > 2 \times N$ and a similar mapping is used.

We now need to describe how to solve the polynomial systems $(A - \lambda_0 I)(A - \lambda_1 I)...(A - \lambda_{2^r-1} I)X = Y$ that are generated at the rth reduction step in nodes labeled $g_{k \times 2^r}$. Using the ideas of Section 2, we note that we need to solve simultaneously the systems $(A - \lambda_i I)X_i = Y$ in different processors and combine the results. The observation here is that these systems can be solved in the subcubes corresponding to labels where the first $p - r$

Figure 1: Old and new methods on n×n grids

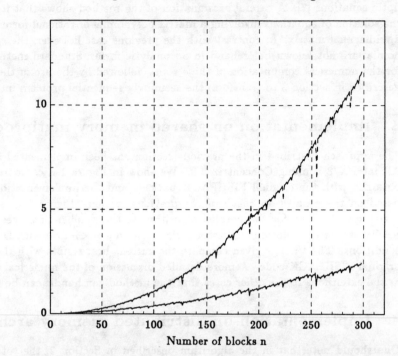

Figure 2: Speed–up with 2, 4, 8 over 1 CE on n×n grids

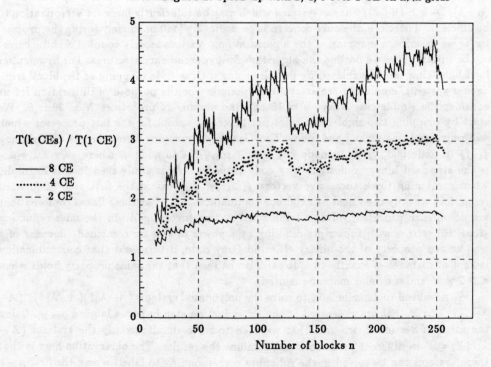

bits are the same as the leading bits of the node $K = g_{k \times 2^r}$. To be more specific, if the binary encoding of K is $k_1 k_2$ where k_2 consists of r bits then we solve the system $(A - \lambda_i I) X_i = F$ in node $k_1 b(i)$ where $b(i)$ represents the binary encoding of i. It can be easily proved that each node will solve one block system except for a small number of the last nodes in the Gray sequence which have no system to solve. This process requires broadcasting of the right hand side in node K to its subcube and then gathering/summing of the data in the manner distributed inner products are usually computed in a hypercube. These operations cost $O(r \times n/N)$ which means that the communication overhead in this second phase is higher than that of the first phase where the right hand sides Y are formed. Hence depending on the relation between communication and arithmetic costs of the particular hypercube system it might be preferable not to distribute the work to all processors but only to those located in a smaller subcube.

References

[1] B. Buzbee, G. Golub, and C. Nielson, *On direct methods for solving Poisson's equation*, *SIAM J. Numer. Anal*, vol. 7, pp. 627-656, (Dec. 1970).

[2] T.F. Chan, Y. Saad, *Multigrid Algorithms on the Hypercube Multiprocessor*, *IEEE Trans. Computers*, Vol. C-35, No. 11, pp. 969-977, (Nov. 1986).

[3] E. Gallopoulos and Y. Saad, *A parallel block cyclic reduction algorithm for the fast solution of elliptic equations*, University of Illinois, CSRD report number 659, submitted, (Apr. 1987).

[4] E. Gallopoulos and Y. Saad, *Block cyclic reduction algorithms for parallel architectures*, University of Illinois, CSRD technical report, In preparation.

[5] L. Johnsson, Y. Saad and M.H. Schultz, *Alternating Direction methods on Multiprocessors*, (to appear, Siam J. Stat. Sci. Comput.) Yale University, Department of Computer Science report YALEU/DCS/RR-382, (Sept. 1985).

[6] Y. Saad, M.H. Schultz, *Data Communication in Hypercubes*, Yale University, Department of Computer Science report YALEU/DCS/RR-428, (Oct. 1985).

[7] P. N. Swarztrauber, *A direct method for the discrete solution of separable elliptic equations*, *SIAM J. of Numer. Anal.*, v. 11, pp. 1136-1150, (Dec. 1974).

[8] P. N. Swarztrauber, *Approximate cyclic reduction*, *SIAM J. Sci. Stat. Comp.*, vol. 8, pp. 199-209, (May 1987).

[9] P. Swarztrauber and R. Sweet, *Efficient Fortran subprograms for the solution of elliptic partial differential equations*, NCAR Technical Note IA-109, Boulder, (July 1975).

[10] R. A. Sweet, *A cyclic reduction algorithm for solving block tridiagonal systems of arbitrary dimension*, *SIAM J. Numer. Anal.*, vol. 14, pp. 707-720, (Sept. 1977).

[11] R. A. Sweet, *A parallel and vector variant of the cyclic reduction algorithm*, *SIAM J. Sci. Stat. Comput.*, (to appear).

CHAPTER 11

Block Tridiagonal Systems on the Alliant FX/8*

Kenneth Dowers[†]
S. Lakshmivarahan[†]
Sudarshan K. Dhall[†]
J. Diaz[†]

Abstract. The problem of solving scalar tridiagonal systems has been extensively studied [1] [2] [3] [4]. This paper examines the solution of tridiagonal systems in which the non-zero elements of the system are mxm matrices, forming a block tridiagonal system of equations.

Two solution methods are examined: Gaussian Elimination in block form (also called Ritchmeyer's algorithm) [2], and a cyclic reduction based algorithm. It is shown that on multi-vector machines such as the Alliant FX/8, which can exploit the parallelism in the cyclic reduction algorithm, the reduction algorithm executes in a shorter time.

1. INTRODUCTION

This paper examines the solution of tridiagonal systems in which the non-zero elements of the system are m x m matrices, forming a block tridiagonal system of equations. This system of equations also appear in the numerical solution of certain partial differential equations.

Two solution methods are examined: Gaussian Elimination in block form (also called Ritchmeyer's algorithm) [2], and a cyclic reduction based algorithm.

Ritchmeyer's algorithm is an algorithm similar to Gaussian elimination; while the algorithm is sequential in nature (due to recurrences), parallelism may be introduced at the block level; i.e, each matrix multiplication, scalar linear system solution, etc., can be done in a parallel way.

The block cyclic reduction algorithm is very similar to the cyclic reduction that is used in the scalar case, but has been generalized. It operates on the same principal of repeatedly reducing the number of equations in the system; however, in this case, the equations are matrix equations.

2. RITCHMEYER'S ALGORITHM

The description closely follows the one given in the book by Gentzsch in [2]. Let

$$A X = K \tag{2.1}$$

be the block tri-diagonal system to be solved where

---

* The work on this report was partially supported by the Energy Research Center, Tulsa, Oklahoma, and the Governer's Council on Science and Technology, State of Oklahoma. We wish to thank Argonne National Laboratory for granting access to their Alliant FX/8. We also wish to thank Dr. D. Sorenson for discussion leading to this report.

† Parallel Processing Institute, School of Electrical Engineering and Computer Science, University of Oklahoma, Norman, OK 73019

$$A = \begin{bmatrix} A_1 & C_1 & 0 & 0 & . & . & 0 \\ B_2 & A_2 & C_2 & 0 & . & . & 0 \\ 0 & B_3 & A_3 & C_3 & . & . & 0 \\ . & . & . & . & . & . & . \\ . & . & . & . & . & . & . \\ . & . & . & . & . & . & . \\ 0 & 0 & . & . & B_{n-1} & A_{n-1} & C_{n-1} \\ 0 & 0 & . & . & 0 & B_n & A_n \end{bmatrix} \tag{2.2}$$

$$X = (X_1^t, X_2^t, \cdots, X_n^t)^t \tag{2.3}$$

$$K = (K_1^t, K_2^t, K_3^t, \cdots, K_n^t)^t \tag{2.4}$$

where t denotes the matrix transpose, $n = 2^s - 1$, for some $s > 0$, each A_i, B_i, and C_i are m x m matrices, and each K_i and X_i are column vectors of length m.

The solution of the algorithm is given in [2] as:

$$L_1 = A_1 \ , \ \ U_1 = L_1^{-1} C_1 \ , \ \ Y_1 = L_1^{-1} K_1 \ ,$$

for i = 2 to n - 1 do
$L_i = A_i - B_i \ U_{i-1}$
$U_i = L_i^{-1} \ C_i$
$Y_i = L_i^{-1} \ (K_i - B_i \ Y_{i-1})$
end for

$$X_n = Y_n$$

for i = n - 1 downto 1 do
$X_i = Y_i - U_i \ X_{i+1}$
end do

As indicated in the algorithm statement, the decomposition of A and the solution of Y may be computed concurrently, reducing the number of steps from three to two. Also, as can be seen in the algorithm statement, the LU decomposition of A contains the recurrence:

$$U_i = (A_i - B_i \ U_{i-1})^{-1} \ C_i$$

This makes the explicit vectorization of Ritchmyer's algorithm at the level of n difficult. However, at the level of the matrix multiplications and inverses, parallelism can be introduced. Hence the algorithm can be said to be parallel at the level of m.

3. CYCLIC REDUCTION

The cyclic reduction based algorithm is based on the parallel algorithm given in [1] for the solution of scalar tridiagonal systems of equations. The algorithm's basic goal is to recursively eliminate the number of equations in the original system until only one equation remains, solve that system, and use that solution to back-substitute until all unknowns are obtained.

The block-tridiagonal cyclic reduction algorithm can be stated as follows (a complete derivation is given in [5]):

$$A_i^{(0)} = A_i \ , B_i^{(0)} = B_i \ , C_i^{(0)} = C_i \text{ and } K_i^{(0)} = K_i$$

Define, for $l \geq 1$,

$$h = 2^{l-1}$$

$$\alpha_i^{(l-1)} = -A_i^{(l-1)} [B_{i-h}^{(l-1)}]^{-1}$$

and

$$\gamma_i^{(l-1)} = -C_i^{(l-1)}$$

Define

$$P_i^{(l)} = (A_i^{(l)} , B_i^{(l)} , C_i^{(l)} , K_i^{(l)})^t$$

where

$$A_i^{(l)} = \alpha_i^{(l)} A_{(i-h)}^{(l-1)}$$
$$B_i^{(l)} = B_i^{(l-1)} + \alpha_i^{(l)} C_{(i-h)}^{(l-1)} + \gamma_i^{(l)} A_{(i+h)}^{(l-1)}$$
$$C_i^{(l)} = \gamma_i^{(l)} C_{(i+h)}^{(l-1)}$$
$$K_i^{(l)} = K_i^{(l-1)} + \alpha_i^{(l)} K_{(i-h)}^{(l-1)} + \gamma_i^{(l)} K_{(i+h)}^{(l-1)}$$

As a boundary condition define

$$A_i^{(l)} = C_i^{(l)} = K_i^{(l)} = 0$$
$$B_i^{(l)} = 1$$

for all $i \le 0$ and $i \ge (n + 1)$.

(a) The Reduction Phase:

```
FOR l = 1 s−1 in steps of 1 DO
   FOR i ε { 2^l , 2×2^l , 3×2^l , · · · , 2^n − 2^l } DO in PARALLEL
      compute P_i^(l)
   End
End
```

(b) The Backward Substitution Phase:

Compute $X_{n/2} = [B_{n/2}^{s-1}]^{-1}$

```
FOR l = (s−1) 1 step −1  DO
   FOR i ε { 2^{l-1} , 3×2^{l-1} , 5×2^{l-1} , · · · , 2^n − 2^{l-1} } DO in PARALLEL
```

$$X_i = [B_i^{(l-1)}]^{-1} (K_i^{(l-1)} - A_i^{l-1} X_{i-2^{l-1}} - C_i^{l-1} X_{i+2^{l-1}})$$

```
   End
End
```

This algorithm is well suited for multi-vector machines; each processor can be assigned an iteration of the innermost loop in the reduction and filling in phase (the loop whose index is i), and the matrix operations involved with iteration can be done using the vector capabilities of each processor. This algorithm has a larger operation count than does Ritchmeyer's algorithm, but has a higher level of parallelism.

4. COMPARISON OF PERFORMANCE

Both algorithms were implemented on the Alliant FX/8, and timings were taken for both algorithms for m = 8, 16, and 32, n = 7, 15, 31, ... , 511. Also, for Ritchmeyer's algorithm, runs were done for m=64, n = 7, ... , 63, and m = 128, n = 7, and 15, to investigate the performance of the algorithm at larger block sizes. Each matrix operation (including the Gaussian Elimination, matrix-matrix multiply, and matrix-vector multiply) was via GAXPY operations.

Table 1b			
Ritchmeyer's Algorithm: n = 7			
m	1 Processor	2 Processors	4 Processors	8 Processors
8	0.0440	0.0472	0.0471	0.0520
16	0.1813	0.1826	0.1755	0.1899
32	0.8660	0.7859	0.7166	0.7497
64	5.7042	4.3680	3.6067	3.6644
128	37.1146	23.9869	17.1806	16.7236

Table 1c			
Ritchmeyer's Algorithm: n = 15			
m	1 Processor	2 Processors	4 Processors	8 Processors
8	0.0939	0.1006	0.1003	0.1100
16	0.3890	0.8112	0.3760	0.4067
32	1.8594	1.6875	1.5377	1.6073
64	12.1904	9.3256	7.7144	7.8373
128	79.4740	51.3155	36.7772	35.8210

Table 1d			
Ritchmeyer's Algorithm: n = 31			
m	1 Processor	2 Processors	4 Processors	8 Processors
8	0.1939	0.2080	0.2073	0.2272
16	0.8065	0.8112	0.7791	0.8434
32	3.8542	3.4969	3.1894	3.3297
64	25.1780	19.2576	15.9305	16.1719

Table 1e			
Ritchmeyer's Algorithm: n = 63			
m	1 Processor	2 Processors	4 Processors	8 Processors
8	0.3937	0.4224	0.4209	0.4604
16	1.6407	1.6512	1.5842	1.7104
32	7.8427	7.1131	6.4848	6.7589
64	51.1280	39.1156	32.3367	32.8437

Table 1f			
Ritchmeyer's Algorithm: n = 127			
m	1 Processor	2 Processors	4 Processors	8 Processors
8	0.7928	0.8498	0.8492	0.9272
16	3.3019	3.3224	3.2010	3.4458
32	15.7905	14.3276	13.0773	13.6180

Table 1g			
Ritchmeyer's Algorithm: n = 255			
m	1 Processor	2 Processors	4 Processors	8 Processors
8	1.5901	1.7037	1.7007	1.8595
16	6.6329	6.6648	6.4118	6.9152
32	31.7218	28.7483	26.2223	27.3378

Table 1h			
Ritchmeyer's Algorithm: n = 511			
m	1 Processor	2 Processors	4 Processors	8 Processors
8	3.0922	3.3247	3.3315	3.6663
16	12.9290	13.0125	12.5291	13.5741
32	62.1100	56.2659	51.2972	53.4700

Table 2b - Cyclic Reduction			
n = 7			
Block Size	1 Processor	2 Processors	4 Processors	8 Processors
8	0.0502	0.0399	0.0296	0.0295
16	0.2272	0.1747	0.1228	0.1222
32	1.0810	0.8395	0.6101	0.6083

Table 2c - Cyclic Reduction			
n = 15			
Block Size	1 Processor	2 Processors	4 Processors	8 Processors
8	0.1324	0.0904	0.0599	0.0514
16	0.6000	0.3999	0.2576	0.2139
32	2.9190	1.9550	1.2592	1.0749

Table 2d - Cyclic Reduction			
n = 31			
Block Size	1 Processor	2 Processors	4 Processors	8 Processors
8	0.3074	0.1927	0.1209	0.0956
16	1.3924	0.8540	0.5221	0.4029
32	6.8427	4.1868	2.5721	2.0179

Table 2e - Cyclic Reduction			
n = 63			
Block Size	1 Processor	2 Processors	4 Processors	8 Processors
8	0.6682	0.3967	0.2423	0.1817
16	3.0260	1.7578	1.0512	0.7723
32	14.9332	8.6687	5.2103	3.9325

Table 2f - Cyclic Reduction			
n = 127			
Block Size	1 Processor	2 Processors	4 Processors	8 Processors
8	1.4008	0.8042	0.4857	0.3525
16	6.3442	3.5754	2.1164	1.5016
32	31.3751	17.6442	10.5049	7.5804

Table 2g - Cyclic Reduction			
n = 255			
Block Size	1 Processor	2 Processors	4 Processors	8 Processors
8	2.8789	1.6183	0.9728	0.6868
16	13.0296	7.1974	4.2334	2.9419
32	64.5060	35.5612	21.0847	14.9184

Table 2h - Cyclic Reduction			
n = 511			
Block Size	1 Processor	2 Processors	4 Processors	8 Processors
8	5.8448	3.2547	1.9489	1.3530
16	26.4476	14.4819	8.4735	5.7878
32	130.9981	71.4264	42.2546	29.4898

4.1. PERFORMANCE OF RITCHMEYER'S ALGORITHM

The algorithm was tested for 1, 2, 4, and 8 processors. As can be seen in table 1, the execution time is essentially independent of the number of processors. This is due largely to the sizes of blocks used. The For instance, at m = 32, the largest vector length obtained in an individual block operation was 32. Since each Alliant processor can do a total of 32 operations in vector mode, it made little difference whether more than one processor was used.

For m = 64, n = 7, ..., 63, and for m = 128, n = 7, 15, some speedup is seen, with the largest speedup being seen at m = 128, as expected, and being about 2.

4.2. PERFORMANCE OF THE ODD-EVEN REDUCTION ALGORITHM

For this algorithm, parallelism was possible at the level of n (each processor being assigned to operate on a different set of three equations). In addition, each matrix or vector operation can be vectorized, since each Alliant processor is also a vector processor. As can be seen in table 2, a substantial speedup is obtained at each increase in the number of processors used. This algorithm was *not* as simple to implement as Ritchmeyer's algorithm.

An interesting point can be made about processor scheduling here. At first, each matrix and vector operation was fully optimized for vectorization and concurrency. Later, as an experiment, the matrix and vector operation subroutines were optimized for vectorization only, and these results were better than for the latter case. The reason for this is that when a routine optimized for concurrency is called, all processors are applied to the problem. What started out as 8 matrix operations operating in parallel became one matrix operation using all 8 processors, which for the block sizes used, does not produce speedup.

5. CONCLUSIONS

1.   For small block sizes and large n's, cyclic reduction shows a marked speedup as the number of processors increases; Ritchmeyer's algorithm does not.

2.   For number of processors above four, and a reasonable n, reduction outperforms Ritchmeyer's algorithm.

3.   In the reduction algorithm, in the case of small blocks, it is better to do all matrix operations in vector mode only, without concurrency.

4.   Vectorization and concurrency may be introduced at the block level for Ritchmeyer's algorithm.

6. REFERENCES

[1]   R.W. Hockney and C.R. Jesshope. *Parallel Computers*, Adam and Hilger, Bristol, England, 1981, Chapter 5, pp. 280 - 298.

[2]   W. Gentzsch. *Vectorization of Computer Programs with Applications to Computational Fluid Dynamics* , Friedr. Vieweg & Sohn, Wiesbaden, Germany, 1984, Chapter 9, pp. 184-185

[3]   K. Dowers, S. Lakshmivarahan, and S. K. Dhall. On the Comparison of the Performance of the Alliant FX/8, VAX 11/780, and IBM 3081 in Solving Linear Tri-Diagonal Systems, IEEE Region 5 Conference, Tulsa, Oklahoma, 1986

[4]   J. J. Dongarra, F. G. Gustavson, and A. Karp. Implementing Linear Algebra Algorithms for Dense Matrices on a Vector Pipeline Machine, SIAM Review, Vol. 26, No. 1, January, 1984

[5]   K. Dowers. On the Solution of Linear Tridiagonal and Block Diagonal Systems of Equations on Multi-Vector Architectures, Master's Thesis, University of Oklahoma, Norman, Oklahoma, 1987.

CHAPTER 12

A Divide and Conquer Algorithm for Computing the Singular Value Decomposition*

E. R. Jessup†

D. C. Sorensen‡

Abstract — A parallel algorithm for computing the singular value decomposition will be presented. The basic step of the algorithm is a divide and conquer step founded on a rank one modification of a bidiagonal matrix. Numerical difficulties associated with forming the product of a matrix with its transpose are avoided and numerically stable formulae for simultaneously computing left and right singular vectors are derived for the divide and conquer steps. A deflation technique is described which together with a robust root finding method assures computation of the singular values to full accuracy in the residuals and also assures the orthogonality of the singular vectors.

1. Introduction

The singular value decomposition (SVD) of a real $m \times n$ matrix A can be written

$$A = U \Sigma V^T,$$

where U and V are both orthonormal matrices, and Σ is a diagonal matrix with non-negative diagonal elements. The columns of U and V are, respectively, the left and right singular vectors of A; the diagonal elements of Σ are its singular values. A standard algorithm for computing the singular value decomposition involves first reducing a matrix A to upper bidiagonal form B using elementary orthogonal transformations [8] to obtain $A = \hat{U} B \hat{V}^T$ and then computing the SVD of $B = \hat{Y} \Sigma \hat{X}^T$. Combining the two results gives

(1.1) $$A = \hat{U} (\hat{Y} \Sigma \hat{X}^T) \hat{V}^T = U \Sigma V^T,$$

where $U = \hat{U} \hat{Y}$ and $V = \hat{V} \hat{X}$.

This paper focuses on the computation of the SVD of the bidiagonal matrix B. The fundamental step of the algorithm presented is a divide and conquer mechanism based on a rank one tearing of the bidiagonal matrix B. In all sections, we consider only the case $m = n$. If $m > n$, the initial reduction may be preceded by a QR factorization

*Work supported in part by the Applied Mathematical Sciences subprogram of the Office of Energy Research, U. S. Department of Energy, under Contract W-31-109-Eng-38. Oral presentations based on this work were given at Gatlinburg X held at Fairfield Glade, TN in Oct. 1987 and at the Third SIAM Meeting on Parallel Processing for Scientific Computing held at Los Angeles, CA in Dec. 1987.

† Department of Computer Science, Yale University, New Haven, Connecticut 06520

‡ Mathematics and Computer Science Division, Argonne National Laboratory, Argonne, Illinois 60439

of A using column pivoting and the $m \times n$ triangular matrix R used in place of A . A similar procedure is appropriate for the case $m < n$. Unless otherwise specified, capital Roman letters represent matrices, lower case Roman letters represent column vectors, and lower case Greek letters represent scalars. A superscript T denotes transpose. All matrices and vectors are real.

2. The Basic Step

The purpose of this paper is to extend the divide and conquer technique first suggested in [4] and developed for parallel computation in [7,9] for symmetric matrices to an algorithm for computing the SVD. In this development we are particularly concerned with avoiding numerical errors associated with explicit formation of the matrix BB^T . Although the techniques of [4,7,9] could be applied directly to this tridiagonal matrix to compute the singular values and left singular vectors, one is then faced with either forming $B^T B$ to obtain the right vectors or computing the right vectors using the left vectors together with one of the appropriate linear relations between the two sets. In the first case, it may be difficult to correctly associate left vectors with right vectors when singular values are clustered; in the second case, it is difficult to obtain a well-conditioned linear system relating the two sets when the original matrix A is ill-conditioned. More detail on these points is given in [10].

The following rank one modification of B

$$(2.1) \quad B = \begin{bmatrix} B_1 & \beta e_k e_1^T \\ 0 & B_2 \end{bmatrix} = \begin{bmatrix} B_1 & 0 \\ 0 & B_2 \end{bmatrix} + \beta \begin{bmatrix} e_k \\ 0 \end{bmatrix} (0, e_1^T)$$

allows implicit formation of BB^T as follows

$$BB^T = \begin{bmatrix} B_1 & \beta e_k e_1^T \\ 0 & B_2 \end{bmatrix} \begin{bmatrix} B_1^T & 0 \\ \beta e_1 e_k^T & B_2^T \end{bmatrix} = \begin{bmatrix} B_1 B_1^T & 0 \\ 0 & B_2(I - e_1 e_1^T)B_2^T \end{bmatrix} + \begin{bmatrix} \beta e_k \\ B_2 e_1 \end{bmatrix} (\beta e_k^T, e_1^T B_2^T) .$$

Now observe that $\hat{B}_2 \hat{B}_2^T = B_2(I - e_1 e_1^T)B_2^T$, where $\hat{B}_2 = B_2 \begin{bmatrix} 0 & 0 \\ 0 & I \end{bmatrix}$ is the bidiagonal matrix B_2 with its first column deleted. Thus, the tearing shown in (2.1) is equivalent to making a rank one update to the symmetric tridiagonal matrix

$$(2.2) \quad BB^T = \begin{bmatrix} B_1 & \beta e_k e_1^T \\ 0 & B_2 \end{bmatrix} \begin{bmatrix} B_1^T & 0 \\ \beta e_1 e_k^T & B_2^T \end{bmatrix} = \begin{bmatrix} B_1 B_1^T & 0 \\ 0 & \hat{B}_2 \hat{B}_2^T \end{bmatrix} + \begin{bmatrix} \beta e_k \\ \alpha e_1 \end{bmatrix} (\beta e_k^T, \alpha e_1^T) ,$$

where $\alpha e_1 = B_2 e_1$. One may now compute the singular value decompositions $B_1 = U_1 \Sigma_1 V_1^T$ and $\hat{B}_2 = \hat{U}_2 \hat{\Sigma}_2 \hat{V}_2^T$ independently and note that

$$(2.3) \quad BB^T = \begin{bmatrix} U_1 \Sigma_1^2 U_1^T & 0 \\ 0 & \hat{U}_2 \hat{\Sigma}_2^2 \hat{U}_2^T \end{bmatrix} + \begin{bmatrix} \beta e_k \\ \alpha e_1 \end{bmatrix} (\beta e_k^T, \alpha e_1^T)$$

$$= \begin{bmatrix} U_1 & 0 \\ 0 & \hat{U}_2 \end{bmatrix} \left(\begin{bmatrix} \Sigma_1^2 & 0 \\ 0 & \hat{\Sigma}_2^2 \end{bmatrix} + \begin{bmatrix} u_1 \\ \hat{u}_2 \end{bmatrix} (u_1^T, \hat{u}_2^T) \right) \begin{bmatrix} U_1^T & 0 \\ 0 & \hat{U}_2^T \end{bmatrix} ,$$

where $u_1 = \beta U_1^T e_k$ and $\hat{u}_2 = \alpha \hat{U}_2^T e_k$. This sequence of computations is equivalent to deleting a column from B and is the basis for the method presented in this paper.

The splitting outlined above might be considered a special case of the general rank one updates to the singular value decomposition described in [2]. It is equivalent to a splitting suggested by Arbenz and Golub in [1].

The approach presented here includes a deflation technique suitable for finite precision arithmetic that avoids working with the squares of the singular values. It also includes a scheme for simultaneous computation of the left and right singular vectors which avoids problems associated with an ill-conditioned bidiagonal matrix B.

3. Deflation

The interior matrix in equation (2.3) is the sum of a diagonal and a rank one matrix. Its eigenvalues can thus be found via the techniques in [3,7]. However, because the squares of the small singular values are not as well separated as the singular values themselves, the deflation rules of [7] concerning nearly equal singular values are likely to be invoked in an inappropriate way. In [10], a set of deflation rules that do not depend on the squared values Σ_1^2 and $\hat{\Sigma}_1^2$ are developed.

Specifically, if the $(n-k) \times (n-k-1)$ matrix $\bar{B}_2$ is defined by $(0, \bar{B}_2) \equiv \hat{B}_2 = B_2 \left[I - e_1 e_1^T \right]$, and if the SVD of $\bar{B}$ is given by $\bar{B}_2 = \bar{U}_2 \bar{\Sigma}_2 \bar{V}_2^T$, then the matrix B may be factored as

$$(3.1) \qquad B = \begin{bmatrix} U_1 & 0 & 0 \\ 0 & \bar{U}_2 & \bar{u} \end{bmatrix} \begin{bmatrix} \Sigma_1 & 0 & u_1 \\ 0 & \bar{\Sigma}_2 & u_2 \\ 0 & 0 & \mu \end{bmatrix} \begin{bmatrix} V_1^T & 0 & 0 \\ 0 & 0 & V_2^T \\ 0 & 1 & 0 \end{bmatrix}.$$

Note that the permutation of (u_1^T, u_2^T, μ) to the last column of the interior matrix is of notational advantage only and need not be carried out during the actual computation.

The SVD of B can then be determined from the SVD of

$$(3.2) \qquad M \equiv \begin{bmatrix} \bar{\Sigma} & u \\ 0 & \mu \end{bmatrix} \equiv \begin{bmatrix} \Sigma_1 & 0 & u_1 \\ 0 & \bar{\Sigma}_2 & u_2 \\ 0 & 0 & \mu \end{bmatrix}.$$

The following lemma provides the key criterion for deflation during the computation of the SVD of M in exact arithmetic.

LEMMA (3.3) If $u^T e_j = 0$, then $\bar{\sigma}_j = e_j^T \bar{\Sigma} e_j$ is a singular value of M and e_j is a left and right singular vector for M corresponding to $\bar{\sigma}_j$.

This lemma shows we can deflate in finite precision when a component of the vector u is small. That is, we can accept e_j as a left and right singular vector corresponding to singular value $\bar{\sigma}_j$ whenever μ_j is smaller than a specified tolerance. It is also possible to deflate when two diagonal entries of $\bar{\Sigma}$ are nearly equal or when one of these diagonal entries is small by using Givens rotations to introduce a zero component in the vector u. These operations are summarized in the following equations:

$|\sigma_k - \sigma_{k+1}|$ small:

$$\begin{bmatrix} c & -s \\ s & c \end{bmatrix} \begin{bmatrix} \sigma_k & 0 & \mu_k \\ 0 & \sigma_{k+1} & \mu_{k+1} \end{bmatrix} \begin{bmatrix} c & s & 0 \\ -s & c & 0 \\ 0 & 0 & 1 \end{bmatrix} = \begin{bmatrix} \hat{\sigma}_k & 0 & 0 \\ 0 & \hat{\sigma}_{k+1} & \tau \end{bmatrix} + cs(\sigma_k - \sigma_{k+1}) \begin{bmatrix} 0 & 1 & 0 \\ 1 & 0 & 0 \end{bmatrix},$$

σ_k small:

$$\begin{bmatrix} c & -s \\ s & c \end{bmatrix} \begin{bmatrix} \sigma_k & \mu_k \\ 0 & \mu \end{bmatrix} = \begin{bmatrix} \hat{\sigma}_k & 0 \\ 0 & \tau \end{bmatrix} - s\sigma_k \begin{bmatrix} 0 & 0 \\ 1 & 0 \end{bmatrix},$$

In both cases, the submatrix on the left is replaced by the first submatrix on the right of the equation with the second term representing the error introduced by making this replacement. The proof of Lemma (3.3) along with others supporting the above comments are given in [10] and provide the basis for deflation rules in finite precision arithmetic. This deflation scheme achieves the desired effect. Namely, the original problem of finding the singular values of M by updating is replaced by the smaller problem of the same form with the diagonal elements of Σ numerically distinct and nonzero and with the elements of u having all components bounded away from zero relative to the size of the largest singular value of the matrix. Note that small singular values are not set to zero in this procedure.

4. Computation of Singular Values and Singular Vectors

After deflation, we need only explicitly compute the singular value decomposition of $M = Y \Sigma X^T$ in (3.2), where $\tilde{\Sigma}$ has distinct, positive elements, and the vector u has only nonzero elements. The squares of the singular values and the left singular vectors are given by the eigendecomposition

$$(4.1) \qquad Y\Sigma Y^T \equiv MM^T \equiv \begin{bmatrix} \tilde{\Sigma}^2 & 0 \\ 0 & 0 \end{bmatrix} + \begin{bmatrix} u \\ \mu \end{bmatrix} (u^T, \mu).$$

The discussion in Section 3 of [7] shows that the singular values of M are the positive roots of

$$(4.2) \qquad 1 + u_1^T (\Sigma_1^2 - \sigma^2 I)^{-1} u_1 - (\mu/\sigma)^2 + u_2^T (\tilde{\Sigma}_2^2 - \sigma^2 I)^{-1} u_2 = 0$$

which is just the secular equation for the updating problem (4.1). This equation can be solved using the method described in [3,7] by setting $\lambda = \sigma^2$ and then taking square roots. Refinements can and probably should be made to the method to compute the positive roots of (4.2) directly. Once a root $\sigma > 0$ of (4.2) has been found, unnormalized vectors y and x in directions of the corresponding left singular vector y_σ and right singular vector x_σ, respectively, can be computed with

$$(4.3)\, y = \begin{bmatrix} (\Sigma_1^2 - \sigma^2 I)^{-1} u_1 \\ -\mu/\sigma^2 \\ (\tilde{\Sigma}_2^2 - \sigma^2 I)^{-1} u_2 \end{bmatrix}, \quad x = M^T y = \begin{bmatrix} \Sigma_1 (\Sigma_1^2 - \sigma^2 I)^{-1} u_1 \\ u_1^T (\Sigma_1^2 - \sigma^2 I)^{-1} u_1 - (\mu/\sigma)^2 + u_2^T (\tilde{\Sigma}_2^2 - \sigma^2 I)^{-1} u_2 \\ \Sigma_2 (\tilde{\Sigma}_2^2 - \sigma^2 I)^{-1} u_2 \end{bmatrix}.$$

Because σ is a root of (4.2), the middle component of x in (4.3) is equal to -1. Therefore, the singular value σ and the corresponding right and left singular vectors x_σ and y_σ can be computed with the following scheme

Procedure 4.4 :

1) Solve (4.2) for σ ;

$$2)\, y \leftarrow \begin{bmatrix} y_1 \\ \eta \\ y_2 \end{bmatrix} = \begin{bmatrix} (\Sigma_1^2 - \sigma^2 I)^{-1} u_1 \\ -\mu/\sigma^2 \\ (\bar{\Sigma}_2^2 - \sigma^2 I)^{-1} u_2 \end{bmatrix} \;\; ; \;\; x \leftarrow \begin{bmatrix} x_1 \\ \xi \\ x_2 \end{bmatrix} = \begin{bmatrix} \Sigma_1 y_1 \\ -1 \\ \bar{\Sigma}_2 y_2 \end{bmatrix} \tilde{} \;;$$

3) $x_\sigma \leftarrow x/\|x\|$ and $y_\sigma \leftarrow y/\|y\|$.

It is worth noting that in the implementation of Procedure 4.4, the differences on the diagonals of $\Sigma_1^2 - \sigma^2 I$ and $\bar{\Sigma}_2^2 - \sigma^2 I$ are maintained and updated directly using the iterative corrections to the approximate root σ in the root finding method. When quadratic convergence of the root finder occurs, these differences have been computed to full relative accuracy, and the computed singular vectors are numerically orthogonal. This accuracy would be lost due to numerical cancellation if, instead, we were to first compute the root σ and then subtract. Determination of x and y in Step 2 of Procedure 4.4 thus consists of simple diagonal scalings. A detailed analysis of the orthogonality of these vectors in finite precision arithmetic is given in [10].

5. Remarks and Conclusions

Limited computational experience with this technique is reported in [10]. There we show that the numerical properties appear to be sound in practice, with orthogonal singular vectors and accurate singular values computed for very ill-conditioned problems. Reasonable parallel performance was observed but speedups were not linear. However, when compared to DSVDC from LINPACK, the performance in both serial and parallel modes were more than satisfactory. The method blends well with block reduction to bidiagonal form as described in [6,10] and hence is rich in matrix-matrix and matrix-vector operations. Additional experience is needed but this method seems well worth considering further.

6. Acknowledgments

We are grateful to Peter Arbenz and Gene Golub for several enlightning discussions. In particular, their paper [1] motivated us to consider the formulation developed in Section 3. We also wish to acknowledge a stimulating discussion during a special session on divide and conquer methods during the Gatlinburg X meeting held at Fairfield Glade in October 1987. A question from Gene Wachspress prompted us to reconsider the computation of right singular vectors so that we ultimately discovered the method described in Section 4.

7. References

[1]   P. Arbenz and G.H. Golub, *On the Spectral Decomposition of Hermitian Matrices Subjected to Indefinite Low Rank Perturbations,* Manuscript NA-87-07, Computer Science Department, Stanford University, 1987.

[2]   J.R. Bunch and C.P. Nielsen, *Updating the Singular Value Decomposition,* Numer. Math. 31, pp.111-129, 1978.

[3]   J.R. Bunch, C.P. Nielsen, and D.C. Sorensen, *Rank-One Modification of the Symmetric Eigenproblem,* Numer. Math. 31, pp. 31-48, 1978.

[4]   J.J.M. Cuppen, *A Divide and Conquer Method for the Symmetric Tridiagonal Eigenproblem,* Numer. Math. 36, pp. 177-195, 1981.

[5]   J.J. Dongarra, J.R. Bunch, C.B. Moler, and G.W. Stewart, *LINPACK Users' Guide,* SIAM Publications, Philadelphia, 1979.

[6]   J.J. Dongarra, S. Hammarling, and D.C. Sorensen, *LAPACK Working Note # 2 Block Reduction to Tridiagonal and Hessenberg Form for the Eigenvalue Problem* ANL-MCS TM No. 99 , September 1987.

[7]   J.J. Dongarra and D.C. Sorensen, *A Fully Parallel Algorithm for the Symmetric Eigenproblem,* SIAM J. Sci. Stat. Comput. 8, pp.139-154, 1987.

[8]   G.H. Golub and C.L. Van Loan, *Matrix Computations,* The Johns Hopkins University Press, 1983.

[9]   I.C.F. Ipsen and E.R. Jessup, *Solving the Symmetric Tridiagonal Eigenvalue Problem on the Hypercube,* Research Report 548, Department of Computer Science, Yale University, 1987.

[10]   E.R. Jessup and D.C. Sorensen, *A Parallel Algorithm for Computing the Singular Value Decomposition of a Matrix* ANL-MCS TM No. 102 , December 1987.

CHAPTER 13

A Multiprocessor Scheme for the Singular Value Decomposition*

Michael Berry[†]
Ahmed Sameh[†]

Abstract. We present a multiprocessor scheme for determining the singular value decomposition of rectangular matrices in which the number of rows is substantially larger or smaller than the number of columns. In this scheme, we perform an initial $\mathbf{QR}$ factorization on the tall matrix (either $\mathbf{A}$ or $\mathbf{A^T}$) using a multiprocessor block Householder algorithm. We then apply a one–sided Jacobi multiprocessor method on the resulting upper triangular $\mathbf{R}$ to yield $\mathbf{RV} = \mathbf{U\Sigma}$, from which the the desired singular value decomposition is obtained. On an Alliant FX/8 computer system with 8 processors, speedups near 5 are obtained for our scheme over an optimized implementation of the Linpack/Eispack routines which perform the classical bi–diagonalization technique. Our scheme performs exceptionally well for rank deficient matrices as well as for those rectangular matrices having clustered or multiple singular values, and may be well suited for applications such as real–time signal processing.

1. Introduction. The singular value decomposition (SVD) is commonly used in determining the solution of unconstrained linear least squares problems. In applications such as real–time signal processing, the solution to these problems is needed in the shortest possible time. Given the growing availability of multiprocessor computer systems, there has been great interest in the development of parallel implementations of the singular value decomposition. We present a multiprocessor scheme for determining the SVD of an m by n matrix A, in which either $m \gg n$ or $n \gg m$, that has demonstrated significant speedups over implementations of the more classical SVD techniques on shared–memory computer systems such as the Alliant FX/8 and Cray X–MP/48.

Without loss of generality, suppose A is a real m by n matrix with $m \gg n$. The singular value decomposition of A can be defined as

$$\mathbf{A} = \mathbf{U\Sigma V^T} , \tag{1}$$

* Research supported by the National Science Foundation under Grant Nos. US NSF MIP–8410110 and US NSF DCR85–09970, the U.S. Department of Energy under Grant No. US DOE DE–FG02–85ER25001, the U.S. Air Force Office of Scientific Research under Grant No. AFOSR–85–0211, and the IBM Donation.

† Center for Supercomputing Research and Development, University of Illinois at Urbana–Champaign, Urbana, IL, 61801

where $U^T U = V^T V = I_n$ and $\Sigma = diag(\sigma_1, \cdots, \sigma_n)$. The orthogonal matrices U and V define the orthonormalized eigenvectors associated with the n eigenvalues of AA^T and A^TA, respectively. The singular values of A are defined as the diagonal elements of Σ which are the nonnegative square roots of the n eigenvalues of AA^T. The multiprocessor scheme we are proposing consists of two stages which are outlined below.

2. Block Householder Reduction. Given the m by n matrix A, where $m \gg n$, we perform a block generalization of Householder's reduction for orthogonal factorization so that

$$A = QR,\tag{2}$$

where Q is m by n and has orthogonal columns. We note that R, an n by n upper–triangular matrix, will be singular for rank deficient matrices A. Block formulations of Householder transformations were proposed by Brenlund and Johnsen [BrJo74], Dietrich [Diet76], and more recently by Bischof and Van Loan [BiVa85]. We adopt the Bischof and Van Loan Method which is based on the observation that the product of k Householder transformations $H_k = P_k \cdots P_2 P_1$ can be written in the form

$$H_k = I - V_k U_k^T,\tag{3}$$

where U_k, V_k are m by k matrices so that $U_k = (u_1, u_2, \cdots, u_k)$ and $V_k = (P_k V_{k-1}, u_k)$. In [BiVa85] it was shown that such a block scheme has the same numerical properties as the classical Householder factorization method.

One major motivation for using this block generalization method on a machine such as an Alliant FX/8 lies in its extensive use of vector–matrix, matrix–vector, and matrix–matrix multiplication modules. As demonstrated in [BGHJ86] and [GaJM86], optimal performance of classical linear algebra algorithms on hierarchical memory systems, such as that of the Alliant FX/8 and *CEDAR* [KDLS85], can be achieved by reformulating the algorithm in terms of matrix–vector and/or matrix–matrix operations (BLAS3). On the Alliant FX/8, the product of Householder transformations, H_k, given in (3) is determined and applied using vector–matrix, matrix–vector, and matrix–matrix multiplication modules which have been exclusively optimized in order to exploit the 8 vector processors and hierarchical memory system. The efficiency of this particular block Householder method is best realized in the speedups obtained over vectorized implementations of the classical Householder factorization [DBMS79]. For a 1000 by 100 matrix A, speedups as high as 9 have been achieved.

3. One–Sided Jacobi Method. To the r by n matrix R, where r is the column rank of A (or row rank for A^T), determined by the block Householder factorization of A, we apply a one–sided Jacobi method based on the one–sided iterative orthogonalization scheme of Hestenes [Hest58] (see also [Kais72] and [Nash75]). Our goal is to determine an orthogonal matrix V as a product of plane rotations so that

$$RV = \tilde{Q} = (\tilde{q}_1, \tilde{q}_2, \cdots, \tilde{q}_n),\tag{4}$$

and

$$\tilde{q}_i^T \tilde{q}_j = \sigma_i^2 \delta_{ij},$$

where the columns of $\tilde{Q}$, $\tilde{q}_i$, are orthogonal, and δ_{ij} is the Kronecker delta. We then may write $\tilde{Q}$ as

$$\tilde{Q} = \tilde{U}\Sigma \text{ with } \tilde{U}^T \tilde{U} = I_n,\tag{5}$$

and hence combining (4) and (5) we obtain

$$R = \tilde{U}\Sigma V^T.\tag{6}$$

From (2), (6) we obtain

$$A = QR = Q\tilde{U}\Sigma V^T = U\Sigma V^T,$$

where $U = Q\tilde{U}$. Thus we can realize the decomposition in (1) after determining the matrix V in (4).

We construct the matrix $\mathbf{V}$ via the (i,j) plane rotation

$$(\mathbf{r_i}, \mathbf{r_j}) \begin{bmatrix} c & -s \\ s & c \end{bmatrix} = (\tilde{\mathbf{r}}_i, \tilde{\mathbf{r}}_j) \quad i < j \ ,$$

so that

$$\tilde{\mathbf{r}}_i^T \tilde{\mathbf{r}}_j = 0, \quad \text{and} \quad \|\tilde{\mathbf{r}}_i\| > \|\tilde{\mathbf{r}}_j\| \ , \tag{7}$$

where $\mathbf{r_i}$ designates the $i-th$ column of $\mathbf{R}$. This is accomplished by choosing

$$c = \left[\frac{\beta+\gamma}{2\gamma}\right]^{1/2} \quad \text{and} \quad s = \left[\frac{\alpha}{2\gamma c}\right] \quad \text{if } \beta > 0 \ , \tag{8}$$

or

$$s = \left[\frac{\gamma-\beta}{2\gamma}\right]^{1/2} \quad \text{and} \quad c = \left[\frac{\alpha}{2\gamma s}\right] \quad \text{if } \beta < 0 \ , \tag{9}$$

where $\alpha = 2\,\mathbf{r_i}^T\mathbf{r_j}$, $\beta = \|\mathbf{r_i}\|^2 - \|\mathbf{r_j}\|^2$, and $\gamma = (\alpha^2 + \beta^2)^{1/2}$. Note that (7) requires the columns of $\tilde{\mathbf{Q}}$ to decrease in norm from left to right, and hence the resulting σ_i to be in monotonic non-increasing order. Several schemes for ordering the (i,j) plane rotations may be used in orthogonalizing the columns of $\mathbf{R}$. One possible orthogonalization scheme (see [BeSa86]) is given in Figure 1. For $n = 8$ we can orthogonalize the pairs $(1,8)$, $(2,7)$, $(3,6)$, $(4,5)$ simultaneously via the first orthogonal transformation $\mathbf{V_1}$ which consists of the direct sum of 4 plane rotations. A sweep consists of $n(n-1)/2$ plane rotations and at the end of any particular sweep s_i we have

$$\mathbf{V}_{s_i} = \mathbf{V_1} \mathbf{V_2} \cdots \mathbf{V_n} \ ,$$

and hence

$$\mathbf{V} = \mathbf{V}_{s_1} \mathbf{V}_{s_2} \cdots \mathbf{V}_{s_t} \ , \tag{10}$$

where t is the number of sweeps required for convergence. This process is iterative since orthogonality between columns established by one transformation may be destroyed in subsequent transformations.

X	7	6	5	4	3	2	1
	X	5	4	3	2	1	8
		X	3	2	1	8	7
			X	1	8	7	6
				X	7	6	5
					X	5	4
						X	3
							X

Figure 1. Orthogonalization Scheme for One–Sided Jacobi Method.

We note that neither row accesses nor column interchanges are required by the one–sided Jacobi method and that the rotations can be applied by the following vector operations,

$$\mathbf{r_i}^{k+1} = c\,\mathbf{r_i}^k + s\,\mathbf{r_j}^k \ ,$$

$$\mathbf{r_j}^{k+1} = -s\,\mathbf{r_i}^k + c\,\mathbf{r_j}^k \ ,$$

where $\mathbf{r_i}$ denotes the $i-th$ column of matrix $\mathbf{A}$, and c, s are determined by either (8) or (9). On multiprocessors, such as the Alliant FX/8, no processor synchronization is required since each processor can be assigned one rotation and hence orthogonalizes one pair of the n columns of the matrix $\mathbf{R}$.

Following the convergence criterion used in [Nash75], we count the number of times

$$\frac{\mathbf{r_i^T r_j}}{(\mathbf{r_i^T r_i})(\mathbf{r_j^T r_j})} \tag{11}$$

falls, in any sweep, below a given *tolerance*. The algorithm terminates when this condition is satisfied $n(n-1)/2$ times after any sweep. Upon termination, the matrix $\mathbf{R}$ has been overwritten by the matrix $\tilde{\mathbf{Q}}$ from (5) and hence the singular values σ_i can be obtained via the n square roots of the diagonal entries of $\mathbf{R^T R}$. The matrix $\mathbf{U}$ in (1), which contains the left singular values of the original matrix $\mathbf{A}$, is readily obtained by scaling the resulting matrix $\mathbf{R}$ (now overwritten by $\tilde{\mathbf{Q}} = \tilde{\mathbf{U}}\Sigma$) by the singular values σ_i followed by the premultiplication of the Householder transformation matrix $\mathbf{Q}$ from (2), i.e., $\mathbf{U} = \mathbf{Q}\tilde{\mathbf{U}}$. The matrix $\mathbf{V}$, which contains the right singular vectors of the original matrix $\mathbf{A}$, is obtained as in (10) as the product of the orthogonal $\mathbf{V_k}$'s. This product is accumulated separately via application of the rotations specified by (8) and (9) to the $n \times n$ identity matrix.

In Figure 2, we present the results of computing (1) for $n \times 128$ matrices having the singular values $\sigma_i = i$, $i = 1, \cdots, 128$, on an Alliant FX/8 by our proposed multiprocessor method (referred to as **HQRJAC**) and by the Linpack routine **DSVDC**, which performs the classical bi–diagonalization SVD algorithm. These experiments were performed using 64–bit double precision floating–point numbers, and the accuracy obtained in the singular values and vectors by both methods were identical. Our version of **DSVDC** was fine–tuned for the Alliant FX/8 through the use of matrix–vector multiplication kernels (BLAS2) in the application and accumulation of Householder transformations. We note that **HQRJAC** consistently outperformed **DSVDC** with a speedup ranging from 4 to 5. As mentioned earlier, **HQRJAC** performs well for rank deficient rectangular matrices and those matrices having clustered or multiple singular values. Although increasing the *tolerance* chosen for the ratio given in (11) can significantly improve the overall performance of **HQRJAC**, we maintained our *tolerance* for these experiments to be 10^{-14} to assure competitive accuracy with **DSVDC**.

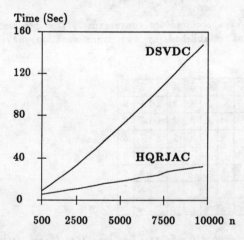

Figure 2. *Computation times for* **HQRJAC** *and* **DSVDC** *on an Alliant FX/8.*

REFERENCES

[BGHJ86]   M. Berry, K. Gallivan, W. Harrod, W. Jalby, S. Lo, U. Meier, B. Phillipe and A. Sameh, "Parallel Numerical Algorithms on the *CEDAR* System," *Lecture Notes in Computer Science*, vol. 237, pp. 25–39, 1986.

[BeSa86]   M. Berry and A. Sameh, "Multiprocessor Jacobi schemes for dense symmetric eigenvalue and singular value decompositions," CSRD Report No. 546, University of Illinois at Urbana–Champaign, 1986.

[BiVa85]   C. Bischoff and C. Van Loan, "The WY representation for products of Householder Matrices," TR 85–681, Department of Computer Science, Cornell University, 1985.

[BLVL85]   R. P. Brent, F. T. Luk, and C. Van Loan, "Computation of the singular value decomposition using mesh connected processors," *J. VLSI and Computer Systems,* vol. 1, no. 3, pp. 242–270, 1985.

[BrLu85]   R. P. Brent and F. T. Luk, "The solution of singular value and symmetric eigenproblems on multiprocessor arrays," *SIAM J. Sci. Stat. Comput.,* vol. 6, pp. 69–84, 1985.

[BrJo74]   O. Bronlund and T. Johnsen, "QR–factorization of partitioned matrices," *Computer Methods in Applied Mechanics and Engineering,* vol. 3, pp. 153–172, 1974.

[Diet76]   G. Dietrich, "A new formulation of the hypermatrix Householder–QR decomposition," *Computer Methods in Applied Mechanics and Engineering,* vol. 9, pp. 273–280, 1976.

[DBMS79]   J. Dongarra, J. Bunch, C. Moler, and G. W. Stewart, *LINPACK Users' Guide,* SIAM, Philadelphia, 1979.

[GaJM86]   K. Gallivan, W. Jalby, and U. Meier, "The Use of *BLAS3* in Linear Algebra on a Parallel Processor with a Hierarchical Memory," CSRD Report No. 610, University of Illinois at Urbana–Champaign, 1986.

[Hest58]   M. R. Hestenes, "Inversion of matrices by biorthogonalization and related results," *J. Soc. Indust. Appl. Math.,* vol. 6, pp. 51–90, 1958.

[Kais72]   H. F. Kaiser, "The JK method: a procedure for finding the eigenvectors and eigenvalues of a real symmetric matrix," *The Computer Journal,* vol. 15, no. 33, pp. 271–273, 1972.

[KDLS86]   D. Kuck, E. Davidson, D. Lawrie, and A. Sameh, "Parallel Supercomputing Today and the *CEDAR* Approach," *Science,* vol. 231, pp. 967–974, 1986.

[Nash75]   J. C. Nash, "A one–sided transformation method for the singular value decomposition and algebraic eigenproblem," *The Computer Journal,* vol. 18, no. 1, pp. 74–76, 1975.

[Same71]   A. Sameh, "On Jacobi and Jacobi–like algorithms for a parallel computer," *Math. Comp.,* vol. 25, pp. 579–590, 1971.

[VanL85]   C. Van Loan, "The block Jacobi method for computing the singular value decomposition," Technical Report No. 85–680, Department of Computer Science, Cornell University, Ithaca, New York, 1985.

CHAPTER 14

A Parallel Algorithm for the Singular Value Decomposition of Rectangular Matrices

William D. Shoaff*
C. David Chan†

Abstract. The classical sequential algorithm for computing the singular value decomposition (SVD) of a matrix has time complexity $O(n^3)$ and therefore can only be used on small matrices. Parallel algorithms for computing the SVD have recently been proposed. The best of these use Jacobi rotations to diagonalize the matrix. However, these Jacobi-SVD algorithms require that the matrix be square, and, when this is not the case, the matrix must be preprocessed. A parallel algorithm is presented that computes the SVD of a rectangular matrix without preprocessing. This algorithm is compared against a selection of other parallel SVD algorithms.

Key Words. Singular value decomposition, parallel algorithms, Jacobi rotation, Givens rotation

1. Introduction. The singular value decomposition (SVD) of an m by n real matrix A, with $m \geq n$, is given by

$$A = U\Sigma V^T,$$

where U is an m by m orthogonal matrix, Σ is an m by n diagonal matrix with diagonal elements $\sigma_1 \geq \sigma_2 \geq \cdots \geq \sigma_n \geq 0$, and V is an n by n orthogonal matrix.

Many algorithms for computing the SVD have recently been proposed [1], [2], [3], [4], [5], [6]. Most of these algorithms require that the matrix A be square, and when this is not the case, some device must be used to obtain a square matrix. For example, columns of zeros can be appended to A, or the QR decomposition of A can be computed and then the SVD of the upper triangular matrix R can be found. Other algorithms require that the matrix A be in triangular or bidiagonal form before the SVD can be computed.

In our approach, which borrows from the ideas of Brent, Luk, Van Loan [4], and Stewart [7], the SVD of A computed directly. The idea is very simple but, when run in a multiprocessor environment, it does require careful synchronization in that the processors must change the programs that they are running at different phases of the algorithm.

---

* Department of Computer Science, Florida Institute of Technology, Melbourne, Florida 32901
† Division of Science and Mathematics, University of Minnesota–Morris, Morris, Minnesota 56267

2. The Algorithm. To describe the algorithm it helps to consider a simple case first. Let

$$A = \begin{bmatrix} a & b \\ c & d \\ e & f \end{bmatrix}.$$

The algorithm proceeds in two phases. The purpose of the first phase is to diagonalize the upper 2 by 2 block of A and move the $(1,1)$ element down to where it can annihilate the element in row 3 column 1. In the second phase the elements from the third row are annihilated.

A phase one step consists of three substeps. First, a rotation matrix S is used to symmetrize the upper 2 by 2 block of A. Then an inner Jacobi rotation J is used to diagonalize the upper 2 by 2 block. Finally, a permutation P is used to interchange the rows one and two and the columns of A. This sequence of transformations can be expressed as

(1) $$A = (SJP)^T A(\hat{J}\hat{P}),$$

where S, J, P are 3 by 3 matrices with identity last row and column, and $\hat{J}$ and $\hat{P}$ are J and P stripped of their last row and column. The result of the update given by equation 1 is a new matrix of the form

$$A = \begin{bmatrix} d' & 0 \\ 0 & a' \\ f' & e' \end{bmatrix}.$$

There are two substeps in a phase two step. Given the matrix A above, an inner Givens rotation is used to zero the $(3,2)$ element, and then the second and third rows of A are permuted. After completion of these steps the matrix A has the form

$$A = \begin{bmatrix} d' & 0 \\ f'' & 0 \\ g'' & a'' \end{bmatrix}.$$

To complete a *forward sweep* in the 3 by 2 case, another phase two step is performed on the first and second row of A. This eliminates the element f'' from A. There is also a *backward sweep* which restores the elements to their original order.

More generally, in the parallel implementation of the algorithm described in Section 3, elements of the matrix A follow paths as shown in Fig. 1. In a forward sweep, the $(1,1)$ element slides down the diagonal and zeros elements in the super and subdiagonal as it passes by. When it reached the last column, it percolates down to the

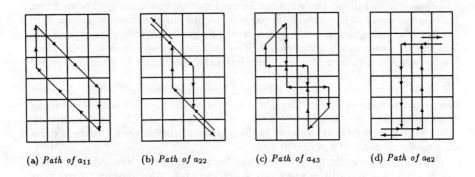

(a) *Path of a_{11}*   (b) *Path of a_{22}*   (c) *Path of a_{43}*   (d) *Path of a_{62}*

FIG. 1. *Paths of Elements*

last row zeroing elements below it as it falls. In a backward sweep, the element in position (m, n) rises along the lower diagonal and bubbles up the first column. This path is shown in Fig. 1 (a). Other diagonal elements slide up the diagonal to position $(1, 1)$, where they turn around and slide down the diagonal until they reach the appropriate column. From there, they percolate down rows until the forward sweep ends. In a backward sweep, they follow a symmetric path back to their original positions, see Fig. 1 (b). Off-diagonal elements in the upper square block follow a roundabout path. For example, in Fig. 1 (c), the element in position $(4, 3)$ moves left, up, left and up to position $(2, 1)$ where it is annihilated. From there, it moves diagonally to position $(1, 2)$, and then back down to position $(3, 2)$ where the forward sweep end. The backward sweep is symmetrical and returns the element to position $(4, 3)$. Elements in the lower rectangle follow paths similar to the path shown in Fig. 1 (d) for the element in position $(6, 4)$.

A curly brace notation is used to denote a phase one step. For example, $\{i, j\}$ indicates a phase one step on the 2 by 2 matrix formed by the intersection of rows i, $i + 1$ and columns $j - 1$ and j. Square brackets denote a phase two step. When $i > j$, $[i, j]$ indicates that the (i, j) element of A is annihilated and rows $i - 1$ and i are exchanged. Otherwise, when $i \le j$, $[i, j]$ indicates that the (j, i) element of A is annihilated and rows j and $j + 1$ are exchanged. Using this notation, the sequential implementation of the algorithm in the 6 by 4 case can be represented as in Fig. 2.

$$
\begin{array}{lll}
\{1,2\} & \{2,3\} & \{3,4\} \quad \| \quad [5,4] \quad [6,4] \\
\{1,2\} & \{2,3\} & \phantom{\{3,4\}} \quad \| \quad [4,3] \quad [5,3] \\
\{1,2\} & & \phantom{\{3,4\}} \quad \| \quad [3,2] \quad [4,2] \\
& & \phantom{\{3,4\}} \quad \| \quad [2,1] \quad [3,1]
\end{array}
$$

$$
\begin{array}{lll}
\{5,4\} & \{4,3\} & \{3,2\} \quad \| \quad [1,2] \quad [1,1] \\
\{5,4\} & \{4,3\} & \phantom{\{3,2\}} \quad \| \quad [2,3] \quad [2,2] \\
\{5,4\} & & \phantom{\{3,2\}} \quad \| \quad [3,4] \quad [3,3] \\
& & \phantom{\{3,2\}} \quad \| \quad [4,5] \quad [4,4]
\end{array}
$$

FIG. 2. *A Forward and Backward Sweep of a 6 by 4 Matrix.*

3. Parallel Implementation. Premultiplying or postmultiplying a matrix A by a rotation matrix changes only two rows or columns, respectively, of A. Thus, it is possible to partition a matrix into disjoint pairs of rows or columns and perform rotations on these pairs independently. Fig. 3 exhibits a parallel implementation of the algorithm in the 6 by 4 case. The numbers to the side are the time steps, and the six numbers following them give the current position of the original six rows. The Roman I or II above a dash indicates a phase one or phase two step to be performed during the time step.

Phase one and phase two steps can be implemented concurrently. Forward and backward sweeps can also execute in parallel. During the first three steps, the first element propagates to the last position on the diagonal and thus phase two of the first forward sweep can begin at step 4. The backward sweep begins at step 7, and the forward sweep ends at step 8, A new forward sweep begins at step 13, and the backward sweep ends at step 14.

There are $m + n - 2$ parallel steps in a forward or backward sweep. Forward sweeps end at steps $m + n - 2, 3m + n - 2, 5m + n - 2, \ldots$, while backward sweeps end at steps $2m + n - 2, 4m + n - 2, 6m + n - 2$, and so on. The time complexity of the algorithm is $O(mS)$, where S is the number of sweeps needed to reduce the off-diagonal elements to a negligible size.

To implement the algorithm assume that both m and n are even and append one row or column of zeros to A when this is not the case. Also assume a mesh connected array of processors as described by Stewart [7] except extended to rectangular

```
                    I
 1.        1  —  2      3      4      5      6
                    I
 2.        2     1  —  3      4      5      6
              I           I
 3.        2  —  3     1  —  4      5      6
                    I           II
 4.        3     2  —  4     1  —  5      6
              I           II          II
 5.        3  —  4     2  —  5     1  —  6
                    II          II
 6.        4     3  —  5     2  —  6     1
              II          II          I
 7.        4  —  5     3  —  6     2  —  1
                    II          I
 8.        5     4  —  6     3  —  1     2
                          I           I
 9.        5     6     4  —  1     3  —  2
                    II          I
10.        5     6  —  1     4  —  2     3
              II          II          I
11.        5  —  1     6  —  2     4  —  3
                    II          II
12.        1     5  —  2     6  —  3     4
              I           II          II
13.        1  —  2     5  —  3     6  —  4
                    I           II
14.        2     1  —  3     5  —  4     6
```

FIG. 3. *Parallel Implementation of Phase One and Phase Two Steps.*

matrices. The processors are labeled

$$
\begin{aligned}
(2i-1, 2j-1), & \quad i = 1, \ldots, m/2, & \quad j = 1, \ldots, n/2, \\
(2i, 2j), & \quad i = 0, \ldots, m/2, & \quad j = 1, \ldots, n/2 - 1, \\
(2i, 2j), & \quad i = 1, \ldots, m/2 - 1, & \quad j = 0, n/2.
\end{aligned}
$$

Each processor (k, l) contains four data registers which hold the $(k, l), (k+1, l), (k, l+1)$, and $(k+1, l+1)$ elements of A. Diagonal processors (k, k) and $(m-k, n-k), k = 1, \ldots, n-1$ compute the rotation parameters of phase one steps and broadcast these to other processors in the same row and column. The processors $(l + 2k, l)$ and $(l + 2k + 1, l - 1), k = 0, \ldots, (m-n-2)/2$ are used to compute the rotations that zero the elements in the l-th column during phase two operations. These processors broadcast their rotation parameters to other processors in the same row. If it is not assumed that the rotation parameters can be broadcast in constant time, then each step must be delayed by one execution cycle and twice the number of steps given above are required to complete a sweep.

4. Numerical Results. To test the algorithm, random m by n matrices whose elements were uniformly distributed in the interval $[-1, 1]$ were generated and the algorithm was run on these matrices. The stopping criteria was that off(A) be reduced to 10^{-12} times its original value. The test for convergence was made at each phase one step of forward and backward sweeps, but not during phase two when the diagonal elements fail to lie on either the main diagonal or the diagonal of the lower square block. The results of these tests are given in Table 1 which show the minimum, average, and maximum number of sweeps needed to compute the SVD of the matrix. These results compare favorably with the results of Brent, Luk, and Van Loan [4] which are given in the last column, labeled USVD, of Table 1.

TABLE 1
Minimum, Average, and Maximum Sweep Using the Algorithm.

m	n	Trials	(Min)Sweeps(Max)	USVD
4	4	1000	(2.50)3.11(4.00)	2.97(4.00)
6	4	1000	(2.83)3.38(4.17)	
6	6	1000	(2.67)3.68(4.67)	3.76(4.87)
8	4	1000	(2.88)3.43(4.13)	
8	8	1000	(3.25)4.07(5.00)	4.21(5.14)
10	6	1000	(3.10)3.96(5.00)	
10	10	1000	(3.50)4.36(5.30)	4.55(5.44)
20	10	1000	(3.85)4.68(5.75)	
20	20	1000	(4.40)5.18(6.00)	5.54(6.01)
30	16	100	(4.77)5.18(5.87)	
30	30	100	(5.13)5.58(6.47)	6.09(6.80)
40	20	100	(4.95)5.43(6.05)	
40	40	100	(5.30)5.97(6.48)	6.40(6.98)
50	26	50	(5.26)5.77(6.04)	
50	50	50	(5.70)6.22(6.72)	6.72(7.34)
80	40	10	(5.89)6.08(6.59)	
80	80	10	(6.31)6.54(6.88)	7.30(7.79)
100	50	10	(6.10)6.39(6.79)	
100	100	10	(6.26)6.82(7.24)	7.56(8.00)
150	76	5	(6.73)6.91(7.01)	
150	150	5	(6.78)6.83(6.89)	7.73(8.03)
200	100	1	(6.97)6.97(6.97)	
200	200	1	(7.33)7.33(7.33)	8.10(8.10)

REFERENCES

[1] M. BERRY AND A. SAMEH, *A multiprocessor scheme for singular value decomposition*, in Third
    SIAM Conference on Parallel Processing for Scientific Computing, 1987.
[2] R. P. BRENT AND F. T. LUK, *The solution of singular value problems using systolic arrays*, in
    SPIE Real Time Signal Processing VII, January 1984, pp. 7–12.
[3] ———, *The solution of singular-value and symmetric eigenvalue problems on multiprocessor
    arrays*, SIAM J. Sci. Statist. Comput., 6 (1985), pp. 69–84.
[4] R. P. BRENT, F. T. LUK, AND C. VAN LOAN, *Computation of the singular value decomposition
    using mesh-connected processors*, J. VLSI Computer Systems, 1 (1985), pp. 242–270.
[5] V. HARI AND V. KREŠIMIR, *On jacobi methods for singular value decompositions*, SIAM J. Sci.
    Statist. Comput., 8 (1986), pp. 741–754.
[6] E. R. JESSUP AND D. C. SORENSEN, *A parallel algorithm for the singular value decomposition*,
    in Third SIAM Conference on Parallel Processing for Scientific Computing, 1987.
[7] G. W. STEWART, *A jacobi-like algorithm for computing the schur decomposition of a nonhermi-
    tian matrix*, SIAM J. Sci. Statist. Comput., 6 (1985), pp. 853–864.

CHAPTER 15

Trace Minimization Algorithm for the Generalized Eigenvalue Problem*

Bill Harrod[†]
Ahmed Sameh[†]

Abstract. A trace minimization algorithm for computing a few of the smallest (or largest) eigenvalues and associated eigenvectors of the generalized eigenvalue problem $Ax = \lambda Bx$ is presented. Here we assume that the matrices A and B are symmetric of order n, with B being positive definite, and that both A and B are so large and sparse that a factorization of either matrix is impractical. In each iteration of this algorithm, first investigated by A. Sameh and J. Wisniewski, we are simultaneously approximating the p desired eigenpairs p$\ll$n by minimizing the trace of a p $\times$ p matrix subject to quadratic constraints. This paper presents an improved computational scheme which incorporates a variety of acceleration strategies, as well as demonstrates the suitability of the algorithm for a multiprocessor with two levels of parallelism, such as the Alliant FX/8.

Introduction. The problem of computing a few of the smallest (or largest) eigenvalues and the eigenvectors of the large, sparse, generalized eigenvalue problem

$$Ax = \lambda Bx \tag{1}$$

where A and B are symmetric matrices of order n, with B being positive definite arises in many applications. We present a method: "trace minimization" [SaWi82], which is suitable for multiprocessors. This method is competitive with the Lanczos algorithm, especially when the matrices A and B are so large and sparse with a sparsity pattern such that the Cholesky factor of either matrix has to be contained in auxiliary storage.

The main idea behind the algorithm depends on the following observation:

* Research supported by the National Science Foundation under Grant Nos. US NSF MIP–8410110 and US NSF DCR85–09970, the U.S. Department of Energy under Grant No. US DOE DE–FG02–85ER25001, and the IBM Donation.

† Center for Supercomputing Research and Development, University of Illinois at Urbana–Champaign, Urbana, IL, 61801

Let $\mathbf{Y}$ denote the set of all $n \times p$, $p < n$, matrices Y such that $Y^T B Y = I_p$. Then

$$\min_{Y \in \mathbf{Y}} \operatorname{tr}(Y^T A Y) = \sum_{i=1}^{p} \lambda_i,$$

where $\operatorname{tr}(A) = \sum_{i=1}^{n} \alpha_{ii}$ denotes the trace of $A = (\alpha_{i,j})$, and

$$\lambda_1 \leq \lambda_2 \leq \cdots \leq \lambda_p < \lambda_{p+1} \leq \cdots \leq \lambda_n$$

are the eigenvalues of (A,B).

Algorithm. Let Y_k be an $n \times p$ matrix approximating the p eigenvectors corresponding to the smallest p eigenvalues of (1), such that

$$Y_k^T A Y_k = \Sigma_k = \operatorname{diag}(\sigma_1^{(k)}, \ldots, \sigma_p^{(k)}),$$

and $Y_k^T B Y_k = I_p$. Now, choose an $n \times p$ correction matrix Δ_k and an $p \times p$ scaling matrix S_k to construct the next iterate

$$Y_{k+1} = (Y_k - \Delta_k)S_k,$$

so that

$$Y_{k+1}^T A Y_{k+1} = \Sigma_{k+1} = \operatorname{diag}(\sigma_1^{(k+1)}, \ldots, \sigma_p^{(k+1)}),$$

$$Y_{k+1}^T B Y_{k+1} = I_p$$

and

$$\operatorname{tr}(Y_{k+1}^T A Y_{k+1}) < \operatorname{tr}(Y_k^T A Y_k). \tag{2}$$

The most time–consuming part of this algorithm is the choice of Δ_k which ensures that (2) holds. One possible choice is the solution of the following constrained optimization problem

$$\min_{Y_k^T B \Delta_k = 0} \operatorname{tr}((Y_k - \Delta_k)^T A(Y_k - \Delta_k)) \tag{3}$$

If A is positive definite, then (3) is reduced to solving the p independent problems:

$$\min_{Y_k^T B d_j = 0} (y_j^{(k)} - d_j^{(k)})^T A(y_j^{(k)} - d_j^{(k)}) \tag{4}$$

where $y_j^{(k)} = Y_k e_j$ and $d_j^{(k)} = \Delta_k e_j$. Using Lagrange multipliers and dropping the subscripts from (4), we see that solving (4) is equivalent to solving the following system of linear equations

$$\begin{bmatrix} A & BY \\ Y^T B & 0 \end{bmatrix} \begin{bmatrix} d \\ l \end{bmatrix} = \begin{bmatrix} Ay \\ 0 \end{bmatrix} \tag{5}$$

where 2l is a vector representing the Lagrange multipliers. Let the orthogonal factorization of BY be given by QR, where Q has p orthonormal columns and R is an upper triangular matrix of order p. let $\tilde{Q}$ be such that $[Q, \tilde{Q}]$ is an orthogonal matrix of order n. Then solving (5) for d, can be found by solving the following positive definite system

$$(\tilde{Q}^T A \tilde{Q})g = \tilde{Q}^T Ay \tag{6}$$

where $d = \tilde{Q}g$. These p independent systems can be solved using the conjugate gradient algorithm, such that we obtain $y - d$ without explicitly solving for g. The algorithm can be terminated once a measure of the error falls below the corresponding measure of the error in the eigenvector.

The trace minimization algorithm may be summarized as follows:

Stage 1 (Initialization)

Choose an n $\times$ s matrix Y of rank s, where s $\geq$ p+1.

Stage 2 (B–orthonormalization)

Obtain $\hat{Y}$ in the space spanned by Y such that $\hat{Y}^T B \hat{Y} = I_s$.

Stage 3 (Forming a section)

Compute the spectral decomposition $\hat{Y}^T A \hat{Y} = U\Sigma U^T$.
Set $\tilde{Y} = \hat{Y}U$. Thus

$$\tilde{Y}^T A \tilde{Y} = \Sigma, \text{ and } \tilde{Y}^T B \tilde{Y} = I_s,$$

where $\Sigma = \text{diag}(\sigma_1, \sigma_2, ..., \sigma_p)$.

Stage 4 (Test for convergence)

Accept $(\sigma_j, \tilde{y}_j)$ as an eigenpair, if a specified error tolerance is satisfied.

Stage 5 (Factorization)

Compute the orthogonal factorization BY = QR.

Stage 6 (Updating)

Solve the s linear systems (6). The resulting s solution vectors form the columns of the new matrix Y; goto Stage 2.

Column j of Y globally converges to the eigenvector x_j corresponding to λ_j for $1 \leq j \leq s$ with an asymptotic rate of convergence less than or equal to $|\lambda_j/\lambda_{s+1}|$.

Accelerating Convergence. The trace minimization method, which has a linear rate of convergence, can be improved tremendously if it is applied to the s eigenvalue problems

$$(A - \nu_j B)x_j = (\lambda_j - \nu_j)Bx_j,$$

where ν_j's are suitable shifts chosen from the Ritz values σ_j, $1 \leq j \leq s$. An appropriate strategy for choosing these shifts is necessary to maintain global convergence and to ultimately achieve cubic convergence. The only change to the method is that (5) is replaced by the following system:

$$\begin{bmatrix} A - \sigma B & BY \\ Y^T B & 0 \end{bmatrix} \begin{bmatrix} d \\ l \end{bmatrix} = \begin{bmatrix} (A - \sigma B)y \\ 0 \end{bmatrix} \qquad (7)$$

We also investigate the trace minimization algorithm applied to the standard eigenvalue problem

$$Ax = \lambda x \qquad (8)$$

where A is a large, sparse, symmetric matrix. In this case, the above algorithm is simplified and polynomial precondtioning may be used to accelerate the convergence of this scheme. For example if $Ax = \lambda x$, then $P_m(A)x = P_m(\lambda)x$, where $P_m(\lambda)$ is a polynomial of degree m chosen so as to enhance the seperation between $\lambda_{n+1-s}, \lambda_{n+2-s}, ..., \lambda_n$ and the rest of the spectrum. For more economical procedure we consider instead the generalized eigenvalue problem $x = \tau P_m(A)x$, where $\tau = 1/P_m(\lambda)$. If $P_m(A)$ is chosen as

$$P_m(\lambda) = \frac{T_m\left(\dfrac{2\lambda - \gamma_0 - \gamma_1}{\gamma_0 - \gamma_1} \right)}{T_m\left(\dfrac{2 - \gamma_0 - \gamma_1}{\gamma_0 - \gamma_1} \right)}$$

where T_m is a Chebyshev polynomials, then we can choose the parameters $\gamma_0 \leq \lambda_1$ and $\lambda_s \leq \gamma_1 \leq \lambda_{s+1}$ to dampen the eigenvalues $\lambda_1, \lambda_2, ..., \lambda_{n-s}$. Leading the algorithm to converge to the largest s eigenvalues $\lambda_{n+1-s}, \lambda_{n+2-s}, ..., \lambda_n$

The following table demonstrates the improved rate of convergence achievable using polynomial preconditioning. The coeffecent matrix is an admittance matrix of order 494 for a bus power system. The matrix can be found in the Harwell/BCS Sparse matrix test collection, the entry is called 494_bus. The Trace Minization method computes the largest 10 eigenvalues, using 12 vectors. The algorithm is terminated when the relative error for all of the eigenvalues is less than or equal to 10^{-6}. The third column refers to the number of matrix_vector products, and the fourth column shows the time consumed on an Alliant FX/8 parallel computer.

degree of polynomial	number of iterations	number of A mult's	time
	31	372	3.84
2	15	336	1.62
4	9	360	1.14

Newton's Method. After the eigenpairs have converged to a given tolerance, we can improve the accuracy of the computed eigenvalues and eigenvectors via Newton's Method (see [PeWi79]). Let $Y \in \mathbf{R}^{n \times p}$ such that

$$Y^T A Y = G = \Sigma + E$$

where $\|E\|$ is small, and $Y^T B Y = I_p$. There exists δY and δG such that

$$A(Y - \delta Y) - B(Y - \delta Y)(G + \delta G) = 0$$

and $(Y - \delta Y)^T B(Y - \delta Y) = I_p$. Ignoring the 2nd order terms, we get

$$A\delta Y - B\delta YG + BY\delta G = R$$

and $Y^T B \delta Y = 0$, where $R = AY - BYG$. Assuming that $Ge_j \approx \gamma_j e_j$, then we have p linear systems

$$\begin{bmatrix} A - \gamma_j B & BY \\ Y^T B & 0 \end{bmatrix} \begin{bmatrix} \delta y_j \\ \delta g_j \end{bmatrix} = \begin{bmatrix} (A - \gamma_j B) y_j \\ 0 \end{bmatrix} \tag{9}$$

where $\delta Y = (\delta y_1, ..., \delta y_p)$ and $\delta G = (\delta g_1, ..., \delta g_p)$. It is interesting to observe that this is identical to the linear systems (7). The following table shows the number of iterations of the Trace Minimization algorithm followed by Newton's method for the 494_bus matrix where $p = 10$ and $s = 12$. The relative error of the eigenvalues is less than or equal to 10^{-15} and the relative error of the eigenvectos is less than or equal to 10^{-10}.

degree of polynomial	number of Trace Min. iterations	number of Newton iterations	number of A mult's
	7	1	93
2	5	1	114
4	4	1	138

REFERENCES

[PeWi79]   G. Peters and J. H. Wilkinson. *Inverse Iteration, Ill-Condtioned Equations and Newton's Method.* **SIAM Review**, Vol. 21, No. 3, pp. 339–360, July, 1979.

[SaWi82]   Ahmed H. Sameh and John Wisniewski. *A Trace Minimization Algorithm for the Generalized Eigenvalue Problem.* **SIAM Journ. Numer. Anal.**, Vol. 19, No. 6, pp. 1243–1259, Dec., 1982.

CHAPTER 16

A Parallel, Hybrid Algorithm for the Generalized Eigenproblem

Shing C. Ma*
Merrell L. Patrick*
Daniel B. Szyld*

Abstract. We present a parallel algorithm for computing all eigenvalues, and their corresponding eigenvectors, in a specified interval for the generalized eigenproblem, $Ax = \lambda Bx$, where A and B are banded, real, symmetric and B is positive definite. Eigenvalues are isolated in parallel, using the Sturm sequence property of leading principal minors of $A - \mu B$. Concurrently, eigenpairs are computed accurately using a superlinear method which combines inverse and Rayleigh quotient iterations. Results obtained from implementation of this algorithm on a shared memory MIMD architecture are presented.

1. Introduction. The solution to the generalized eigenproblem

$$(1) \qquad Ax = \lambda Bx$$

is required in many applications, such as, in vibration mode superposition analysis [3, p. 558].

We assume that A and B are real and symmetric $n \times n$ matrices, and B is positive definite. This implies that all the eigenvalues are real. We further assume that A and B are banded since our algorithm is designed, and therefore most efficient, for such matrices. Given an interval, we compute all the eigenvalues contained in it, as well as their corresponding eigenvectors. Alternatively, the lowest m eigenvalues can be computed for a given m.

When $B = I$, we have the standard eigenproblem

$$(2) \qquad Ax = \lambda x.$$

For real, symmetric and tridiagonal A, a well known solution method for problem (2) is based on the Sturm sequence property used with the sectioning method [1][4][6]. Given $\mu \in \mathbb{R}$, the Sturm sequence provides the number of eigenvalues less than μ. The Sturm sequence of problem (2) where A is tridiagonal can be computed using a second order linear recurrence [13, pp. 300-314], or a more stable non-linear recurrence [2]. We can determine the number of eigenvalues in any interval simply by evaluating the Sturm sequences at its boundary points. The sectioning method repeatedly "slices" intervals containing eigenvalues into equal size subintervals until they are sufficiently small to approximate the eigenvalues at their midpoints.

The parallel algorithm for solving problem (2) where A is tridiagonal due to Lo, Philippe and Sameh [6], called TREPS2, uses the sectioning method to isolate eigenvalues, i.e., successively partitions the given interval to the point where all eigenvalues lie in different subintervals. Once an eigenvalue has been isolated the Zeroin method due to Dekker [5], which is a root finding procedure, is used to accurately compute the eigenvalue. The eigenvectors corresponding to

---

* Department of Computer Science, Duke University, Durham, North Carolina 27706.

computed eigenvalues are obtained by the inverse iteration and the modified Gram-Schmidt procedures. The algorithm TREPS2 was much faster than using a combination of EISPACK's BISECT, TINVIT and TQL2.

In this paper, we present a parallel algorithm which solves problem (1) with an innovative scheme for extracting eigenvalues. Results obtained from the implementation of this algorithm on a parallel computer are also presented.

2. The Algorithm. Our algorithm SECTRQ (Sectioning method with inverse and Raleigh quotient iterations) can be divided into two stages : eigenvalue isolation and eigenvalue extraction. The sectioning method is used to isolate distinct eigenvalues, i.e., intervals containing more than one eigenvalues are repeatedly partitioned into several subintervals until all distinct eigenvalues lie in different intervals. The Sturm sequence for problem (1) can be computed using a modified elimination procedure with a special pivoting scheme [13, pp. 437-439] which was used in [7][8][10].

If multiple or clustered eigenvalues exist, the sectioning process will continue until the subintervals containing these eigenvalues are sufficiently small to approximate them at the intervals' midpoints. The eigenvectors corresponding to these eigenvalues are computed using the inverse iteration method and orthogonalization techniques, as described in [9]. Typically, very few iterations are required if the eigenvalues are sufficiently accurate [14].

Once an eigenvalue has been isolated, we use a superlinear algorithm [12] to accurately compute the eigenpair. Let $\delta_k = (\gamma - \eta, \gamma + \eta)$ be an interval containing an isolated eigenvalue, λ_k, for some k. The algorithm is a combination of inverse and Rayleigh quotient iterations of the form

$$(A - \mu B)y_{s+1} = Bz_s \tag{3}$$
$$\omega_{s+1} = (y_{s+1}^T B y_{s+1})^{-1/2} \tag{4}$$
$$z_{s+1} = \omega_{s+1} y_{s+1} \tag{5}$$

where $\mu = \gamma$ in the case of the inverse iteration and $\mu = \mu_s = z_s^T A z_s$ for the Rayleigh quotient iteration. The inverse iteration converges linearly to the eigenvalue closest to γ, while the Rayleigh quotient iteration exhibits cubic convergence. However, the latter may converge to different eigenvalues depending on the initial vector z_0 and often to λ_i far from γ. In each step of our algorithm either $\mu = \gamma$ or $\mu = \mu_s$ are used in equation (3). The decision is based on a criterion that guarantees superlinear convergence of the algorithm to an eigenvalue in δ_k. This criterion is briefly discussed below (see also [14]). The criterion used guarantees that the method will converge to the eigenvalue in δ_k and not to another, i.e., it will converge within δ_k.

2.1. The Switching Criterion. Let λ_k be the isolated eigenvalue in δ_k and x_k be its corresponding eigenvector. At each step of the inverse iteration method, there are two residuals that can be computed, $q_s = Az_s - \gamma B z_s$ with respect to the shift γ, and $r_s = Az_s - \mu_s B z_s$ with respect to the Rayleigh quotient. They are related by the equation $r_s = q_s + (\gamma - \mu_s) B z_s$. Since $z_s \to x_k$ and $\mu_s \to \lambda_k$, $r_s \to 0$ as $s \to \infty$. We use the B^{-1}-norm to study the convergence of q_s and obtain $\| r_s \|_{B^{-1}}^2 = \| q_s \|_{B^{-1}}^2 - (\gamma - \mu_s)^2$. This equality is important, since it shows that

$$\| r_s \|_{B^{-1}} \leq \| q_s \|_{B^{-1}}, \quad |\gamma - \mu_s| \leq \| q_s \|_{B^{-1}} \tag{6}$$

and that

$$\| q_s \|_{B^{-1}} \to |\gamma - \lambda_k|, \text{ as } s \to \infty. \tag{7}$$

It turns out that the convergence sequence is actually monotonically decreasing, i.e.,

$$\| q_{s+1} \|_{B^{-1}} \leq \| q_s \|_{B^{-1}} . \tag{8}$$

This inequality is crucial in our criterion for switching to the Rayleigh quotient iteration.

From equations (7,8) we obtain

$$|\gamma - \lambda_k| \leq \| q_s \|_{B^{-1}} . \tag{9}$$

The importance of equation (8) is that it guarantees that the bound (9) never deteriorates as the iteration continues.

The inequality (9) can be rewritten as the inclusion theorem

$$(10) \qquad \gamma - \| q_s \|_{B^{-1}} \leq \lambda_k \leq \gamma + \| q_s \|_{B^{-1}} .$$

From this follows that when looking for an eigenvalue in the interval $\delta_k = (\gamma - \eta, \gamma + \eta)$ by the inverse iteration method using the shift γ, λ_k can be guaranteed to lie in δ_k as soon as

$$(11) \qquad \| q_s \|_{B^{-1}} < \eta.$$

From that point on, by equations (6,8), both $\| q_{s'} \|_{B^{-1}} \leq \eta$ and $\| r_{s'} \|_{B^{-1}} \leq \eta$ for $s' \geq s$.

Equation (11) is the switching criterion. Further analysis can be found in [12]. Equation (8) and the inclusion theorem (10) guarantee that the criterion (11) will be satisfied after certain number of inverse iterations. At that point we know from equation (7) that $\mu_s \in \delta_k$.

We point out that from equations (3-5) $\| q_s \|_{B^{-1}} = \omega_s$. Since ω_s is computed in every iteration, we obtain $\| q_s \|_{B^{-1}}$ and thus $\| r_s \|_{B^{-1}}$ without further cost and even without computing q_s or r_s. The switching criteria (11) can thus be rewritten as

$$(12) \qquad \omega_s < \eta$$

and this is actually how it is implemented in our method. If λ_k is "isolated enough", then μ_s is guaranteed to converge to λ_k. Although it is reasonable to expect that the steps of the inverse iteration would have diminished the component of z_0 in the directions other than x_k, this is not taken for granted in our algorithm as we monitor the Rayleigh quotient μ_s at each iteration. At some point if μ_s falls outside δ_k, the method reverts to the inverse iteration with shift γ using the latest iterate of the Rayleigh quotient iteration as the initial vector. From our experience, we rarely need to revert to the inverse iteration.

3. Parallel Implementation. In this section we discuss how SECTRQ is parallelized and implemented on a shared memory MIMD architecture. The strategy we used to parallelize SECTRQ is similar to that of [4][6][7]. Our algorithm has been implemented on the FLEX/32 Multicomputer at NASA Langley Research Center, Hampton, VA, which has eighteen processors. Assume that p processors are used.

We use the sectioning method to isolate the eigenvalues. Let $I = (l, u)$ be the interval we partition into $k + 1$ subintervals. Assuming that we know the number of eigenvalues less than l and u, the k Sturm sequence evaluations at $l + \frac{u-l}{k+1}$, $l + 2\frac{u-l}{k+1}, \ldots, l + k\frac{u-l}{k+1}$ can be evaluated in parallel. In our algorithm, we conveniently used $k = p$, so processor i evaluates the Sturm sequence at $l + i\frac{u-l}{p+1}$.

Once the eigenvalues are isolated, i.e., each eigenvalue lies in a different interval, we accurately determine the eigenvalues and their corresponding eigenvectors by the superlinear method which combines inverse and Rayleigh quotient iterations. Clearly, the eigenvalues in different intervals can be computed independent of one another. In our implementation, the intervals containing the isolated eigenvalues are stored in a queue in shared memory. Whenever, a processor is available, it works on the next interval, if any, in the queue.

4. Numerical Experiments. In this section, we present numerical results of several experiments of increasing degrees of difficulties.

Experiment 1 : Matrix A is a real, symmetric tridiagonal matrix obtained from discretization of Poisson's equation in one dimension, and $B = I$. The matrix size $n = 400$, and the given interval is $(0, 0.385)$ which contains 80 eigenpairs. Figure 1 shows the execution times when different methods are used to extract eigenvalues which have been isolated by the sectioning method. Note that using the inverse and Rayleigh quotient iterations combination to extract eigenvalues is much faster than using either the Zeroin or the bisection methods. This result is representative of all our experiments.

Experiment 2 : Matrices A and B are obtained from piecewise linear finite element [11] discretization of the Sturm-Liouville problem in one dimension

$$-\frac{d}{dx}\left(p(x)\frac{du}{dx}\right) + q(x)u = \lambda u$$

where $u = u(x)$, $0 < x < \pi$ and $u(0) = 0$, $u'(\pi) = 0$ and $p(x) > 0$. Here, both A and B are symmetric tridiagonal and positive definite matrices. We use $p(x) = 2$, $q(x) = 1$ and the given interval is $(0, 2000)$ which contains 32 eigenpairs. Figure 2 shows the effect of varying mesh sizes on the execution times when $p = 18$. For this problem, the execution time appears to increase linearly with the order of the matrices.

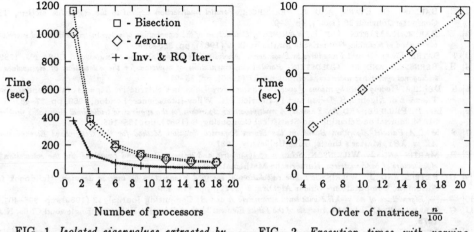

FIG. 1. *Isolated eigenvalues extracted by different methods.*

FIG. 2. *Execution times with varying mesh sizes.*

Experiment 3 : The matrices for this experiment are used to analyze a very long flexible space structure, where A has bandwidth of 35 and B is a diagonal matrix. The degrees of freedom $n = 486$, and the given interval is $(0, 65000)$ which contains 18 eigenpairs. Figures 3 and 4 show the execution times and speedups, respectively. By using SECTRQ, the execution time is substantially reduced when multiple processors are used. Even though only 18 eigenpairs are computed, the speedup achieved is over 10 for $p = 18$.

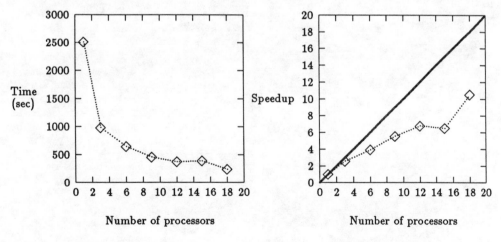

FIG. 3. *Execution time.*

FIG. 4. *Speedup.*

5. Concluding Remarks. To accurately determine an isolated eigenvalue, the superlinear method which combines inverse and Rayleigh quotient iterations is substantially faster than the traditional procedures which are widely used, namely the Zeroin or the bisection methods. From our test runs, the speedups obtained are quite good, especially when the number of eigenpairs desired is much greater than the number of processors. For $Ax = \lambda Bx$, each Sturm sequence evaluation entails a triangular factorization. Since the time required for a triangular factorization is proportional to the square of the bandwidth of the matrix involved, our algorithm is most appropriate for problems with matrices of small to medium bandwidths.

REFERENCES

[1] R. BARLOW, D. EVANS, AND J. SHANEHCHI, *Parallel multisection applied to the eigenvalue problem*, The Computer Journal, 26 (1983), pp. 6–9.

[2] W. BARTH, R. MARTIN, AND J. WILKINSON, *Calculation of the eigenvalues of a symmetric tridiagonal matrix by the method of bisection*, Numerische Mathematik, 9 (1967), pp. 386–393.

[3] K. BATHE, *Finite Element Procedures in Engineering Analysis*, Prentice-Hall, Inc., Englewood Cliffs, NJ, 1982.

[4] H. BERNSTEIN AND M. GOLDSTEIN, *Parallel implementation of bisection for the calculation of eigenvalues of tridiagonal symmetric matrices*, Computing, 37 (1986), pp. 85–91.

[5] T. DEKKER, *Finding a zero by means of successive linear interpolation*, in Constructive Aspects of the Fundamental Theorem of Algebra, B. Dejon and P. Henrici, eds., Wiley-Interscience, London, 1969, pp. 37–48.

[6] S. LO, B. PHILIPPE, AND A. SAMEH, *A multiprocessor algorithm for the symmetric tridiagonal eigenvalue problem*, SIAM Journal on Scientific and Statistical Computing, 8 (1987), pp. 155–165.

[7] S. MA, *A Parallel Algorithm Based on the Sturm Sequence Solution Method for the Generalized Eigenproblem, $A\bar{x} = \lambda B\bar{x}$*, Master's thesis, Duke University, 1987.

[8] R. MARTIN AND J. WILKINSON, *Solution of symmetric and unsymmetric band equations and the calculation of eigenvectors of band matrices*, Numerische Mathematik, 9 (1967), pp. 279–301.

[9] G. PETERS AND J. WILKINSON, *The calculation of specified eigenvectors by inverse iteration*, Handbook for Automatic Computation - Linear Algebra, 2 (1971), pp. 418–439.

[10] ———, *Eigenvalues of $A\bar{x} = \lambda B\bar{x}$ with band symmetric A and B*, Computing Journal, 12 (1969), pp. 398–404.

[11] G. STRANG AND G. FIX, *An Analysis of the Finite Element Method*, Prentice-Hall, Inc., Englewood Cliffs, N.J., 1973.

[12] D. SZYLD, *Criteria for combining inverse and Rayleigh quotient iterations*, SIAM Journal on Numerical Analysis, (1988). To appear.

[13] J. WILKINSON, *The Algebraic Eigenvalue Problem*, Clarendon Press, Oxford, 1965.

[14] ———, *Inverse iteration in theory and in practice*, Symposia Mathematica, 10 (1972), pp. 361–379.

Accelerating with Rank–One Updates
Timo Eirola and Olavi Nevanlinna

Consider iterations for solving $Ax = b$ (A is n by n nonsingular) which are based on a splitting $A = M - N$. We discuss rank–one updates to improve $\text{inv}(M)$ as an approximation to $\text{inv}(A)$ during the iteration. The update kills and reduces singular values of $N\text{inv}(M)$ and thus speeds up the convergence. The basic algorithm terminates after at most n sweeps, and if full n sweeps are needed, then $\text{inv}(A)$ has been computed. Suitability for parallel processing and the effect of restricted memory length for the updates are also discussed.

CHAPTER 17

Preconditioned Conjugate Gradient Methods for General Sparse Matrices on Shared Memory Machines*

Edward Anderson[†]
Youcef Saad[†]

Abstract. This paper considers how to implement some preconditioners for conjugate gradient methods on a multiprocessor. The emphasis is on linear systems $Ax = f$ for which the coefficient matrix A is large, sparse, and nonsymmetric. Our first approach is to parallelize the standard preconditioners based on an incomplete factorization $A \approx LU$. We develop a reordering technique based on the structure of A and discuss how it can be used to introduce parallelism in both the ILU(0) factorization and the triangular solves at each iteration. As an alternative, we consider polynomial preconditioners which approximate A^{-1} by a polynomial in A. These methods offer a high degree of parallelism and their performance can be enhanced by a simple diagonal or block diagonal scaling. Numerical experiments are presented which show that for a small number of processors, the standard ILU(0) preconditioner parallelizes reasonably well and may be better than polynomial preconditioning for general unstructured problems.

1. Introduction. Preconditioned conjugate gradient methods represent a good alternative to direct methods for large sparse linear systems, such as those arising from three–dimensional models. The performance of these methods can be greatly improved by the use of preconditioning, in which the system $Ax = b$ is replaced by an auxiliary system $Q^{-1}Ax = Q^{-1}b$ for some preconditioning matrix Q. The auxiliary system will exhibit a better convergence rate using the conjugate gradient method if the eigenvalues of $Q^{-1}A$ are more favorably distributed than those of A. In practice, we would like $Q^{-1}A$ to be closer to the identity matrix in some sense so that its eigenvalues should be

* This work was supported by the National Science Foundation under contracts NSF MIP–8410110 and NSF DCR85–09970, by the Department of Energy under contract DE–FG02–85ER25001, by the Air Force under contract AFSOR–85–0211 and by an IBM donation.

† Center for Supercomputing Research and Development, University of Illinois, Urbana, Illinois 61801.

clustered about 1. Preconditioning then consists either of finding a matrix Q which is an approximation to A such that systems of the form $Qy = z$ are easy to solve, or of directly finding Q^{-1}, which should be an approximation of A^{-1}.

Incomplete factorizations of the coefficient matrix A are one technique for approximating A by an easily invertible matrix Q. A procedure similar to the usual LU–decomposition is used to find $A = LU + E$, where L is a sparse lower triangular matrix, U is a sparse upper triangular matrix, and E is an error matrix, which is not explicitly computed. (1), (3) In the ILU(0) factorization considered here, no fill–in is allowed in the L and U factors so L and U have the same nonzero structure as the lower and upper triangular parts of A. We choose as the preconditioning matrix $Q = LU$. At each iteration, solving $Qy = z$ for y involves two triangular solves: $L\hat{y} = z$ and $Uy = \hat{y}$. In the following section, we show how to parallelize this time–consuming portion of the computation by means of a preliminary analysis of the structure of the triangular factors. Section 3 contains a brief discussion of how this same analysis can be used to parallelize the ILU(0) factorization as well, and some numerical results are presented.

The alternative approach of approximating A^{-1} directly is used in polynomial preconditioning, where $Q^{-1}A$ is replaced by a polynomial in A of the form $s(A){\cdot}A$. Polynomial preconditioning is particularly well–suited to structured problems, where A has only a few nonzero diagonals and multiplication by A or a polynomial in A is highly vectorizable, but here we consider its use for general sparse matrices. Section 4 outlines one technique for choosing the polynomial s and contains a comparison of polynomial preconditioning to ILU(0) preconditioning for problems on which either method can be used.

Our code development and experiments have been done on an Alliant FX/8, a shared memory multiprocessor with 8 computational elements (CEs), each with vector hardware. Our test matrices have been chosen from the nonsymmetric matrices in the Boeing Computer Services/Harwell test collection. All matrices are assumed to be stored by rows in the sparse A–JA–IA format, where

A = nonzeros of the coefficient matrix, stored by rows

JA = column numbers of each element of A

IA = index into each row of A and JA. The kth row of the matrix is found in locations IA(k) to IA(k+1) of A.

The iterative method on which the preconditionings have been used is the restarted version of GMRES described in (6) in which the dimension of the Krylov space is fixed at 10. GMRES stops when the residual is reduced to ϵr_0, where r_0 is the initial residual, and for our experiments $\epsilon = 1 \times 10^{-8}$.

2. Parallelizing the solution of triangular systems.
Solving a lower triangular system of equations $Lx = b$ by the usual forward elimination algorithm is generally regarded as a sequential operation. The ith row equation is solved for x_i:

$$x_i = l_{i,i}^{-1}(b_i - l_{i,1}x_1 - \cdots - l_{i,i-1}x_{i-1})$$

and if L is dense, the $l_{i,j}$'s are nonzero and each of the components $x_1, \cdots, x_{i-1}$ must be known in order to compute x_i. However, if L is sparse, some of the $l_{i,j}$'s will be zero and it may not be necessary to determine all of the preceding $i-1$ components of x before solving for x_i. In fact, for many triangular systems the structure of the coefficient matrix allows a reordering of the rows into equivalence classes, where the rows in each class are independent and can be solved in parallel.

The equivalence class for the ith row of a lower triangular system can be defined by

$$lclass(i) = \begin{cases} 1, & \text{if } l_{ij}=0 \text{ for all } j<i \\ 1 + \max_{j<i}\{lclass(j) : l_{ij}\neq 0\}, & \text{otherwise} \end{cases}$$

This definition places each row equation in the first possible class in which it can be solved, which is the next class after the last required x–component has been computed. Since each row equation in a class depends only on the components of x found in previous classes, the rows within a class may be solved concurrently. Similar ideas are suggested in (2), (7), and (8).

A parallel algorithm for the solution of a lower triangular system is then obtained by solving the row equations in increasing order of their equivalence class numbers, distributing the rows in each class among the processors. This technique is equivalent to reordering the rows and unknowns of the system to create a block lower triangular system with diagonal matrices for the diagonal blocks, and then performing a block forward solve. It should be noted that simply reordering the calculations may not be an effective way to obtain parallelism for every sparse system. For example, a lower bidiagonal matrix with a dense nonzero subdiagonal has a string of dependencies $x_1 \rightarrow x_2 \rightarrow \cdots \rightarrow x_n$, so each equivalence class will contain only one row. However, good speedups are obtained on a small number of processors for most unstructured matrices or for structured matrices which do not contain any completely dense subdiagonals.

3. Parallelizing the ILU(0) factorization. The ILU(0) preconditioner is of interest in this context because it can be parallelized by the same means used to parallelize the solution of the sparse lower triangular systems. Here the fundamental unit of computation is the factorization of one row of a sparse matrix A, which consists of subtracting multiples of previously determined rows of U from the current row in order to eliminate the nonzeros to the left of the diagonal. The multiples are stored to form a new row of L and the remaining elements form a new row of U in the approximate factorization $A \approx LU$. If the rows of A are organized into equivalence classes as before, then the rows in each class can be factored independently using only rows of U found in previous classes.

The parallel implementation of the ILU(0) preconditioner therefore consists of the following stages:

(1) Find the classes of independent rows.

(2) Form the sparse factors L and U of the ILU(0) factorization by the parallel scheme.

(3) Solve the preconditioned system by GMRES, where the triangular solves at each iteration have been parallelized.

We have included the time for (1) with the GMRES time. The sequential method consists of a sequential ILU(0) factorization plus the solution by GMRES using standard forward and back solves for the triangular systems at each iteration; step (1) is not required. A comparison of the parallel method on 8 CEs of an Alliant FX/8 to the sequential method on 1 CE for some nonsymmetric matrices from the BCS/Harwell test collection is shown in Table 1. Each stage of the computation shows a respectable speedup of about 5 to 6.

Matrix	ILU(0) time			GMRES time			Total time		
	Seq'l	Par'l	Ratio	Seq'l	Par'l	Ratio	Seq'l	Par'l	Ratio
jpwh_991	0.089	0.017	5.25	1.16	0.25	4.68	1.25	0.26	4.72
orsirr_1	0.100	0.018	5.43	3.73	0.72	5.21	3.84	0.73	5.22
orsreg_1	0.212	0.035	6.06	9.18	1.72	5.34	9.39	1.75	5.35
sherman5	0.324	0.061	5.33	17.15	3.34	5.13	17.47	3.40	5.14

Table 1: Times in seconds and speedups for ILU(0) + GMRES(10)

4. Polynomial preconditioning. An interesting alternative to incomplete factorization methods is the use of polynomial preconditioning. The principle of these methods is to replace the linear system $Ax = b$ by a system of the form $s(A)Ax = s(A)b$, where s is a low degree polynomial such that $s(A)$ approximates A^{-1}. Since the only operations required by this approach are matrix–vector multiplications and vector–vector primitives, the degree of parallelism is very high. The principal problem then is how to obtain a good polynomial s.

One approach to finding an acceptable s is to choose the least squares polynomial with respect to some norm. A first run with the nonpreconditioned GMRES algorithm is made. This is nothing but a projection technique on a Krylov subspace $K_m \equiv span[r_0, Ar_0, \ldots, A^{m-1}r_0]$, where r_0 is the initial residual and m is the number of steps. The result is some approximate solution to the problem and a Hessenberg matrix that represents the projection of the matrix A in the Krylov subspace. The eigenvalues of this Hessenberg matrix will provide good approximations to some of the outermost eigenvalues of A. A rough convex hull of the spectrum is then constructed. The polynomial is chosen so as to minimize $\| 1 - t \cdot s(t) \|_w$ over all polynomials of degree k, where $\|\cdot\|_w$ is an L_2 norm associated with a Chebyshev weight w. The implementation details of this approach can be found in (4) and (5).

We have implemented a polynomial preconditioned GMRES algorithm coupled with a diagonal or block diagonal scaling. Results for the nonsymmetric test matrices we considered were mixed, as can be seen in Table 2. The performance of polynomial preconditioning was comparable to ILU(0) preconditioning in some cases but up to four times slower in others.

5. Conclusions. From the results presented here, we can conclude that for a small number of processors, the ILU(0) preconditioner does parallelize reasonably well. For general sparse nonsymmetric matrices, ILU(0) is even better than polynomial preconditioning when either method can be used. The advantage to users of iterative methods is that this method is robust, efficient, and well understood and requires only knowledge of the structure of the coefficient matrix A. These conclusions probably do not hold for larger number of processors, in which this implementation of ILU(0)

Matrix	Rows	Seq'l ILU(0)	Parallel ILU(0)	Polynomial
jpwh_991	991	1.25	0.26	0.31
orsirr_1	1030	3.84	0.74	2.88
orsreg_1	2205	9.39	1.75	5.77
sherman5	3312	17.47	3.40	4.20

Table 2: Preconditioning + GMRES(10), time in seconds

preconditioning would be limited by the inherent parallelism in the structure of A. In this case, polynomial preconditioning, combined with some form of diagonal or block diagonal scaling, would again have to be considered.

REFERENCES

(1) H. C. Elman, *Iterative Methods for Large, Sparse, Nonsymmetric Systems of Linear Equations*, PhD thesis, Yale University Research Report #229, 1982.

(2) A. Greenbaum, *Solving Sparse Triangular Linear Systems Using Fortran with Parallel Extensions on the NYU Ultrcomputer Prototype*, Ultracomputer Note #99, Courant Institute, April 1986.

(3) J. A. Meijerink and H. A. van der Vorst, *An Iterative Solution Method for Linear Systems of which the Coefficient Matrix is a Symmetric M–Matrix*, Mathematics of Computation, 31 (1977), pp. 148–162.

(4) Y. Saad, *Least squares polynomials in the complex plane with applications to solving sparse nonsymmetric linear systems*, SIAM J. Numer. Anal., 24 (1987), pp. 155–169.

(5) Y. Saad, *Practical use of polynomial preconditionings for the conjugate gradient method*, SIAM J. Sci. Stat. Comp., 6 (1985), pp. 865–881.

(6) Y. Saad and M. H. Schultz, *GMRES: a Generalized Minimal Residual method for solving nonsymmetric linear systems*, SIAM J. Sci. Stat. Comp., 7 (1986), pp. 856–869.

(7) J. H. Saltz, *Automated Problem Scheduling and Reduction of Synchronization Delay Effects*, ICASE Report no. 87–22, July 1987.

(8) O. Wing and J. W. Huang, *A Computational Model of Parallel Solution of Linear Equations*, IEEE Transactions on Computers, v. C–29 no. 7, July 1980, pp. 632–638.

CHAPTER 18

A Comparison of Some Vectorized ICCG Linear Solvers and LINPACK Banded Solvers for a Problem of Two Phase Flow in a Centrifuge*

Eugene L. Poole[†]
Jeffrey W. Frederick[‡]

Abstract. A problem in fluid dynamics, two phase fluid flow in a centrifuge, is used to compare the direct solution of linear systems using LINPACK routines with the iterative solution of linear systems using Incomplete Choleski Conjugate Gradient (ICCG). Two different approaches to vectorize ICCG are compared, using the diagonal ordering, which yields shorter vector lengths, $O(w)$, for a $w \times h$ grid, and using multi-color orderings which yield long vector lengths, $O(wh)$. The key consideration is the greater degree of vectorization for the multi-color approach versus the somewhat better convergence results for the diagonal ordering. Results are given for runs on a MicroVax, Convex C-1, Cray-2, and Cray-XMP.

1. Introduction. A problem in fluid dynamics, two phase fluid flow in a centrifuge, is used to compare the direct solution of linear systems using LINPACK routines with the iterative solution of linear systems using the Incomplete Choleski Preconditioned Conjugate Gradient method (ICCG). Two different approaches to vectorize ICCG are compared, using the diagonal ordering, which yields shorter vector lengths, $O(w)$, for an $w \times h$ grid, and using multi-color orderings, which yields long vector lengths, $O(wh)$. Two key considerations are the greater degree of vectorization for the multi-color approach versus the somewhat better convergence results for the diagonal ordering.

The methods used were applied to a fluid dynamics problem which models the separation of two immiscible liquids, e.g. oil and water, in a continuous process centrifuge. The computational region is a rectangular grid. Finite difference equations are derived from the conservation and momentum equations and pressures and velocities are calculated at each step in a time marching routine. See Harlow and Amsden [1975] for a general description of the numerical model used. A more detailed description of the physical problem is given by Frederick [1988]. This paper

---

* This work was supported in part by Pittsburgh Supercomputing Center Grant PSCA245

† Awesome Computing Inc., 1390 Big Bethel Rd., Hampton, Va. 23666

‡ University of Virginia, Dept. of Mechanical Engineering, Charlottesville, Va. 22903

addresses the main computational load, the solution of a new Poisson equation for pressures, at each time step. Several thousand time steps are typically required where at the kth time step equation (1),

$$A^k x^k = b^k \qquad (1)$$

must be solved. Since the pressures, x^{k-1} from the previous time step serve as a good initial guess for the pressures at the current time step, the iterative approach should benefit as the steady state solution is approached.

2. Description of Methods. The incomplete Choleski conjugate gradient method has been studied by many authors as an effective iterative method for solving large, sparse, symmetric linear systems of equations both on scalar and vector computers. (See for example, Van der Vorst [1982,1983,1986] Lichnewsky [1983], Jordan [1984]). Figure 1 shows the standard ICCG algorithm.

Choose x^0 ; Set $r^0 = b - A\,x^0$

Solve $M\,\hat{r}^0 = r^0$; Set $p^0 = \hat{r}^0$

Loop $k = 0, 1, \ldots,$ kmax

$$\alpha_k = \frac{(\hat{r}^k,\ r^k)}{(p^k,\ A\,p^k)}\ ;\ x^{k+1} = x^k + \alpha_k\,p^k\ ;\ r^{k+1} = r^k - \alpha^k\,A\,p^k$$

if $\|r^{k+1}\| \leq$ tolerance then stop

Solve: $M\,\hat{r}^{k+1} = r^{k+1}$

$$\beta_k = \frac{(\hat{r}^{k+1},\ r^{k+1})}{(\hat{r}^k,\ r^k)};\ p^{k+1} = \hat{r}^{k+1} + \beta_k p^k$$

Figure 1. Preconditioned Conjugate Gradient Algorithm

The main computational steps at each iteration in the algorithm are a matrix-vector multiplication, three vector add-multiply operations, two inner products, and the preconditioning step. Since all but the preconditioning step are easily vectorized primary consideration is given to the preconditioning step, $M\hat{r} = r$, where M is a symmetric, positive definite matrix. For ICCG, $M = LDL^T$ is chosen to be an incomplete factorization of A where L is lower triangular and in this case, has the same non-zero structure as the lower triangular part of A. For this problem, A is symmetric positive definite and has five diagonals including the main diagonal. Various approaches are taken to vectorize the factorization and, more importantly, the forward and back solutions necessary to solve the triangular systems at each iteration. The diagonal ordering approach, orders the grid points by diagonals and so the equations along successive diagonals beginning in the lower left corner of the rectangular grid and proceeding to the upper right corner are solved in vector loops. Using this approach, the maximum vector length encountered in the forward and back solves is the $min(w, h)$ for a grid having w grid points per row and h rows. This approach must be modified when the matrix structure is more complicated and often the diagonals along which computations may occur in a vector loop with no dependencies become even shorter and may be skewed thus further reducing the vector performance. For such problems or whenever one dimension of the rectangular region is small the vector lengths are short and the cost of the preconditioning step can be significantly more than the matrix-vector multiplication even though the same number of operations are performed. The motivation for this approach is to achieve the same rate of convergence for the resulting ICCG method as for natural ordered ICCG on the same problem since algebraically the two methods are equivalent. See Ashcraft [1987] for a more detailed discussion of this approach and some results on square regions.

The other approach considered is multi-color orderings of the grid points where the equations in the linear system are reordered in such a way to achieve vector lengths which are $O(\frac{wh}{p})$ in the forward and back solves. Here p is the number of colors used to order the equations. For the problems considered here, the Red-Black ordering $(p = 2)$ of the unknowns is used. For the Poisson problem all of the equations associated with the odd numbered grid points are solved first followed by the equations for the even numbered grid points. Since odd grid points are only coupled to even grid points in the finite difference equations and vice versa for the even grid points the odd equations can be solved in a single vector loop with no dependencies. In this case the preconditioning step costs about the same as the matrix-vector multiplication. When the equations are more complicated the number of colors is increased. and grid points are ordered, or colored, to group grid points that are not coupled by the equations. See Poole [1987] for a more detailed discussion of multi-color orderings applied to a general class of problems on rectangular regions. The main drawback for this method has been that the convergence of the ICCG method is not as good as for the diagonal and/or natural ordering. However, using a modification based on Eisenstat [1981] and described in Poole [1987] the red-black preconditioning for the Poisson problem can be implemented with essentially the same cost per iteration as conjugate gradient without preconditioning. This is because the matrix-vector multiply at each iteration is eliminated being combined with the preconditioning step. The same modification can be carried out on the diagonally ordered ICCG algorithm but the savings is not nearly so great since we lose the matrix multiply which vectorized very well and must instead use the short vector forward and back solves.

3. Results. Results were obtained for two test cases, on grids of dimension 13×20 and 19×38 using the diagonal ordering, red-black ordering for the standard ICCG method and the Eisenstat modification applied to red-black ICCG. The LINPACK routines, SPBFA and SPBSL, were also used to factor and solve the Poisson equations. The LINPACK routines used are the standard routines accessible on many computers including the Cray-2 and Cray-XMP. There are some versions of these routines which have been modified for the Cray-2 or Cray-XMP which may perform better. Table 1 gives results for the two test cases obtained on the Cray-XMP/48 at the Pittsburgh Supercomputing Center. The total CPU time includes the time to set up the new matrices and to calculate new velocities using the new pressures at each time step.

260 Unknowns: 13 × 20 Grid Size				
Method Used	Total Time (sec)	Eqn Solution Time (sec)	Number of Computations	Mflops
LINPACK	204.1	129.4	621,240,000	4.8
Diagonal	106.9	31.9	1,138,847,060	32
Red-Blk	91.0	20.1	1,310,842,000	65
Red-Blk/Eisen	83.3	12.4	893,250,704	72

722 Unknowns: 19 × 38 Grid Size				
Method Used	Total Time (sec)	Eqn Solution Time (sec)	Number of Computations	Mflops
LINPACK	709.3	531.2	3,380,280,000	6.4
Diagonal	269.2	104.3	3,944,178,650	45.9
Red-Blk	235.9	70.3	5,734,040,248	81
Red-Blk/Eisen	206.3	41.7	3,918,777,930	94

Table 1. Cray XMP/48 Results for 10,000 Time Steps

The number of computations given is the total number of floating point operations required for the solution of the linear systems for 10000 time steps and is used to calculate the Mflop rates given also in Table 1. The equation solution time is the dominant factor when the LINPACK routines are used for the solution of the linear systems since they did not vectorize well on the Cray. For all three ICCG solvers the linear equation solution time was significantly reduced so that it was no longer the dominant computation at each time step. Note that for both problems the Red/Black Eisenstat ICCG algorithm was more than twice as fast as the diagonal ordered ICCG algorithm and nearly twice as fast as the standard ICCG algorithm in the time required to solve the linear systems.

Figure 2 shows the advantage of the diagonal ordering in terms of the number of iterations required to solve the linear systems at each time step. The values plotted are the total number of ICCG iterations required to solve 100 linear systems (i.e. 100 time steps). For both problems the convergence criteria used for the ICCG methods was to stop when the inner product of the residual r was less than 10^{-6}. Note that after the first 100 time steps the number of iterations required to solve the linear systems drops dramatically.

Figure 3 shows the number of floating point operations required by each method for each 100 time steps. Here the direct method requires the fewest number of floating point operations although the difference is small after the first 200 time steps. The savings afforded by the Eisenstat modification of the basic ICCG algorithm is

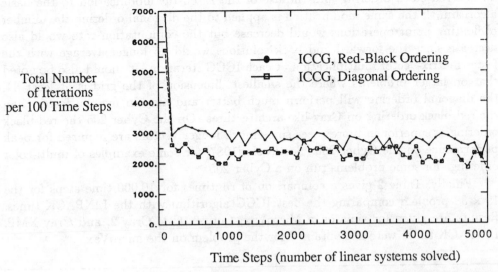

Figure 2. Comparison of Convergence Rates for ICCG using Red-Black and Diagonal Grid Orderings

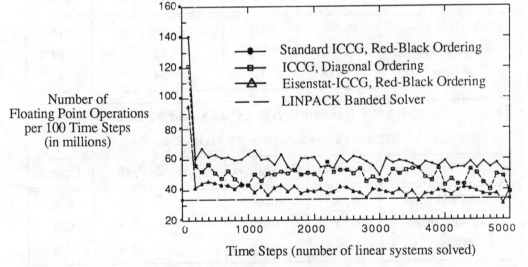

Figure 3. Comparison of Amount of Computations per 100 Linear System Solutions for 5000 Time Steps

evident in that even though the Eisenstat ICCG algorithm requires more iterations than diagonal ordered ICCG, it requires fewer floating point operations. The standard ICCG implementation requires $O(29n)$ operations per iteration whereas the Eisenstat implementation requires $O(19n)$ operations per iteration where n is the dimension of the linear system ($n = w \times h$).

The results show that for these problems the rate of convergence advantage of the diagonal ordering approach is more than offset by the higher computation rate for the red-black ordering and the use of the Eisenstat modification to the basic algorithm. If the same modification is applied to the diagonal ordering the number of floating point operations would decrease but the computation rate would also decrease since the forward and back solutions would no longer average with the faster matrix-vector multiplication at each ICCG iteration. It should also be noted that on larger problems where the smallest dimension of the grid is at least 64, the diagonal ordering will perform much better and may in fact be faster than the red-black ordering on Cray-like architectures. On the Cyber 205 the red-black ordering is superior in more cases since much longer vectors are required for peak performance on the Cyber 205. See Poole [1987] for some examples of multi-color orderings for some problems run on a Cyber 205.

Finally, Table 2 gives a comparison of runtimes for 10,000 time steps for the 19×38 problem comparing the best ICCG algorithm with the LINPACK times. Results are given for runs on a MicroVax, Convex C-1, Cray-2, and Cray-XMP. LINPACK times were not obtained for the problem on the microVax.

Red-Black Ordering, Eisenstat Modification 722 Unknowns: 19×38 Grid Size				
	MicroVax	Convex C-1	Cray XMP/48	Cray-2
Total CPU time	46,358	3396	206	278
Eqn soln time	6069	625	48	84
Mflops	.148	7.6	94	47

LINPACK Banded Solver SPBFA and SPBSL 722 Unknowns: 19×38 Grid Size				
	MicroVax	Convex C-1	Cray XMP/48	Cray-2
Total CPU time	N/A	14,360	709	900
Eqn soln time	N/A	11,530	531	654
Mflops	N/A	.3	6.4	5.2

Table 2. Computer Comparison for 10,000 Time Steps

For both test problems the use of a vectorized iterative solver significantly reduced the equation solution time compared to the time required by the LINPACK solver. The overall time to solve the test problems was reduced dramatically requiring only minutes on a Cray to solve each problem.

4. Conclusions. The choice of linear equation solvers may dramatically effect execution times on vector computers. In this case the ICCG iterative method was shown superior to a commonly used linear solver for a relatively small problem. The positive effect of decreased iterations from using the solutions at the previous time step as the initial guess for the ICCG solution of the linear systems was demonstrated. The implementation of ICCG using red-black ordering was shown superior to diagonal ordering for these problems and the advantage of using a more computationally efficient implementation of ICCG was demonstrated. Finally, in choosing an iterative method or even competing implementations of an iterative method, one must consider not only rate of convergence, rate of computation, and amount of computation but the combined effect of all three factors on overall execution time.

REFERENCES

Ashcraft, C. and Grimes, R. [1987]. *On Vectorizing Incomplete Factorization and SSOR Preconditioners*, SIAM J. Sci. Stat. Comput. To appear.

Frederick, J. [1988]. *Two Phase Flow in a Centrifuge*, to appear, Phd. Dissertation, University of Virginia.

Harlow, F. H. and Amsden, A. A. [1975]. *Numerical Calculation of Multiphase Fluid Flow*, J. Comp. Phys., Vol. 17, No. 1, pp. 19-52.

Jordan, T. [1984]. *Conjugate Gradient Preconditioners for Vector and Parallel Processors*, in Elliptic Problem Solvers, Birkhoff and Schoenstadt (Eds.), Academic Press, N.Y.

Lichnewsky, A. [1983]. *Some Vector and Parallel Implementations for Linear Systems Arising in PDE Problems*, Presented at the SIAM Conference on Parallel Processing for Scientific Computing, Norfolk, Va, November.

Poole, E. and Ortega, J. [1987]. *Multicolor ICCG Methods for Vector Computers*, SIAM J. Numer. Anal., Vol. 24, No. 6, pp. 1394-1418.

van der Vorst, H. [1982]. *A Vectorizable Variant of Some ICCG Methods*, SIAM J. Sci. Stat. Comput., pp. 350-356.

van der Vorst, H. [1983]. *On the Vectorization of Some Simple ICCG Methods*, First Int. Conf. Vector and Parallel Computation in Scientific Applications, Paris.

van der Vorst, H. [1986]. *The Performance of Fortran Implementations for Preconditioned Conjugate Gradients on Vector Computers*, Parallel Computing 3, pp. 49-58.

CHAPTER 19

Incomplete Domain Decomposition Preconditioners for the Conjugate Gradient Method

Gérard Meurant*

Abstract. The aim of this paper is to present incomplete Domain Decomposition preconditioners that can be used on parallel computers with a large number of processors in connection with the Conjugate Gradient method for solving symmetric linear systems. We summarize the techniques which have been presented in [8] and we present new ideas to further enhance the efficiency of our preconditioners. Finally, we show some numerical results that demonstrate the usefulness of the preconditioners described in this paper.

1. Introduction. In the last years, there has been a great deal of interest in Domain Decomposition (DD) methods for solving elliptic PDEs. This type of method was originally developed by Soviet mathematicians for solving large problems on computers with small memories, but recently a new interest has been found in these methods for application on parallel computers. The new trend in DD methods is to use this general framework to construct preconditioners for the Conjugate Gradient method for solving symmetric linear systems. Several interesting papers have recently appeared addressing this problem: Bjorstad & Widlund [1], Bramble, Pasciak & Schatz [2], Golub & Mayers [7], Chan & Resasco [3]. A conference dedicated to Domain Decomposition methods was organized in Paris last year [6]. In this paper, we present a class of preconditioners which blends together the techniques of Domain Decomposition and that of Incomplete Block Factorization introduced in Concus, Golub & Meurant [4]. Our methods are purely algebraic, working only with the matrix of the linear system regardless of the underlying PDE. This helps solving a larger class of problems than the methods relying on bilinear forms and facilitates the possible generalizations to unsymmetric problems. However, there is a price to pay for this freedom and the preconditioners we developed are not spectrally equivalent to the original operator but, as the numerical experiments will show, we obtain very good convergence rates, even for very difficult problems with strongly discontinuous coefficients. The main point is that the DD methods presented in this paper give both efficiency and parallelism.

Section 2 introduces the model problem we are solving. In Section 3 we briefly present the tools we are using to construct the preconditioners. Section 4 describes an exact DD solver and then, Section 5 shows how to derive an incomplete decomposition which was first described in [8]. We conclude in Section 6 with some numerical experiments on the CRAY X-MP/416 for some difficult problems.

---

* CEA, Centre d'Etudes de Limeil-Valenton, BP 27, 94190 Villeneuve St Georges, France

2. The Model Problem. The problem we want to solve is a linear elliptic PDE,

$$-\frac{\partial}{\partial x}\left(a(x,y)\frac{\partial u}{\partial x}\right) - \frac{\partial}{\partial y}\left(b(x,y)\frac{\partial u}{\partial y}\right) + cu = f, \quad \text{in } \Omega \subset R^2, \quad u|_{\partial\Omega} = 0 \quad \text{or} \quad \frac{\partial u}{\partial n}|_{\partial\Omega} = 0,$$

Ω being a rectangle. With standard finite differences schemes (5 point or 9 point) and row–wise ordering, we obtain a block tridiagonal linear system :

$$A\,x\,=\,b,$$

with

$$A = \begin{pmatrix} D_1 & A_2^T & & & \\ A_2 & D_2 & A_3^T & & \\ & \ddots & \ddots & \ddots & \\ & & A_{n-1} & D_{n-1} & A_n^T \\ & & & A_n & D_n \end{pmatrix}.$$

D_i is point tridiagonal strictly diagonally dominant , A_i is diagonal (5 point scheme) or tridiagonal (9 point scheme). A is a positive definite symmetric M-matrix. From now on, we will only consider the 5 point scheme.

3.Tools. The first tool we use is straightforward. We are looking at how to solve tridiagonal linear systems with sparse right hand sides. Suppose T is a tridiagonal matrix and

$$b = \begin{pmatrix} 0, & \cdots, & 0, & b_n \end{pmatrix}^T.$$

We want to compute x_n in the solution of $Tx = b$. If T is factored as $T = L\,D\,L^T$ with L lower triangular and D diagonal, then it is obvious that $x_n = D_{n,n}^{-1}\,b_n$.

In the same way, if $b = (b_1, \quad 0, \quad \cdots, \quad 0)^T$ and we want to compute x_1, it is more efficient to factor T as $T = L^T \Delta L$ with Δ diagonal, then $x_1 = \Delta_{1,1}^{-1} b_1$.

The second technique, which was developed in Concus, Golub & Meurant [4]concerns approximating the inverse of a tridiagonal matrix. Let T be a diagonally dominant tridiagonal matrix, then T^{-1} is full, but the elements decrease away from the diagonal (see Concus & Meurant [5]). So, we approximate T^{-1} by $trid(T^{-1})$, a tridiagonal matrix whose elements are the same as the corresponding ones of T^{-1}.

The third tool is the block (incomplete) factorization INV of Concus, Golub & Meurant [4]. Suppose A is a block tridiagonal matrix, then the block Cholesky factorization of A can be written as

$$A = (\Delta + L)\,\Delta^{-1}\,(\Delta + L^T)$$

$$\Delta = \begin{pmatrix} \Delta_1 & & & \\ & \ddots & & \\ & & \ddots & \\ & & & \Delta_n \end{pmatrix}, L = \begin{pmatrix} 0 & & & \\ A_2 & 0 & & \\ & \ddots & \ddots & \\ & & A_n & 0 \end{pmatrix}$$

$$\begin{cases} \Delta_1 = D_1, \\ \Delta_i = D_i - A_i\,\Delta_{i-1}^{-1}\,A_i^T. \end{cases}$$

To construct an incomplete decomposition, we simply replace the inverse with a sparse approximation,

$$\Delta_i = D_i - A_i\,trid(\Delta_{i-1}^{-1})\,A_i^T.$$

It follows that all the Δ_i are tridiagonal matrices.

4. An exact DD solver. To develop a DD method, we partition the domain Ω into strips $\Omega_i, i = 1, \ldots, k$ and we renumber the unknowns in such a way that the components of x related to the subdomains appear first and then the ones for the interfaces. The notations are detailed in [8]. With this (block) ordering, the system can be written as

$$
\begin{pmatrix}
B_1 & & & & & & C_1 & & & & \\
& B_2 & & & & & E_2 & C_2 & & & \\
& & B_3 & & & & & E_3 & \ddots & & \\
& & & \ddots & & & & & \ddots & C_{k-1} & \\
& & & & B_k & & & & & E_k & \\
C_1^T & E_2^T & & & & & B_{1,2} & & & & \\
& C_2^T & E_3^T & & & & & B_{2,3} & & & \\
& & \ddots & \ddots & & & & & \ddots & & \\
& & & C_{k-1}^T & E_k^T & & & & & B_{k-1,k} &
\end{pmatrix}
\begin{pmatrix}
x_1 \\ x_2 \\ \vdots \\ x_k \\ x_{1,2} \\ x_{2,3} \\ \vdots \\ x_{k-1,k}
\end{pmatrix}
=
\begin{pmatrix}
b_1 \\ b_2 \\ \vdots \\ b_k \\ b_{1,2} \\ b_{2,3} \\ \vdots \\ b_{k-1,k}
\end{pmatrix},
$$

with each B_i related to a subdomain Ω_i,

$$
B_i =
\begin{pmatrix}
D_i^1 & (A_i^2)^T & & & \\
A_i^2 & D_i^2 & (A_i^3)^T & & \\
& \ddots & \ddots & \ddots & \\
& & A_i^{m_i-1} & D_i^{m_i-1} & (A_i^{m_i})^T \\
& & & A_i^{m_i} & D_i^{m_i}
\end{pmatrix},
\quad i = 1, \ldots, k.
$$

D_i^j and $B_{i,j}$ are point tridiagonal, m_i is the number of mesh lines in Ω_i.

The matrices C_i and E_i have a very special structure,

$$
C_i =
\begin{pmatrix}
0 \\ \vdots \\ 0 \\ C_i^{m_i}
\end{pmatrix},
\qquad
E_i =
\begin{pmatrix}
E_i^1 \\ 0 \\ \vdots \\ 0
\end{pmatrix}.
$$

$C_i^{m_i}$ and E_i^1 are diagonal.

To derive an exact DD solver, we eliminate $x_1, \ldots, x_k$ to get a reduced system involving only the unknowns for the interfaces.

$$
\begin{pmatrix}
B_{1,2}' & F_2^T & & & \\
F_2 & B_{2,3}' & F_3^T & & \\
& \ddots & \ddots & \ddots & \\
& & F_{k-2} & B_{k-2,k-1}' & F_{k-1}^T \\
& & & F_{k-1} & B_{k-1,k}'
\end{pmatrix}
\begin{pmatrix}
x_{1,2} \\ x_{2,3} \\ \vdots \\ x_{k-2,k-1} \\ x_{k-1,k}
\end{pmatrix}
=
\begin{pmatrix}
b_{1,2}' \\ b_{2,3}' \\ \vdots \\ b_{k-2,k-1}' \\ b_{k-1,k}'
\end{pmatrix}
$$

It is easy to see that we have the following formulas for $i = 1, \ldots, k-1$,

$$
B_{i,i+1}' = B_{i,i+1} - C_i^T B_i^{-1} C_i - E_{i+1}^T B_{i+1}^{-1} E_{i+1},
$$

$$
F_i = -C_i^T B_i^{-1} E_i,
$$

$$
b_{i,i+1}' = b_{i,i+1} - C_i^T B_i^{-1} b_i - E_{i+1}^T B_{i+1}^{-1} b_{i+1}.
$$

To simplify these expressions, we use 2 factorizations for the matrices B_i coresponding to a subdomain, a block LU one (top–down) which can be written as

$$
B_i = (\Delta_i + L_i)\, \Delta_i^{-1}\, (\Delta_i + L_i^T),
$$

$$\begin{cases} \Delta_i^1 = D_i^1, \\ \Delta_i^j = D_i^j - A_i^j \left(\Delta_i^{j-1}\right)^{-1} \left(A_i^j\right)^T. \end{cases}$$

and a block UL one (bottom–up),

$$B_i = (\Sigma_i + L_i^T) \, \Sigma_i^{-1} \, (\Sigma_i + L_i),$$

$$\begin{cases} \Sigma_i^{m_i} = D_i^{m_i}, \\ \Sigma_i^j = D_i^j - \left(A_i^{j+1}\right)^T \left(\Sigma_i^{j+1}\right)^{-1} A_i^{j+1}. \end{cases}$$

With these notations, we have the following result,

Theorem

For $i = 1, \ldots, k-1$,

$$B_{i,i+1}' = B_{i,i+1} - \left(C_i^{m_i}\right)^T \left(\Delta_i^{m_i}\right)^{-1} C_i^{m_i} - \left(E_{i+1}^1\right)^T \left(\Sigma_{i+1}^1\right)^{-1} E_{i+1}^1.$$

If

$$G_i^1 = \left(\Sigma_i^1\right)^{-1} E_i^1,$$

$$G_i^l = -\left(\Sigma_i^l\right)^{-1} A_i^l \, G_i^{l-1}, \; l = 2, \ldots, m_i,$$

then,

$$F_i = -C_i^{m_i} \, G_i^{m_i}.$$

The proof, which use the tool #1, can be found in [8].

5. Domain Decomposition Preconditioners. From this factorization we can derive an approximation that will give a Domain Decomposition Preconditioner. To do this, we approximate the inverses in the same way we were doing in INV.

Here, we are interested in choosing the preconditioner M in order to have good convergence and parallelism. We want also to be able to generalize to unsymmetric problems; that is why we are considering an algebraic approach.

Consider first the 2 strips case.

$$\begin{pmatrix} B_1 & 0 & C \\ 0 & B_2 & E \\ C^T & E^T & B_{1,2} \end{pmatrix} \begin{pmatrix} x_1 \\ x_2 \\ x_{1,2} \end{pmatrix} = \begin{pmatrix} b_1 \\ b_2 \\ b_{1,2} \end{pmatrix}.$$

As a preconditioner we take

$$M = L \begin{pmatrix} M_1^{-1} & 0 & 0 \\ 0 & M_2^{-1} & 0 \\ 0 & 0 & M_{1,2}^{-1} \end{pmatrix} L^T, \quad \text{with} \quad L = \begin{pmatrix} M_1 & 0 & 0 \\ 0 & M_2 & 0 \\ C^T & E^T & M_{1,2} \end{pmatrix}.$$

Then

$$M = \begin{pmatrix} M_1 & 0 & C \\ 0 & M_2 & E \\ C^T & E^T & M_{1,2}^* \end{pmatrix},$$

$$M_{1,2}^* = M_{1,2} + C^T M_1^{-1} C + E^T M_2^{-1} E.$$

To have M being an approximation to A, it is natural to take M_i as an approximation to B_i and $M_{1,2}$ as an approximation to $B_{1,2} - C^T B_1^{-1} C - E^T B_2^{-1} E$, i.e. the Schur complement.

Then, $M_{1,2}^* \approx B_{1,2}$. The problem is how to choose $M_1, M_2, M_{1,2}$?

For M_1, M_2 we select INV block preconditioners, but many other choices are possible. With this particular choice, we have for M_1 a block LU approximation :

$$M_1 = (\Delta_1 + L_1) \, \Delta_1^{-1} \, (\Delta_1 + L_1^T)$$

$$\begin{cases} \Delta_1^1 = D_1^1, \\ \Delta_1^j = D_1^j - A_1^j \ trid((\Delta_1^{j-1})^{-1}) \ (A_1^j)^T. \end{cases}$$

and for M_2 a block UL aproximation :

$$M_2 = (\Sigma_2 + L_2^T) \ \Sigma_2^{-1} \ (\Sigma_2 + L_2)$$

$$\begin{cases} \Sigma_2^{m_2} = D_2^{m_2}, \\ \Sigma_2^j = D_2^j - (A_2^{j+1})^T \ trid((\Sigma_2^{j+1})^{-1}) \ A_2^{j+1}. \end{cases}$$

Then, we set

$$M_{1,2} = B_{1,2} - (C_1^{m_1})^T \ trid((\Delta_1^{m_1})^{-1}) \ C_1^{m_1} - (E_2^1)^T \ trid((\Sigma_2^1)^{-1}) \ E_2^1.$$

With suitable hypothesis, it is possible to show that the preconditioner M just constructed is positive definite.

It is easy to generalize this construction to k strips.

$$M = L \begin{pmatrix} M_1^{-1} & & & & & & & \\ & M_2^{-1} & & & & & & \\ & & \ddots & & & & & \\ & & & M_k^{-1} & & & & \\ & & & & M_{1,2}^{-1} & & & \\ & & & & & \ddots & & \\ & & & & & & M_{k-1,k}^{-1} \end{pmatrix} L^T,$$

$$L = \begin{pmatrix} M_1 & & & & & & & & \\ & M_2 & & & & & & & \\ & & \ddots & & & & & & \\ & & & \ddots & & & & & \\ & & & & M_k & & & & \\ C_1^T & E_2^T & & & & M_{1,2} & & & \\ & C_2^T & E_3^T & & & H_2 & M_{2,3} & & \\ & & \ddots & & & & \ddots & \ddots & \\ & & & C_{k-1}^T & E_k^T & & & H_{k-1} & M_{k-1,k} \end{pmatrix}$$

In the lower right corner of L is the factor of an incomplete block Cholesky decomposition of the reduced system.

$$M_{1,2} = B_{1,2} - (C_1^{m_1})^T \ trid((\Delta_1^{m_1})^{-1}) \ C_1^{m_1} - (E_2^1)^T \ trid((\Sigma_2^1)^{-1}) \ E_2^1.$$

$$M_{i,j} = B_{i,j} - (C_i^{m_i})^T \ trid((\Delta_i^{m_i})^{-1}) \ C_i^{m_i} - (E_j^1)^T \ trid((\Sigma_j^1)^{-1}) \ E_j^1$$

$$-H_i \ trid(M_{i-1,j-1}^{-1}) \ H_i^T.$$

$$G_i^1 = diag((\Sigma_i^1)^{-1}) \ E_i^1),$$

$$G_i^j = -diag((\Sigma_i^j)^{-1}) \ A_i^j \ G_i^{j-1}),$$

$$H_i = -(C_i^{m_i})^T G_i^{m_i}.$$

where $diag$ defines a diagonal approximation. Then, H_i is diagonal.

M_i is chosen as an INV block LU or UL approximation of B_i. But, one can choose for M_i a DD preconditioner in the orthogonal (i.e. column) direction. This gives a preconditioner using boxes. This is more costly than directly defining a preconditioner on a mesh with boxes but the former procedure is simpler, avoiding handling the problem of corners.

Whatever is the approximation, we can solve independently for the M_i's i.e. for each subdomain, but we have a block recursion for the reduced system i.e. the interfaces. We call this method **INVDD**.

There are many other possibilities :

1) take $H_i = 0$, $\forall i$; then everything is parallel as there is no more recursion within the interfaces (**INVDDH**).

2) take only "some" $H_i = 0$, as needed by the number of available processors (**INVDDS**).

3) form the reduced system and apply the same method recursively.

Notice that these approaches are purely algebraic and are feasible for any diagonally dominant block tridiagonal M-matrices regardless of their origins.

These DD preconditioners are easily generalized to unsymmetric problems.

Moreover, one can use different approximations (in place of INV) for the subdomains, like FFT-based preconditioners or point preconditioners. Modified (i.e. zero row sums) preconditioners are also possible.

6. Numerical results. We solve the following problem with discontinuous coefficients

$$-div(\lambda \nabla u) = f \quad \text{in } \Omega =]0, 1[\text{ x }]0, 1[$$

$$u|_{\partial\Omega} = 0$$

$$\lambda = 1000 \text{ in }]\tfrac{1}{4}, \tfrac{3}{4}[\text{ x }]\tfrac{1}{4}, \tfrac{3}{4}[,$$
$$\lambda = 1 \text{ elsewhere.}$$

$$h = \frac{1}{101}, \quad \|r^k\| \leq 10^{-6} \|r^0\|.$$

We are interested in looking at the number of iterations of the Conjugate Gradient method as a function of subdomains (which means the number of processors we can use).

# of subdomains	INVDD	INVDDH	INVDDS
2	22	22	22
4	24	24	24
8	26	26	26
16	26	26	26
24	30	30	30
32	31	35	33
40	32	45	41
50	32	54	41

We can see that for INVDD there is a very slow increase in the number of iterations with respect to the number of subdomains. The number of iterations for INVDDH is the same as for INVDD up to 30 subdomains and then increases more rapidly because the subdomains become too narrow and we are neglecting the coupling between the interfaces. For INVDSS we take one every two $H_i = 0$ so for a large number of subdomains, the results are in between INVDD and INVDDH.

The number of iterations for the INV preconditioner of [4] is 22 but it should be mentioned that the work per iteration is about half the one for DD preconditioners.

We solve several other problems with strongly discontinuous coefficients which behave the same as the previous one. Unfortunately, up to now, we were not able to get results using the four processors of the CRAY X-MP/416 with microtasking under COS 1.15. But some results obtained on a related method suggest that we could hope reaching about 500 Mflops for large problems and speed up greater than 3.9 (at least for INVDDH). With multitasking, the speed up is around 3.5, but these results have to be confirmed.

7. Conclusions. The Domain Decomposition methods presented in this paper offer a great deal of parallelism when used with the Conjugate Gradient method. But they are also efficient because the increase of the number of iterations with the number of subdomains is very small. Moreover, they can be extended to unsymmetric problems.

References

[1] P. BJORSTAD & O.B. WIDLUND, *Iterative methods for the solution of elliptic problems on regions partioned into substructures.* SIAM J. on Numer. Anal. v 23, n 6, (1986) pp 1097–1120.

[2] J.H. BRAMBLE, J.E. PASCIAK & A.H. SCHATZ, *The construction of preconditioners for elliptic problems by substructuring. I.* Math. of Comp. v 47, n 175, (1986) pp 103–104.

[3] T. CHAN & D. RESASCO, *A domain decomposition fast Poisson solver on a rectangle.* Yale Univ. report YALEU/DCS/RR 409 (1985).

[4] P. CONCUS, G.H. GOLUB & G. MEURANT, *Block preconditioning for the conjugate gradient method.* SIAM J. Sci. Stat. Comp., v 6, (1985) pp 220–252.

[5] P. CONCUS & G. MEURANT, *On computing INV block preconditionings for the conjugate gradient method.* BIT v 26 (1986) pp 493–504.

[6] R. GLOWINSKI, G.H. GOLUB, G. MEURANT & J. PERIAUX, *Proceedings of the first international symposium on domain decomposition methods for partial differential equations.* SIAM (1987).

[7] G.H. GOLUB & D. MAYERS, *The use of preconditioning over irregular regions.* In "Computing methods in applied science and engineering VI", R. Glowinski & J.L. Lions Eds, North–Holland (1984).

[8] G. MEURANT, *Domain decomposition vs block preconditioning.* In [6].

PART II

Numerical Methods

CHAPTER 20

Parallel Multivariate Numerical Integration

E. de Doncker[*]
J. Kapenga[*]

Abstract. An adaptive algorithm for numerical integration over an N-dimensional cube will be discussed, together with its implementation in a portable manner on a range of MIMD shared memory architectures.

We shall investigate the influence of various elements of the algorithm on the performance. A technique for a-priori performance analysis, on the basis of the serial code, will be applied and justified at the hand of performance results.

1. Introduction. In view of the significant increase of the number of integrand evaluations needed with higher dimensions, numerical multivariate integration may be troublesome or totally unfeasible when the function is not smooth or the dimension is high. This situation is favorable to a parallel adaptive approach, because of the fairly large granularity of the subdivision step in adaptive quadrature algorithms.

Good performance results were reported in [10] from a parallel version of the two-dimensional integrator *Triex* [2], and by Genz [5], from a parallel version of his integrator *Adapt* for the N-dimensional cube. Genz made use of compiler directives for his parallelization. We applied a parallel work pool management scheme based on the use of monitors [7,11,12].

In the present paper we shall outline an integration algorithm for the N-dimensional cube (section 2), together with its parallelization (section 3). It will then be our goal, in section 4, to examine factors which may limit performance expectations of the parallel code.

2. Algorithm. The algorithm (*Cubex*) incorporates a global adaptive extrapolation strategy similar to that in some of the (1-dimensional) *Quadpack* integrators [17]. For many cases of local singular integrand behaviour as well as for smooth functions, the extrapolation technique (by means of the *epsilon* algorithm [19]) is justified by the form of the N-dimensional quadrature error functional expansion [4,13,14]. Genz provided extrapolation based integration algorithms [3,4] which were local adaptive in nature.

The present algorithm comprises successive *levels* (or *stages*) each of which produces, in a global adaptive way [15], a particular integral result upon completion of the stage, and extrapolates on the sequence of the integral results currently available.

The global adaptive error reduction process within each stage consists of successive subdivision steps where cubes are subdivided and the subcubes obtained are integrated over, until a global stage accuracy requirement is fulfilled. In the beginning of a stage, all subcubes are *active* (or *large* with respect to that level). At each step of the stage, the active cube with the largest error estimate is

* Western Michigan University, Computer Science Department, Kalamazoo MI 49008. `dedonker@anl-mcs.arpa`

TABLE 1

stage #	integration result	extrapolation result
1	0.9768662	
2	0.9828943	
3	0.9883845	1.0444184
4	0.9924385	1.0038805
5	0.9951894	1.0009978
6	0.9969797	1.0000000

selected for subdivision. A subdivision may yield large and/or small subcubes, which are added to the active or the inactive pool respectively. At the end of the stage, if the extrapolated result does not satisfy the accuracy requirement, the active and the inactive pool are merged in order to start the next stage with active regions only.

Local elements of the algorithm are the quadrature rules by Genz and Malik [6] for integration over the subcubes and an analysis of the integrand over the cubes by calculating fourth divided differences as in the integrators *Adapt* [6] and *Half* [18]. The integrand analysis provides an ordering on the different directions with respect to how badly behaved the function is in each. This enables *Adapt* and *Half* to employ, as type of subdivision, a bisection of the selected cube perpendicular to the direction in which the integrand varies most locally.

In order to use the above bisection method in our adaptive extrapolation strategy, a cube resulting from a subdivision in the k-th stage, $k = 1, 2, ...,$ is considered *large* with respect to level k when its volume exceeds $2^{-Nk}V_0$ (where V_0 is the volume of the original cube) and moreover: the direction of its smallest axis does not coincide with that of its worst integrand behaviour, or the length of its smallest axis exceeds 2^{-k} times the length of the corresponding axis of the original cube.

Table 1 gives the results of the integration of $f(x,y) = 0.49(xy)^{-0.3}$ over the unit square. The sequence of the adaptive integration results available at the end of stages 0, 2, ..., 5 and the corresponding extrapolation results are given. Stage 0 corresponds to the integration over the original cube.

3. Parallel Algorithm. According to [11,12], we represent the algorithm by its *Control* and *Work* parts [10] as follows.
Control:
        Initialize and Create Slave Processes;
        While the user-requested accuracy is not achieved do
                Work;
                Finalize stage computations;
        Enddo.
Work:
        While the active set acceptance criterion is not satisfied do
                Select one or more cubes for subdivision;
                Get space for the subcubes about to be generated;
                Subdivide and integrate;
                Update global variables and stop in case of termination conditions;
                Add subcubes to pools;
        Enddo;
The algorithm is executed in parallel by P processes one of which, designated as the *master* process, executes the control. *Work* is executed by all processes (*master* and *slaves*) in parallel.

We found that the algorithm parallelizes well as a task pool algorithm on MIMD machines with shared memory. For the parallelization we have developed a set of macros layered over the monitor macros of Lusk and Overbeek [11,12], as described in [1,10]. Central to the parallelization is the acquisition of tasks from the active task pool and the maintenance of the pool by the co-existing processes.

Since all synchronization efforts are hidden in the macros, the application programs, including *Cubex*, are naturally portable with the macro sets.

So far we ran the program on a Sequent Balance and an Encore Multimax at Argonne Labs. With the current version we obtained speedups up to about 8. Generally, the more time consuming the integrand evaluations, the better the speedups obtained.

4. Performance modeling. We examined how the algorithm complies with the conditions given in [10], to obtain P speedup when using P processes, $P \leq NE$ (= the effective speedup of the machine). We shall formulate the conditions here to model a speedup $s(P)$ obtained with P processes. We have $s(P) = T(1)/T(P)$ where T represents the execution time as a function of the number of processes used. The goal of the study is not to provide a complex model for a detailed prediction or an accurate a-posteriori analysis of the performance of the parallel code (as for example in [8]). It is, however, the purpose, to effect an a-priori detection of possible limitations on the performance of a parallelized algorithm, based on a performance analysis of a serial version of the code.

If not satisfied, the conditions (2) to (5) below reflect the loss of parallelism through various causes in standard portions of the code of an adaptive task partitioning scheme like the one under study. These are starvation loss, saturation loss and braking loss, defined by Möller-Nielsen and Staunstrup [16], and singularity loss [10], which is the loss caused by processes doing unneeded work through selecting an unimportant task from the pool before some other process has finished its work and added an important task.

Assume

$$(1) \qquad\qquad s(P) \leq E$$

with E the effective parallelism of the system, dictated by hardware and possibly system imposed restrictions on the speedup. For example, by default a maximum of about 10 processors is assigned by the Sequent provided parallel programming library.

In order that the initial and terminal computations (to be carried out by one process) are not too time-consuming we require

$$(2) \qquad\qquad TI + T_c + T_t < \frac{T}{s(P)}$$

where T is the total execution time, TI is the initialization time, T_c represents the process creation time for P processes and T_t is the problem termination time.

In the present version of *Cubex*, the initialization includes the integration (by the master process) over the original cube. Elimination of the start-up phase of the serial algorithm involving the integration over the original cube and the first few subdivisions, would reduce the fraction of the execution time where only one or a small number of processes are active and the others are waiting for new tasks to be put in the pool. However, examination of the times in equation (2) for a number of integration problems revealed that in general TI as well as T_t is insignificant; on the Sequent Balance and the Encore Multimax, T_c may be considerable if the integration problem is small and the number of processes high. The times of equation (2) are given for two examples as part of Tables 2 and 3 below. Example 1 concerns the integral of $f(x, y) = 1/\sqrt{xy}$ over the unit square and example 2 that of $f(x, y, z, u, v) = cos(x + y + z + u + v)$ over the 5-dimensional unit cube. The timings in Tables 2 and 3 were obtained from a serial run of the program for a tolerated relative error of 10^{-8} on the Encore (using the clock routine *gettmr* and are given in microseconds

In order that the end of stage computations are not too costly we require

$$(3) \qquad\qquad \sum_{i=1}^{n}(TFC + TF_i) < \frac{T}{s(P)},$$

where TF_i is the *i-th* stage finalization time and TFC is the overhead for the synchronization primitives in the stage finalization. In the current version of *Cubex*, the slave processes wait while the master process returns to the control module, performs the extrapolation (with termination testing) and merges the the two pools. End of stage and total stage times are given as part of Tables 2 and 3 below, for the problems considered. We found the contribution from TFC to be insignificant. The left hand sides of the equations (2) and (3) are small for these examples.

Equation (3) does not account for the time that the master process waits to be released from the Argonne *askfor* macro at the end of a stage (before it returns to the control module), as to allow the other processes that are still working, to finish their subdivision step. This time may be nearly as big as that of a complete pass through the do-loop in the work module. Typically, some but not all processes are actually doing work in this time span. It may be noted that several different implementations are possible for the part of the algorithm to proceed from one stage to another. When a level is completed, the first processes to leave the askfor monitor (as opposed to the master process in the present version) can be delegated tasks involving the in-between calculations and updating of the data structures. The remaining processes can either wait for the others, or start integrating again as soon as new work is accessible in the pool. However, we noted that equation (3) is usually satisfied with the current version of the integrator, in particular because the number of stages is generally small (especially in higher dimensions and for integrands with a uniform type of behaviour over the whole integration region).

For the subdivision step to have $s(P)$ parallelism we require

$$(4) \qquad \sum_{i=1}^{n} \frac{T_i}{TPO_i}(TPC_i + TPI_i) < \frac{T}{s(P)},$$

where TPI_i is the time of the code at the i-th stage partitioning, embedded in critical sections, TPO_i the time not in critical sections and TPC_i the overhead of the synchronization primitives in the subdivision step. Tables 2 and 3 also give critical times spent getting work (T_{ask}), getting space (T_{sp}), updating global variables (T_{up}) and inserting the new subcubes into the pools (T_{add}). As can be observed, the times for adding work back into the pools may become large, when the number of subdivisions is large. Note that, roughly, $TPC_i + TPI_i \approx T_{ask} + T_{sp} + T_{up} + T_{add}$ and $TPO_i \approx T_i - TF_i - (TPC_i + TPI_i)$. Using these, condition (4) gives an upperbound for the speedup which is in agreement with the observed speedups of the parallel code for these problems (about 5 and 7 for examples 1 and 2 respectively).

Because of the demonstrated importance of the relative efficiency of the section of code involved, we have been concentrating on inserting work more efficiently. We implemented a first modification by using different monitors for adding work back into the two different pools, since a process adding a *small* cube hardly interferes with manipulations on the *active* set. This improved the performance significantly. The parallel management of data structures other than the two interleaved linked lists used on the work pools by the integrator at present, is investigated in [9] and is promising. Several other general task management options have been implemented and are under consideration.

As a guard against singularity and starvation loss (so that the processes would not perform many unneeded subdivisions), it is requested that

$$(5) \qquad \sum_{i=1}^{n} \frac{T_i}{T} \frac{s(P)}{NI_i} < 1,$$

where T_i is the total time for stage i and NI_i is the number of *active* cubes at the beginning of stage i, which will be partitioned in the course of the stage.

For example, the integration of the function $f(x, y, z) = 1/(x+y+z)^2$ over the three-dimensional unit cube suffers from singularity loss due to the point singularity in the origin. For a serial run of this problem when a relative accuracy of 10^{-4} is requested, it emerges that the number of useful tasks in the work pool at the beginning of the stages is not more than 5. Only the subcubes near the origin are actually subdivided. For this example, the left hand side of equation (5) is about $0.24s(P)$. Consequently, a speedup of not more than about 4 should be expected, which is confirmed by running the parallel code.

5. Conclusions. We outlined a global adaptive extrapolation algorithm for the N-dimensional cube and described its a-priori performance analysis based on the serial code. Although good speedups are obtained in general, the major factors, we found, that may affect the parallel performance are critical times spent updating global data structures as will as singularity loss in the case of point singularities.

Acknowledgment. We thank the Argonne Laboratories Mathematics and Computer Science Division and the Advanced Computing Research Facility for excellent access to the machines and for their valuable support. We also thank Ian Robinson for helpful discussions in the initial development of the algorithm Cubex.

TABLE 2

$T = 5100834$ $TI = 54424$ $T_t = 32$

Stage#i	T_i	TF_i	T_{ask}	T_{sp}	T_{up}	T_{add}
1	21298	91	108	61	256	226
2	74255	588	272	154	885	1121
3	287527	785	962	594	3473	10989
4	697466	2248	2243	1485	8307	45075
5	1474005	3999	4405	3940	16524	180638
6	2491813	1023	6884	5957	30858	524887

TABLE 3

$T = 51234808$ $TI = 79135$ $T_t = 33$

Stage#i	T_i	TF_i	T_{ask}	T_{sp}	T_{up}	T_{add}
1	1399300	93	1012	753	3536	7771
2	49756233	583	30301	22176	112969	6083231

REFERENCES

[1] E. DE DONCKER AND J. KAPENGA, *Parallelization of adaptive integration methods*, in NUMERICAL INTEGRA-TION; RECENT DEVELOPMENTS, SOFTWARE AND APPLICATIONS, P. Keast and G. Fairweather, eds., Reidel, 1987, pp. 207–218.

[2] E. DE DONCKER AND I. ROBINSON, *An algorithm for automatic integration over a triangle using nonlinear extrapolation*, ACM Transactions on Mathematical Software, 10 (1984), pp. 1–16.

[3] A. GENZ, *An adaptive multidimensional quadrature procedure*, Computer Physics Communications, 4 (1972), pp. 11–15.

[4] ———, *THE APPROXIMATE CALCULATION OF MULTIDIMENSIONAL INTEGRALS USING EXTRAP-OLATION METHODS*, PhD thesis, Univ. of Kent at Canterbury, 1975.

[5] ———, *The numerical evaluation of numerical integrals on parallel computers*, in NUMERICAL INTEGRA-TION; RECENT DEVELOPMENTS, SOFTWARE AND APPLICATIONS, P. Keast and G. Fairweather, eds., Reidel, 1987, pp. 219–229.

[6] A. GENZ AND A. MALIK, *An adaptive algorithm for numerical integration over an n-dimensional rectangular region*, Journal of Computational and Applied Mathematics, 6 (1980), pp. 295–302.

[7] C. HOARE, *MONITORS: an operating system structuring concept*, Comm. ACM, 17 (Oct. 1974), pp. 549–557.

[8] H. JORDAN, *INTERPRETING PARALLEL PROCESSOR PERFORMANCE MEASUREMENTS*, Tech. Rep., Department of Electrical Engineering and Computer Engineering, Univ. of Colorado, 1985. CSDG 85-1.

[9] J. KAPENGA AND E. DE DONCKER, *Concurrent management of priority queues*, in this conference, 1988.

[10] ———, *A parallelization of adaptive task partitioning algorithms*, Parallel Computing, (1988). to appear.

[11] E. LUSK AND R. OVERBEEK, *IMPLEMENTATION OF MONITORS WITH MACROS: A PROGRAMMING AID FOR THE HEP AND OTHER PARALLEL PROCESSORS*, Tech. Rep., Argonne National Labora-tories, 1983. MCS ANL-83-97.

[12] ———, *USE OF MONITORS IN FORTRAN: A TUTORIAL ON THE BARRIER, SELF-SCHEDULING DO-LOOP AND ASKFOR MONITORS*, Tech. Rep., Argonne National Laboratory, 1984. Report MCS ANL-84-51.

[13] J. LYNESS, *An error functional expansion for n-dimensional quadrature with an integrand function singular at a point*, Mathematics of Computation, 30 (1976), pp. 1–23.

[14] J. LYNESS AND E. DE DONCKER, *On quadrature error expansions. part i*, Journal of Computational and Applied Mathematics, 17 (1987), pp. 131–149.

[15] M. MALCOLM AND R. SIMPSON, *Local versus global strategies for adaptive quadrature*, ACM Transactions on Mathematical Software, 1 (1975), pp. 129–146.

[16] P. MØLLER-NIELSEN AND J. STAUNSTRUP, *EXPERIMENTS WITH A MULTIPROCESSOR*, Tech. Rep., Aarhus University, Aarhus, 1984. CS PB-185.

[17] R. PIESSENS, E. DE DONCKER-KAPENGA, C. ÜBERHUBER, AND D. KAHANER, *QUADPACK, A SUBROU-TINE PACKAGE FOR AUTOMATIC INTEGRATION*, Springer Series in Computational Mathematics, Springer-Verlag, 1983.

[18] P. VAN DOOREN AND L. DE RIDDER, *Algorithm 2, an adaptive algorithm for numerical integration over an n-dimensional cube*, Journal of Computational and Applied Mathematics, 2 (1976), pp. 207–217.

[19] P. WYNN, *On a device for computing the $e_m(s_n)$ transformation*, Mathematical Tables and Aids to Computing, 10 (1956), pp. 91–96.

A High–Performance FFT Algorithm for Vector Supercomputers

David H. Bailey

Many traditional algorithms for computing the fast Fourier transform (FFT) on conventional computers are unacceptable for advanced vector and parallel systems because they employ nonunit, power–of–two memory strides. This paper presents a technique for computing the fast Fourier transform that avoids such strides and appears to be near–optimal for a variety of current vector and parallel computers. Performance results of a program based on this technique are presented. Notable among these results is that a Fortran implementation of this algorithm on the Cray–2 runs up to 75% faster than Cray's assembly–coded library routine.

CHAPTER 21

The Granularity of Homotopy Algorithms for Polynomial Systems of Equations*

S. Harimoto[†]
Layne T. Watson[†]

Abstract. Polynomial systems consist of n polynomial functions in n variables, with real or complex coefficients. Finding zeros of such systems is challenging because there may be a large number of solutions, and Newton-type methods can rarely be guaranteed to find the complete solution list. There are homotopy algorithms for polynomial systems of equations that are globally convergent from an arbitrary starting point with probability one, are guaranteed to find all the solutions, and are robust, accurate, and reasonably efficient. There is inherent parallelism at several levels in these algorithms. Several parallel homotopy algorithms with different granularities are studied on several different parallel machines, using actual industrial problems from chemical engineering and solid modelling.

1. Introduction. Solving nonlinear systems of equations is a central problem in numerical analysis, with enormous significance for science and engineering. A very special case, namely small polynomial systems of equations, occurs frequently enough in solid modelling, robotics, computer vision, chemical equilibrium computations, chemical process design, mechanical engineering, and other areas to justify special algorithms. Polynomial systems are unique in that they have many solutions, of which several may be physically meaningful, and that there exist homotopy algorithms guaranteed to find all these meaningful solutions. The very special nature of polynomial systems and the power of homotopy algorithms are often not fully appreciated, perhaps because globally convergent probability-one homotopy methods are not widely known.

These globally convergent homotopy algorithms for polynomial systems have inherent parallelism at several levels. The purpose of the present paper is to study several different parallel homotopy algorithms for polynomial systems, corresponding to different decomposition and communication strategies and with different granularities, on shared memory machines.

---

*This work was supported in part by AFOSR Grant 85–0250.

† Department of Computer Science, Virginia Polytechnic Institute & State University, Blacksburg, VA 24061.

2. Homotopy algorithm. Let E^p denote p-dimensional real Euclidean space, and let $F : E^p \to E^p$ be a C^2 (twice continuously differentiable) function. The general problem is to solve the nonlinear system of equations $F(x) = 0$. The fundamental mathematical result behind the homotopy algorithm is

Proposition 1. *Let $F : E^p \to E^p$ be a C^2 map and $\rho : E^m \times [0,1) \times E^p \to E^p$ a C^2 map such that*
1) the Jacobian matrix $D\rho$ has full rank on $\rho^{-1}(0)$;
and for fixed $a \in E^m$
2) $\rho(a, 0, x) = 0$ has a unique solution $W \in E^p$;
3) $\rho(a, 1, x) = F(x)$;
4) the set of zeros of $\rho_a(\lambda, x) = \rho(a, \lambda, x)$ is bounded.
Then for almost all $a \in E^m$ there is a zero curve γ of $\rho_a(\lambda, x) = \rho(a, \lambda, x)$, along which the Jacobian matrix $D\rho_a(\lambda, x)$ has full rank, emanating from $(0, W)$ and reaching a zero $\bar{x}$ of F at $\lambda = 1$. Furthermore, γ has finite arc length if $DF(\bar{x})$ is nonsingular.

The general idea of the algorithm is apparent from the proposition: just follow the zero curve γ of ρ_a emanating from $(0, W)$ until a zero $\bar{x}$ of $F(x)$ is reached (at $\lambda = 1$). Of course it is nontrivial to develop a viable numerical algorithm based on that idea, but at least conceptually, the algorithm for solving the nonlinear system of equations $F(x) = 0$ is clear and simple. A typical form for the homotopy map is

$$\rho_W(\lambda, x) = \lambda F(x) + (1 - \lambda)(x - W),$$

which has the same form as a standard continuation or embedding mapping. However, there are two crucial differences. In standard continuation, the embedding parameter λ increases monotonically from 0 to 1 as the trivial problem $x - W = 0$ is continuously deformed to the problem $F(x) = 0$. The present homotopy method permits λ to both increase and decrease along γ with no adverse effect; that is, turning points present no special difficulty. The second important difference is that there are never any "singular points" which afflict standard continuation methods. The way in which the zero curve γ of ρ_a is followed and the full rank of $D\rho_a$ along γ guarantee this.

3. Polynomial systems. Suppose that the components of the nonlinear function $F(x)$ have the form

$$F_i(x) = \sum_{k=1}^{n_i} a_{ik} \prod_{j=1}^{n} x_j^{d_{ijk}}, \quad i = 1, \ldots, n.$$

The ith component $F_i(x)$ has n_i terms, the a_{ik} are the (real) coefficients, and the degrees d_{ijk} are nonnegative integers. The total degree of F_i is $d_i = \max_k \sum_{j=1}^{n} d_{ijk}$. For technical reasons it is necessary to consider $F(x)$ as a map $F : C^n \to C^n$, where C^n is n-dimensional complex Euclidean space. Define $G : C^n \to C^n$ by

$$G_j(x) = b_j x_j^{d_j} - a_j, \quad j = 1, \ldots, n,$$

where a_j and b_j are nonzero complex numbers and d_j is the (total) degree of $F_j(x)$, for $j = 1, \ldots, n$. Define the homotopy map

$$\rho_c(\lambda, x) = (1 - \lambda) G(x) + \lambda F(x),$$

where $c = (a, b)$, $a = (a_1, \ldots, a_n) \in C^n$ and $b = (b_1, \ldots, b_n) \in C^n$. Let $d = d_1 \cdots d_n$ be the *total degree* of the system. The fundamental homotopy result is:

Theorem. *For almost all choices of a and b in C^n, $\rho_c^{-1}(0)$ consists of d smooth paths emanating from $\{0\} \times C^n$, which either diverge to infinity as λ approaches 1 or converge to solutions to $F(x) = 0$ as λ approaches 1. Each geometrically isolated solution of $F(x) = 0$ has a path converging to it.*

Define $F'(y)$ to be the homogenization of $F(x)$:

$$F'_j(y) = y_{n+1}{}^{d_j} F_j(y_1/y_{n+1}, \ldots, y_n/y_{n+1}), \qquad j = 1, \ldots, n.$$

Define a linear function $u(y_1, \ldots, y_{n+1}) = \xi_1 y_1 + \xi_2 y_2 + \cdots + \xi_{n+1} y_{n+1}$ where $\xi_1, \ldots, \xi_{n+1}$ are nonzero complex numbers, and define $F'' : C^{n+1} \to C^{n+1}$ by

$$F''_j(y) = F'_j(y), \qquad j = 1, \ldots, n,$$
$$F''_{n+1}(y) = u(y) - 1.$$

The significance of $F''(y)$ is given by

Theorem. *If $F'(y) = 0$ has only a finite number of solutions in CP^n, then $F''(y) = 0$ has exactly d solutions (counting multiplicities) in C^{n+1} and no solutions at infinity, for almost all $\xi \in C^{n+1}$.*

The import of the above theory is that the nature of the zero curves of the projective transformation $F''(y)$ of $F(x)$ is as follows: There are exactly d (the total degree of F) zero curves, they are monotone in λ, and have finite arc length. The homotopy algorithm is to track these d curves, which contain all isolated (transformed) zeros of F.

4. Computational results. Polynomial systems arise in such diverse areas as solid modelling, robotics, chemical engineering, mechanical engineering, and computer vision. A small problem has total degree $d < 100$ and a large problem has $d > 1000$. Given that d homotopy paths are to be tracked, there are two extreme approaches for the parallel homotopy algorithm.

The first extreme, with the coarsest granularity possible, is to assign one path to each processor, with a master processor controlling the assignment of paths to the slave processors, keeping as many slaves busy as possible, and post-processing the answers computed by the slaves.

The second extreme, with the finest granularity, is to track all d paths on a single processor, distributing the numerical linear algebra, polynomial system evaluation, Jacobian matrix evaluation, and possibly other tasks amongst the other processors. The algorithm at this granularity is a major modification of the serial algorithm. A possible advantage is that the load could be balanced better, resulting in an overall speedup over a coarser grained algorithm.

Several parallel versions of the homotopy algorithm for polynomial systems with different granularities were tried, and results are shown in Tables 1 and 2. The problem number refers to an internal numbering scheme used at General Motors Research Laboratories. These problems are all real engineering problems that have arisen at GM and elsewhere. The times shown are for the serial algorithm (no superscript), a coarse-grained algorithm (superscript 1), and a fine-grained algorithm (superscript 2).

Table 1. Execution time (secs).

Problem number	total degree	80286/ 80287	iPSC–32[1]	VAX 11/750	VAX 11/780	IBM 3090/200	SUN-2 /50	SUN-3 /160	CRAY X-MP
102	256	16257	645	2438	1248	77	10545	8041	251
103	625	34692	1616	5260	2656	163	22634	17126	535
402	4	255	54	41	18	2	158	111	5
403	4	84	19	14	6	1	54	38	1
405	64	3669	335	703	334	14	2958	2429	84
601	60	9450	257	1707	796	78	7417	5903	249
602	60	28783	2795	4332	2054	124	21897	16831	462
603	12	1200	243	325	152	9	1339	1060	37
803	256	—	11527	29779	16221	667	130311	77113	2523
1702	16	1655	163	216	112	7	986	658	21
1703	16	1657	162	216	112	7	984	658	21
1704	16	1628	108	216	112	6	1005	667	21
1705	81	14336	378	1884	999	55	8907	6313	204
5001	576	—	11786	49736	27815	1997	—	237685	5148

Table 1 (continued). Execution time (secs).

Problem number	total degree	Balance 21000	Balance 21000[1]	Balance 21000[2]	Elxsi 6400	Elxsi 6400[1]	Elxsi 6400[2]
102(4)	256	4436	682	1120	508	63	442
103(4)	625	9316	1428	2379	1081	127	936
402(2)	4	72	18	37	10	7	11
403(2)	4	23	6	12	3	3	5
405(2)	64	1199	171	605	124	26	107
601(2)	60	3499	822	1321	392	105	289
602(2)	60	6775	983	2899	769	135	558
603(2)	12	577	133	334	63	20	64
803(8)	256	33746	6094	9961	6991	759	3045
1702(4)	16	399	81	140	50	15	37
1703(4)	16	399	81	142	50	15	37
1704(4)	16	399	68	129	50	11	34
1705(4)	81	3507	494	1053	426	53	267
5001(8)	576	97550	17829	32049	17449	1829	9051

Table 1 (continued). Execution time (secs).

Problem number	total degree	Alliant $FX/8$	Alliant $FX/8^1$	Alliant $FX/8^2$	Encore Multimax
102(4)	256	362	52	215	1022
103(4)	625	769	108	457	2157
402(2)	4	6	2	4	21
403(2)	4	2	1	1	5
405(2)	64	96	16	55	287
601(2)	60	245	35	126	836
602(2)	60	793	148	406	2317
603(2)	12	47	13	32	133
803(8)	256	4459	711	1642	10428
1702(4)	16	36	9	16	91
1703(4)	16	36	9	16	92
1704(4)	16	35	7	15	91
1705(4)	81	308	46	134	800
5001(8)	576	11579	1765	4736	28969

Table 2. Efficiency: [(serial time)/(parallel time)]/(number of processors used).

Problem number	total degree	iPSC-32^1	Balance 21000^1	Balance 21000^2	Elxsi 6400^1	Elxsi 6400^2	Alliant FX/8^1	Alliant FX/8^2
102(4)	256	.76	.81	.99	.81	.29	.87	.42
103(4)	625	.65	.82	.98	.85	.29	.89	.42
402(2)	4	.94	1.00	.97	.36	.40	.72	.83
403(2)	4	.88	.96	.96	.25	.28	.65	.80
405(2)	64	.33	.88	.99	.48	.58	.75	.87
601(2)	60	1.11	.53	1.32	.37	.68	.88	.97
602(2)	60	.31	.86	1.17	.57	.69	.67	.98
603(2)	12	.38	.54	.86	.32	.49	.45	.74
803(8)	256	—	.69	.68	.92	.29	.78	.34
1702(4)	16	.60	.62	.71	.33	.34	.48	.54
1703(4)	16	.60	.62	.70	.33	.34	.47	.54
1704(4)	16	.89	.73	.77	.45	.37	.67	.57
1705(4)	81	1.15	.89	.83	.80	.40	.84	.58
5001(8)	576	—	.68	.61	.95	.24	.82	.31

REFERENCES

[1] A. P. MORGAN AND L. T. WATSON, *Solving nonlinear equations on a hypercube*, in Super and Parallel Computers and Their Impact on Civil Engineering, M. P. Kamat (ed.), ASCE Structures Congress '86, New Orleans, LA, 1986, pp. 1–15.

[2] A. P. MORGAN AND L. T. WATSON, *Solving polynomial systems of equations on a hypercube*, in Hypercube Multiprocessors 1987, M. T. Heath (ed.), SIAM, Philadelphia, PA, 1987, pp. 501–511.

[3] A. P. MORGAN AND L. T. WATSON, *A globally convergent parallel algorithm for zeros of polynomial systems*, J. Parallel Distributed Comput., to appear.

[4] W. PELZ AND L. T. WATSON, *Message length effects for solving polynomial systems on a hypercube*, Parallel Comput., to appear.

[5] L. T. WATSON, *Numerical linear algebra aspects of globally convergent homotopy methods*, SIAM Rev., 28 (1986), pp. 529–545.

[6] L. T. WATSON, S. C. BILLUPS, AND A. P. MORGAN, *Algorithm 652: HOMPACK: A suite of codes for globally convergent homotopy algorithms*, ACM Trans. Math. Software, 13 (1987), pp. 281–310.

CHAPTER 22

A Parallel Homotopy Method for Solving a System of Polynomial Equations*

Shui-Nee Chow[†]
Lionel M. Ni[‡]
Yun-Qui Shen[†]

Abstract. Solving a system of polynomials is an important and fundamental technique to a large class of scientific and engineering applications. Traditional methods are very time-consuming and are not able to find all the solutions. The probability one homotopy method is theoretically able to find all the roots. In numerical implementation of the homotopy method, each root is obtained by following a distinct homotopy curve. Tracking the curve very closely may guarantee the finding of all roots, but it is very computationally expensive. A loose following of the curve can expedite the computation, but some roots will be missing due to the merge of curves and the undesirable divergence may occur. An automated and efficient numerical implementation of the homotopy method is proposed in this paper. The proposed method can expedite the computation process and can automatically detect and correct curve merging. Potential parallelism of the proposed method is fully exploited to allow the implementation in parallel processors.

1. Introduction. To solve a system of n polynomials with n complex variables is represented as

$$P(Z) = \begin{bmatrix} P_1(z_1, z_2, \cdots, z_n) \\ P_2(z_1, z_2, \cdots, z_n) \\ \cdots \\ P_n(z_1, z_2, \cdots, z_n) \end{bmatrix} = 0 \tag{1}$$

where Z is an n-dimensional complex vector $[z_1, \cdots, z_n]$. Let the degree of P_i be d_i for $1 \le i \le n$. Clearly, if $n=1$, then there are d_1 complex roots including multiplicity roots. In general, (1) has at most $d = d_1 \times d_2 \times \cdots \times d_n$ isolated roots. However, traditional methods are not able to solve all roots. The popular IMSL package can only find one root based on the MINPACK implementation of M.J.D. Powell's hybrid algorithm and P. Wolfe's secant method. A good initial guess is very important to the successful handling of the problem in above methods.

Solving all roots of any system of polynomials has been almost impossible until the advent of the *probability one homotopy method* by Chow, Mallet-Paret, and Yorke [1,2]. Theoretically, the homotopy method is able to find all the complex roots with probability one. The homotopy method is globally convergent and numerically stable and accurate. Since the homotopy method is global convergence, there is

* This research was supported in part by the DARPA AMSP Project, in part by the State of Michigan REED Project, and in part by the NSF grant DMS-8401719.

†Department of Mathematics, Michigan State University, East Lansing, Michigan 48824.

‡ Department of Computer Science, Michigan State University, East Lansing, Michigan 48824.

no need of a good initial estimate. The homotopy method is briefly described below. For a recent survey of this method, please refer to [3].

Consider the solving of another system of n polynomials, $Q(Z)$,

$$Q(Z) = \begin{bmatrix} z_1^{d_1} - b_1 \\ \cdots \\ z_n^{d_n} - b_n \end{bmatrix} = 0 \tag{2}$$

where $b_1, b_2, \cdots, b_n$ are n complex constants. Each polynomial, $z_i^{d_i} - b_i = 0$, has exactly d_i distinct complex roots and can be easily obtained. Clearly, $Q(Z)$ has $d = d_1 \times d_2 \times \cdots \times d_n$ distinct complex roots.

Define a homotopy function, $H(Z, t)$,

$$H(Z, t) = \begin{bmatrix} H_1(Z, t) \\ \cdots \\ H_n(Z, t) \end{bmatrix} = (1-t)Q(Z) + tP(Z) + t(1-t)R(Z) \tag{3}$$

where t is a real parameter varying from 0 to 1 and $R(Z)$ is defined as

$$R(Z) = \begin{bmatrix} \sum_{j=1}^{n} c_{j1} z_j^{d_1} \\ \cdots \\ \sum_{j=1}^{n} c_{jn} z_j^{d_n} \end{bmatrix} \tag{4}$$

where c_{ij} for $1 \le i, j \le n$ are n^2 complex constants.

It can be seen from (3) that $H(Z, 0) = Q(Z)$ and $H(Z, 1) = P(Z)$. In [2], it was shown that for almost every choice of the constants b_i and c_{ij}, for $1 \le i, j \le n$, $P(Z) = 0$ can be solved by tracing the solution curves of the following equation.

$$H(Z, t) = 0 \tag{5}$$

Thus, the choice of b_i's and c_{ij}'s is random.

The following properties of the homotopy curve obtained in [2] are essential in tracing the curve.

(P1) The solution set of $H(Z, t) = 0$ has d disjoint smooth homotopy curves for $t \in [0, 1)$.

(P2) Any homotopy curve is a function of t for $t \in [0, 1)$.

(P3) The homotopy curve does not go to infinity in the middle way. The term $R(Z)$ in the homotopy function guarantees this property [2].

(P4) If $P(Z)$ has exactly d roots including multiplicity, there are exactly d end points of these curves at $t = 1$. If the number of roots is less than d, the rest of the curves will go to infinity when t is approaching 1.

2. Parallelism in the Homotopy Method. The homotopy method has an inherent parallelism in which the tracing of all d homotopy curves can be performed in parallel and in an asynchronous fashion. Furthermore, all these curves can be independently traced. If a multiprocessor has d processors, a maximum speedup of d may be achieved. When the number of processors, m, is much greater than the number of curves (or possible roots) d, the time spent in tracing each curve may be expedited by allocating many processors to trace each curve.

Each homotopy curve is obtained by solving (5). We start from $t = 0$ and gradually increase t until t becomes 1. Let $\{t_0, t_1, \cdots, t_k\}$ be the *sampling points*, where $t_0 = 0$ and $t_k = 1$. In tracing a homotopy curve, one concerns the reaching of the final solution at $t = 1$. We will develop techniques to reduce the number of sampling points without sacrificing the accuracy of the final solution.

2.1 Tracing the Homotopy Curve. Let $s_i = t_i - t_{i-1}$ be the *stepping distance* from the sampling point t_{i-1} to the next point t_i. Initially, we have $t_0 = 0$. The value of $Z(t_0)$, which defines the initial point of the curve, is known by choosing one of the roots obtained in solving (2). Now we want to find $Z(t_1)$ for $t_1 = t_0 + s_1$. In general, given $Z(t_{i-1})$, we have to find $Z(t_i)$. Newton's iterative method will be used. Let $Z^{(j)}(t_i)$ be the value after the j-th iteration and $Z^{(0)}(t_i) = Z(t_{i-1})$. We have

$$\left[\frac{\partial(H_1, H_2, \cdots, H_n)(Z^{(j)}(t_i), t_i)}{\partial(z_1, z_2, \cdots, z_n)} \right] (Z^{(j+1)}(t_i) - Z^{(j)}(t_i)) = -H(Z^{(j)}(t_i), t_i) \quad \text{for } j = 0, 1, 2, \cdots \tag{6}$$

Theoretical results show that for a suitable s_i, $Z^{(j+1)}(t_i) \rightarrow Z(t_i)$ when $j \rightarrow \infty$. In practice, Newton's method converges quadratically and $Z^{(j+1)}(t_i)$ is considered to match $Z(t_i)$ when

$$|Z^{(j+1)}(t_i) - Z^{(j)}(t_i)| < \varepsilon \quad \text{for a small } \varepsilon \tag{7}$$

A numerical approach to evaluate (6) and (7) usually defines an upper bound on the number of iterations allowed. If $Z(t_i)$ can not be obtained after the predetermined number of iterations, it is considered divergent. Parallelism that can be exploited in tracing a curve is to evaluate many potential sampling points at the same time. The stepping distance, s_i, thus can be chosen as the largest allowable value. The total number of sampling points, k, then can be reduced.

2.2 Curve Merging. In order to speedup the computation, one may wish to have a greater stepping distance. In a sequential computer, the stepping distance is usually chosen to be a constant (it is $s = s_i$ for all i). If the value of s is large, two undesirable facts may happen. One is that (6) may not converge. In this case, a smaller value of s has to be used.

A serious problem of having a large s is the merge of curves. Since the tracing of all curves is based on the same equation (5) but with different initial values, a greater value of s may merge one curve to another curve and can still satisfy the convergence condition (7). However, some roots will be missed due to curve merging, which is very difficult to detect. The choice of s is problem dependent. A very small value of s may guarantee the finding of all complex roots, but is very time-consuming. A large s may cause curves merged together and lose some solutions.

2.3 Dynamic Stepping Interval. To expedite the finding of a homotopy curve, m possible sampling points are evaluated concurrently from a known point $Z(t_{i-1})$ if there are m processors allocated. Let $t_{i-1} + \Delta_j$ (for $1 \leq j \leq m$) be the j-th possible sampling point. These m possible stepping distances, Δ_j's, are chosen from a *dynamic stepping interval* (DSI), $(I_l, I_h]$, defined below.

$$I_l = \frac{2}{m+1} \bar{s} \quad \text{and} \quad I_h = \frac{2m}{m+1} \bar{s} \tag{8}$$

The *range* of DSI is defined as $I_h - I_l = 2(m-1)\bar{s}/(m+1)$, where $\bar{s}$ is initially chosen to be the previous stepping distance $\bar{s} = s_{i-1} = t_{i-1} - t_{i-2}$. All m Δ_j's are uniformly selected among the dynamic stepping interval. Thus, we have

$$\Delta_j = (I_l + \theta) + \frac{(j-1)(I_h - I_l - \theta)}{m-1} = j\bar{s}\frac{2}{m+1} + \theta\frac{m-j}{m-1} \quad \text{for } 1 \leq j \leq m \tag{9}$$

where θ is a small number and is dependent on the arithmetic accuracy of the machine.

Each processor then is evaluating $Z(t_{i-1} + \Delta_j)$ on an assigned Δ_j based on (6). If all processors have obtained the results, i.e., (7) is satisfied, then s_i is taken to be Δ_m. If some processors have results and some don't, then s_i is taken to be Δ_v such that Δ_{v+1} is the smallest Δ_j's that caused divergence. Note that the evaluation based on Δ_{v+r} for some positive integer $r > 1$ may still converge. This convergence may be caused due to the merge to another curve.

If the processor whose evaluation was based on Δ_1 reported a failure of convergence check, then a new DSI has to be defined. In this case, $\bar{s}$ is reduced to $\bar{s}/m$. Equation (9) is applied again to repeat the evaluation. Clearly, for a sufficient small I_l which is closing to t_{i-1}, there exists a Δ_1 to satisfy the convergence condition.

The DSI is a function of the previous stepping distance s_{i-1}. If all Δ_j's cause convergent results based on t_{i-1}, the next stepping distance, s_i, will be $2ms_{i-1}/m+1$. In the best case, the lower bound of the number of sampling points, k, required can be easily derived as

$$k = \left\lceil \log(1 + \frac{m-1}{2ms_0}) / \log(\frac{2m}{m+1}) \right\rceil \tag{10}$$

The above procedure has tried to reduce the possibility of curve merging. However, it cannot guarantee curve-merging free.

2.4 Merge Checking Points. Curve merging will degrade the performance of the homotopy method. However, there is no guarantee that curve merging won't occur unless the stepping distance is taken to be a sufficient small value at the cost of increased processing time. In addition to reducing the probability of the occurrence of curve merging, our algorithm tries to detect curve merging and to correct curve merging during the computation in case it happens.

A series of c *merge checking points* are defined to be

$$0=t_0<t_1^*<t_2^*<\cdots<t_c^*<1 \tag{11}$$

Note that t_c^* must be chosen to be close to 1 to avoid curve merging between t_c^* and 1. The tracing of each homotopy curve is required to evaluate $Z(t_q^*)$ for $1\leq q\leq c$. In order to do so, the determination of Δ_j's defined in (9) has to be modified such that $\Delta_j=t_q^*-t_{i-1}$ for some j if $t_q^*-t_{i-1}\in(I_l,J_h]$. Formally, we have

$$\Delta_j'=t_q^*-t_{i-1} \quad \text{if} \quad t_{i-1}+\Delta_{j-1}<t_q^*\leq t_{i-1}+\Delta_j \tag{12}$$

Δ_j' will replace Δ_j among those m potential stepping distances.

The merge checking is done by comparing its $Z(t_q^*)$ with all recorded values of $Z(t_q^*)$ obtained by other curves. If there is no match, the process simply records its values and continues its computation. If there is a match, the process first has to notify those processes running these matched curves to rollback its computation and to indicate the location of matched merge checking point.

2.5 Rollback Evaluation. Rollback computation has to reevaluate sampling points between two merge checking points, say t_{q-1}^* and t_q^*. The merge checking points only provide the boundary of the interval, $(t_{q-1}^*,t_q^*]$, that the actual merge point resides. To avoid curve merging again, one has to reduce the stepping distance. This may be achieved by reducing the size of stepping interval. With m processors, the following method is proposed.

If $m>1$, the dynamic stepping interval is chosen to be $(s/2, s]$, where $s=s_0$ initially. The j-th ($1\leq j\leq m$) possible stepping distance, Δ_j, is

$$\Delta_j = s(1-\frac{m-j}{2m}) \quad \text{for} \quad 1\leq j\leq m \tag{13}$$

If curve merging occurs again at t_q^*, then the above procedure is applied again with s reduced to $s/2$. Clearly, for a sufficient small s, curve merging will be eventually resolved. In practice, the initial value of s_0 is usually chosen to be a small value. Thus, curve merging is usually resolved in one pass.

3. Performance Evaluation. Given a system of polynomials with n roots, a parallel processor with n processors can achieve n times speedup. If there are $n\times m$ processors available, then the finding of each root can be further expedited by having m processors working together to solve each root. However, the actual speedup is usually less than $n\times m$ because these processors have to communicate to ensure no curve merging. Also the actual speedup is problem dependent. The following discussion demonstrates the speedup improvements of the single polynomial equation example shown below.

$$P(z) = z^3+(4+2i)z^2+(4+5i)z+1+3i=0 \tag{14}$$

The constant in (2) for this particular example was chosen at random. Here we have $z^3-b=0$, where $b=0.5801\times e^{4.9505i}$. Three roots of this example are $z1=-2-i$, $z2=-1-i$, and $z3=-1$.

An ideal speedup measurement is defined to be the ratio as the number of sampling points calculated sequentially to the number of worst case sampling points obtained in parallel. Without running the algorithms on real parallel processors, the overhead incurred due to processor synchronization and communication is very difficult to estimate and thus is ignored here. Note that the actual speedup is dependent on the architecture of the parallel processor and the parallel programming techniques. The actual speedup should be less than the ideal case.

In the following discussions, we only consider the ideal speedup with a given initial value of s_0. The first row indicated by *sequential* in Table 1 shows the number of sampling points calculated base on a sequential processor (automatic rollback computation is used to resolve curve merging) with respect to different choices of s_0. The rest of rows in Table 1 shows the speedup for different number of processors with $n=3$ and m is between 1 and 4. A "*" in the table indicates that the rollback computation is involved. In these experiments, one pass of rollback computation is able to resolve curve merging. The lower bound is calculated based on (10).

As the number of processors, m, increases, the speedup, in general, increases. This is because the difference between two adjacent Δ_j's is small and the probability that the selected next sampling point will cause curve merging is small. However, further increase m may not improve the speedup due to (10) and little effect on a too small difference between Δ_j's. The initial value of s_0 has little impact of the system performance because it is problem dependent. As a rule, a small s_0 should be used due to the possible involvement of rollback computation. It is interesting to note that the speedup for the case of $s_0=0.04$ and $n\times m=3\times2=6$ is 6.21 which is greater than 6. This is because our proposed method can dynamically adjust the stepping interval to reduce the possibility of curve merging.

4. Concluding Remarks. The homotopy method is the only known approach that is theoretically able to find all the roots of a system of polynomials. However, to speedup the usual numerical implementation of the homotopy method cannot not guarantee the finding of all roots. This paper is the first attempt to systematically make the homotopy method useful by introducing the concepts of merge checking and dynamic stepping interval. In addition to guarantee the finding of all roots, this is also the first paper trying to speedup the computation process by parallelizing the homotopy method in order to take advantage of various existing multiprocessors.

The actual speedup or performance improvement of the proposed approach is dependent on the problem itself. The concept of dynamic stepping interval is introduced to allow each problem dynamically adjusting its stepping distance. The concept of merge checking points is introduced to guarantee the finding of all roots and to reduce the amount of rollback computation.

processors	sampling points	$s_0=0.025$	$s_0=0.03$	$s_0=0.035$	$s_0=0.04$	$s_0=0.045$
sequential		120	102	107*	87*	73*
	curve-1	40	34	39*	31*	30*
	curve-2	40	34	29	31*	23
$n \times m = 3 \times 1$	curve-3	40	34	39*	25	30*
	lower bound	40	34	29	25	23
	speedup	3.00	3.00	2.74	2.81	2.43
	curve-1	21*	21*	23*	14	18*
	curve-2	9	9	12	8	10
$n \times m = 3 \times 2$	curve-3	20*	18*	23*	8	16*
	lower bound	9	8	8	7	7
	speedup	5.71	4.86	4.65	6.21	4.06
	curve-1	23*	19*	10	10	15*
	curve-2	7	7	7	7	6
$n \times m = 3 \times 3$	curve-3	23*	16*	7	7	15*
	lower bound	7	7	6	6	6
	speedup	5.22	5.37	10.7	8.7	4.87
	curve-1	10	7	19*	9	8
	curve-2	7	6	6	6	5
$n \times m = 3 \times 4$	curve-3	7	9	14*	6	5
	lower bound	6	6	6	5	5
	speedup	12	11.33	5.63	9.67	9.13

Table 1.   The number of sampling points required and the corresponding ideal speedup for various s_0's and number of processors, where the "*" indicates a rollback computation involved.

REFERENCES

[1] S.N. Chow, J. Mallet-Paret and J.A. Yorke, *Finding zeroes of maps: homotopy methods that are constructive with probability one*, Math. Comp., (32), 1978, pp.887-899.

[2] S.N. Chow, J. Mallet-Paret and J.A. Yorke, *A homotopy method for locating all zeroes of a system of polynomials*, Functional Differential Equations and Approximation of Fixed Points, H.-O. Peitgen and H.O. Walther, eds., Springer-Verlag Lecture Notes in Math., 730, 1979, pp.77-88.

[3] L.T. Watson, *Numerical linear algebra aspects of globally convergent homotopy methods*, SIAM Review, December 1986, pp.529-545.

CHAPTER 23

A Parallel Nonlinear Integer Programming Algorithm Based on Branch and Bound and Simulated Annealing*

Ken W. Bosworth[†]
G. S. Stiles[‡]
Rick Pennington[‡]

Abstract. The parallel algorithm we present consists of a three phase optimization process, with two phases of this process based on *deterministic* branch and bound, with differing heuristics, and with one phase based on a modified simulated annealing (*nondeterministic*) optimization procedure. The resulting algorithm yields *probable near best solutions*, together with confidence statistics on those solutions. Implementation of the optimization procedure requires at least three processing "units" (one for each phase) which run in a highly independent, *concurrent* fashion, and which can handle a moderate flow of inter-unit messages without bottlenecks or excessive "idling" occuring. Each of the three units can consist of groups of linked processors, running concurrently. Implementation of the algorithm on a transputer-based multiprocessor network is indicated.

1. Introduction. Very large scale nonlinear integer programming problems, or NLIP's, arise in the optimal design of distributed database computer and communication networks. The scale and complexity of these problems precludes their efficient solution using traditional deterministic branch and bound algorithms in a sequential processing environment. A stopgap solution is to resort to nondeterministic branch and bound algorithms, but the computation of sufficient reliability statistics for a verification of near optimal solutions remains a time consuming process. The recent development of multiprocessing environments has made possible the efficient "solution" of these large scale problems. We propose a NLIP nondeterministic parallel algorithm utilizing such an environment in this paper.

2. Problem formulation and terminology. The NLIP we attack is the following. Let $x \in X := X_1 \times X_2 \times \ldots \times X_n$, where X_i is a finite integer subset. $x = (x_1, \ldots, x_n)$ is termed the decision or design vector or "state", and $x_i \in X_i$ the ith decision variable. At hand is a cost functional, $c(x)$, consisting of a sum of nonlinear terms (products and boolean indicators of the decision variables). We

---

* This work supported by a grant from CCINR, Drexel University, Phil. Pa.

[†] Mathematics Dept., Utah State University, Logan, Utah, 84322-3900.

[‡] Electrical Engineering Dept., Utah State University, Logan, Utah, 84322-4120.

also have present feasibility constraints on the design, of the form $G(x) \leq b$, where $b \in Z^m$, and $G : X \mapsto Z^m$ is a nonlinear mapping. The NLIP consists then of

$$\begin{aligned} \min \quad & c(x) \\ \text{s.t.} \quad & x \in X \\ & G(x) \leq b. \end{aligned}$$

Two further key assumptions on the NLIP are *cost monotonicity* and *hereditary infeasibility*. To explain these concepts, we need some additional terminology.

We assume the ordering of the subsets X_i entering the Cartesian product definition of X is fixed. A state x is called a fully assigned state. A partial assignment of the entries in x of the form $x^i := (x_1, \ldots, x_i, \emptyset, \ldots, \emptyset)$ is termed a partially assigned state (at level i). A fully assigned state x is also denoted x^n when necessary, and an empty state assignment is x^0. Given x^i , let $c_i(x^i)$ denote the partial cost *computable* knowing *only* x^i (terms in $c(x)$ involving decision variables $x_j, j > i$, are evaluated as 0). We also note: $c_n(x^n) := c(x)$. Knowing x^i, some components of G may be fully computable. We speak of a partially assigned state x^i as being *infeasible* when some computable component of $G(x^i)$, G_k , exceeds b_k. If x^i and x^j are partially assigned states, x^j is said to be a *descendent* of x^i if $j > i$ and if $x^i = (x_1^i, \ldots, x_i^i, \emptyset, \ldots, \emptyset)$ and $x^j = (x_1^j, \ldots, x_i^j, \ldots, x_j^j, \emptyset, \ldots, \emptyset)$ implies $x_k^i = x_k^j$, $1 \leq k \leq i$.

Cost monotonicity is the requirement that whenever x^j is a descendent of x^i, then $c_j(x^j) \geq c_i(x^i)$. *Hereditary infeasibility* is the requirement that whenever x^j is a descendent of x^i, and x^i is infeasible, then so is x^j.

3. Branch and bound.
Given the NLIP above, together with the cost monotonicity and hereditary infeasibility assumptions, (deterministic) *branch and bound* is a natural means for finding a minimum cost feasible solution.

Branch and Bound Algorithm

Step 0 Initialization: set active list, AL , to $\{x^0\}$. Set best solution to date, x_{opt} , to $\emptyset$, and best cost to date, c_{opt} to ∞.

Step 1 Using some heuristic function, select $x^k \in AL$. If $AL = \emptyset$, exit.

Step 2 Find all descendents x^{k+1} of x^k. Discard any descendents which are infeasible, and go to Step 2 if no descendents remain.

Step 3 Compute partial costs, $c_{k+1}(x^{k+1})$, of remaining descendents. Discard any descendents with $c_{k+1}(x^{k+1}) \geq c_{opt}$. Go to Step 1 if no descendents remain.

Step 4 If $k + 1 = n$, set $x_{opt} = \tilde{x}^n$ and $c_{opt} = c_n(\tilde{x}^n)$, with $\tilde{x}^n$ that descendent with minimal cost. Discard remaining descendents and go to Step 1. Else, insert descendents into AL and go to Step 1.

At the conclusion of **branch and bound**, either $x_{opt} = \emptyset$ and $c_{opt} = \infty$, in which case NLIP is overconstrained ($\nexists$ feasible states $x \ni G(x) \leq b$), or $x_{opt} = \tilde{x}^n, c_{opt} = c_n(\tilde{x}^n)$ is a minimal cost feasible solution pair for NLIP.

We note that the selection heuristic in **Step 1** is left unspecified, and hence our *algorithm* really encompasses an entire class of branch and bound algorithms. The terminology *branch and bound* derives from the expansion of "active nodes" in a *decision tree* in **Step 2** (branching), and the elimination of "inactive nodes" from the tree in **Step 2** (pruning by infeasibility) and **Step 3** (pruning by bounding). The decision tree is the tree with *root node* x^0 , with *branches* $\{(x^0, x^1, \ldots, x^n) \mid x^{i+1}$ is a descendent of x^i $i = 0, 1, \ldots, n-1\}$, and with *leaves* x^n.

It should be noted that the bounding procedure, **Step 3**, is often modified to yield a more efficient algorithm. One has a (lower) bounding function $lb(\cdot)$ defined on the collection of all feasible partial states which is such that for each partial feasible state x^k and all x^n , feasible final descendents of x^k , we have $c_k(x^k) \le lb(x^k) \le c_n(x^n)$. **Step 3** is modified by replacing $c_{k+1}(x^{k+1})$ by $lb(x^{k+1})$.

When working on a NLIP , we often find that variables x^i have varying import in the cost and feasibility computations We term the ith decision variables *cost* type variables, if $\partial c / \partial x^i$ is large or possess a large variation over X_i , and if we *usually* have $G(x) \le b \Rightarrow G(\tilde{x}) \le b$, where x differs from $\tilde{x}$ only in the ith position. We term the ith decision variables *feasibility* type varibles if we *usually* have $G(x) \le b \not\Rightarrow G(\tilde{x}) \le b$, where x differs from $\tilde{x}$ only in the ith position. These types are defined subjectively, and hence provide only a "fuzzy" grading of the decision variables. We assume that one can determine (by experiments or analysis) a grading of the decision variables (strongly cost type, strongly feasibility type, etc.).

One major efficiency problem with branch and bound is the explosion in size of the active list. One means of controlling this problem is to "prune the tree" at early levels (small depths i). One can prune by infeasibility or by partial cost bounding. If the decision variables ordering in the definition of X is such that X_i for small i are feasibility type variables, and X_i for large i are cost type variables, then infeasibility pruning is likely to occur early in the tree. On the other hand, regardless of the variable ordering, the partial costs generally grow very slowly in the early levels of the tree, due to the product/boolean terms involved in the cost function not being computable until all the variables in these terms are known. Thus, pruning by cost bounding will not be possible, as a rule, until deep in the tree.

The above simple observations suggest the following approach to branch and bound. Order the decision variables in level from "strongly feas. type" to "strongly cost type ". Determine some level, k_{thresh} , below which the decision variables are predominantly "cost type". Maintain two active lists, AL_1 , AL_2. AL_1 contains partial states x^k with $1 \le k < k_{thresh}$, and AL_2 contains partial states x^k with $k \ge k_{thresh}$. AL_1 is expanded via a *depth first* selection heuristic (with min. partial cost used as a tiebreaker), and is pruned predominantly via infeasibility. Whenever this expansion produces a feasible state at level k_{thresh}, it is inserted to AL_2. AL_2 is expanded via a selection heuristic based on *expected final costs* , given the partial state x^k. This list is pruned by both infeasibility and cost bounding.

In a sequential processing environment, such a branch and bound procedure has as problems: how to allocate processing time and memory between the two list expansion processes, and how to make use of new fully assigned state information. The first two problems are obvious. To see the third, note that each new fully assigned feasible state found, x^n, whether or not it is a new "best cost" solution, contains information on the level by level accruement of costs. This information can be incorporated in expected cost heuristic used in managing AL_2. One must decide how much processing effort to invest in updating the expected cost heuristic. In a multiprocessing environment allowing concurrent independent processes, these problems are alleviated.

4. Modified Simulated Annealing. We refer to [1,2] for discussions on the philosophy behind, and various implementations of *simulated annealing*. We present a modified simulated annealing algorithm which is tailored to the assumed structure of our NLIP, and the above assumed ordering of the decision variables.

Modified Simulated Annealing

Step 0 Initialization: Input annealing schedule $\{T_i\}_{i=0,...,N}$, define $T_{-i} := 1$ for
$\quad\quad i = 1,...,N$, and input probabilities $\{p_k\}_{k=1,...,n}$. Input fully assigned feas.
$\quad\quad$ state $x = (x_1,...,x_n)$ and set $x_{opt} = x$, $c_{opt} = c(x)$. Set $i = 0$.

Step 1 If $i < -N$ or $i > N$, exit.

Step 2 Draw n uniform random variates $\mu_k \in [0,1]$, $k = 1,...,n$.
$\quad\quad$ If $\mu_k \geq p_k T_i$, don't flag k.
$\quad\quad$ If $\mu_k < p_k T_i$, flag k.

Step 3 If some k is flagged, go to Step 4.
$\quad\quad$ Else select a uniform random variate $\mu \in [0,1]$ and $k \ni \frac{k}{n} \leq \mu < \frac{k+1}{n}$.

Step 4 Perform *depth first search* for new feasible state $\tilde{x}$, with
$\quad\quad \tilde{x}_k \in X_k$, and *first choice* of $\tilde{x}_k = x_k$, if k not flagged,
$\quad\quad \tilde{x}_k \in X_k \setminus \{x_k\}$, if k flagged.
$\quad\quad$ If no path exists past a depth k, reject $\tilde{x}$, set $i \leftarrow i - 1$; go to Step 1.

Step 5 Compute and record the new feasible state's cost, $c(\tilde{x})$, in a stats bin.

Step 6 If $c(\tilde{x}) \leq c(x)$, set $c_{opt} = c(\tilde{x})$, $x_{opt} = \tilde{x}$, $x = \tilde{x}$, $i = 0$ and go to Step 1.

Step 7 Draw a uniform random variate $\mu_i \in [0,1]$.
$\quad\quad$ If $\mu_i \geq \exp[(-c(x) + c(\tilde{x}))/T_i]$, reject $\tilde{x}$, set $i \leftarrow i + 1$, and go to Step 1.
$\quad\quad$ Else, accept $\tilde{x}$, set $x \leftarrow \tilde{x}$, $i \leftarrow i + 1$ and go to Step 1.

We note that the temperature schedule is monotone decreasing from $T_0 = 1$ and the probabilities, $\{p_k\}$, are monotone increasing in k. **Steps 2** and **4** produce perturbed states $\tilde{x}$ which are usually, but not always, nearby x (as leaves on a tree). **Steps 4,6,** and **7** make the algorithm "self scheduling".

5. Parallel algorithm. We assume the NLIP is as above. Using the INMOS T-800 transputer chip, we form a multiprocessing environment consisting of 3 groups of processors, managed by a COMPAQ host running an OCCAM parallel processing shell. E.g., using 8 chips, we dedicate 3 chips each to groups 1 and 2, and 2 chips to group 3. In essence, groups 1 and 2 handle the branch and bound algorithm outlined above, and group 3 runs the modified simulated annealing algorithm. Only limited information flow between the groups is allowed.

Group 1 has the task of performing depth first search on an active list, AL_1 ,consisting of partial states up to level $k_{thresh-1}$. One processor in the group acts as a list manager, and the remaining processors are its slaves. The list manager feeds out partial states to its slaves for expansion, and receives feasible states back from the same for insertion into AL_1. When a partial state at depth k_{thresh} is received, it is sent to group 2's list manager. Lastly, updated best costs to date are received by the list manager (from group 2) for use in pruning.

Group 2 is given the task of performing branch and bound on an active list, AL_2 , consisting of partial states at level k_{thresh} or more. As in group 1, group 2 has one processor acting as list manager, and the remaining processors are slaves. The expansion heuristic is a minimum expected final cost, based on an average of partial cost histories over all fully assigned feasible states found to date. This heuristic is produced by a heuristic manager in group 3, and is passed, along with best costs to date, to the list manager in group 2. The manager of group 2 passes final feasible nodes to the heuristic manager of group 3, and relays best costs to group 1.

Group 3 carries out two functions: computation of statistics, and modified simulated annealing. The first of these tasks is assigned to the heuristic manager. It receives final feasible states from group 2 and from the simulated annealing process. An average of the ensemble partial cost histories is computed, and passed to group 2 for use as an expected final cost predictor. The heuristic manager also records and relays the best cost to date found by either the branch and bound or simulated annealing process, and keeps a list of the best solutions received to date from group 2. The remaining processors in group 3 carry out independent versions of modified simulated annealing. Each receives its seed from the heuristic manager's list of best group 2 solutions. The stats bin of each simulated annealing process contains cost vs. frequency information; this information and the optimal solutions found by each annealer are send to the heuristic manager for preservation and relaying.

The processing would begin with the loading of the entire program onto the network (each transputer receives only its relevant code). The algorithm terminates when either the group 1 and 2 processors empty their active lists, in which case the process is deterministic, or, when sufficient statistics have been compiled by group 3. In this case, the best solution to date is a probable near optimum, with reliability statistics given by the cost vs. frequency records of group 3.

Preliminary simulations are encouraging. It has been demonstrated that the expected cost heuristic is very efficient when sufficient cost histories are known. The modified simulated annealing is robust, and tends not to waste effort on infeasible perturbations. Thus, it is expected that the modified simulated annealing will be able to provide the necessary large sampling of feasible states needed to produce a robust expected cost heuristic. The above indicated 8 transputer network is currently under construction, and is expected to be operational by mid-1988.

REFERENCES

[1] Kirkpatrick, S., Gelatt, C.D., and M.P. Vecchi, *Optimization by Simulated Annealing*. Science, **220** (1983), pp. 671-220.

[2] Kirkpatrick, S. *Optimization by Simulated Annealing: Quantitative Studies*. Jour.Stat. Physics, **34** (1984), pp. 975-986.

Implementation of the Acceptance-Rejection Method on Parallel Processors: A Case Study in Scheduling

William Celmaster[*]

Abstract. The Acceptance-Rejection method (AR) is often used in connection with the generation of nonuniform random distributions. We examine the problem of generating, via AR, vectors of nonuniform numbers on various kinds of vector and parallel processors. Vector and SIMD implementations are generally quite inefficient compared to MIMD implementations. Various scheduling models are presented and the theoretical predictions are verified by tests done on a BBN Butterfly Parallel Processor.

There has been a substantial amount of research done in the area of numerical algorithms for parallel processors. Much of the focus of this activity has centered around the techniques of linear algebra and explicit solutions of differential equations. These kinds of problems tend to have a homogeneity which renders them vectorizable or equivalently, parallelizable on Single Instruction Multiple Data (SIMD) computers.

Unfortunately, it turns out that many physical problems have some inherent inhomogeneity. For example, certain portions of the physical domain may become nonlinear, increasing the computational demands on that region. On a SIMD machine, these kinds of inhomogeneity can be accommodated by forcing the solver program to consider the most general case for each point in the region. Thus, one strategy would be to use a nonlinear solver for the entire region rather than just for the region of nonlinearity. The resulting algorithm might be completely parallelizable, but extremely inefficient for solving the problem of interest.

In this paper, we are motivated by another area of numerical analysis- probability and statistics- in which a very simple algorithm, the Acceptance-Rejection (AR) algorithm, is often used in connection with the generation of nonuniform random distributions.[1] This algorithm can be repeated N times in order to produce N random numbers. Since the repetitions are all independent we would expect that a parallel machine could

---

[*]Bolt Beranek and Newman, Advanced Computers Inc. , 10 Fawcett Street, Cambridge, MA 02238

perform them in parallel. It turns out that it is very difficult to efficiently parallelize the N iterations of AR on a vector or SIMD machine but it is easy to efficiently do the parallelization on a Multiple Instruction Multiple Data (MIMD) machine such as the BBN Parallel Processor.

The significance of this example is that it exemplifies an important aspect of numerical analysis which has hitherto received inadequate attention by the parallel processing community. Many serial algorithms for solving equations, approximating functions, etc. involve testing, branching and iteration. For randomly distributed input data, the execution time of these algorithms follows a discrete probability distribution in much the same way that the AR algorithm follows a discrete probability distribution. Such algorithms are efficient on serial processors and can efficiently be performed in parallel on MIMD processors. However, for SIMD or Vector processors, less efficient techniques must often be used.

As an illustration of the Acceptance-Rejection algorithm, consider the following program for generating a pair of random numbers (X, Y) uniformly on a unit circle.

a.   V is initialized to a number greater than 1.

b.   [Choose a pair of random numbers between 0 and 1 and accept them if they lie within the unit circle. Otherwise repeat.]

   DO WHILE V>1

   1)   Generate U_1, U_2 each uniform on [-1,1].

          (Assume that efficient algorithms exist for generating uniform distributions such as these.)

   2) $U_1^2 + U_2^2 \rightarrow V$

   END DO

c.   [Divide the accepted pair by its magnitude.]

   $(U_1/\sqrt{V}, U_2/\sqrt{V}) \rightarrow (X,Y)$

The heart of this program is the DO-WHILE loop, which continues as long as the pair is "rejected" (V < 1) but terminates when the pair is "accepted" (V > 1). Hence the name "acceptance - rejection". For our purposes, the two significant parameters of this problem are L, the single-loop-iteration execution time and the acceptance probability "p" which is easily computed to be $\pi/4 = 0.79$.

In general, an analysis of the Acceptance-Rejection algorithm can be made by considering the following simple model algorithm which we will refer to as ARM (Acceptance Rejection Model).

Algorithm ARM

R = 1.0

DO WHILE (R>p)

WAIT (L)

R = RAND ()

END DO

The input variable p represents the acceptance probability and the wait-time L represents the compute-time of the non-branching part of the algorithm. It turns out that this performance model also represents the behavior of iterative algorithms whose termination time is based on certain convergence criteria.

On a MIMD parallel machine such as the BBN Butterfly Parallel Processor, each processor is capable of executing its own instructions independently of the other processors. Consider the problem of generating N numbers, R, on P processors.

For the case N>>P the most effective approach would seem to be that known as self-scheduling or dynamical task allocation. In that method, tasks wait in a task queue until one of the processors is free. The free processor then pops the task from the queue. If one were to assume that this allocation did not involve any overhead, one would conclude that processors would continually be kept busy doing useful work until there were no more tasks in the queue. This situation has been investigated on the BBN Butterfly Processor.

This is illustrated in Figure 1 produced which was by produced by the event-logging utility, GIST, on the Butterfly machine. Each random number is generated by a task, denoted by the event pair 1-2. If the task-queue is not refilled until the last value of R is obtained, then processors which finish earlier than the last are effectively idle as is readily seen in Figure 1. In that case, the inefficiency is measured by computing the total amount of idle time. A theoretical estimate of the straggler time, for the example of generating numbers on the unit circle, leads to a prediction that the speedup of P processors relative to 1 processor is, for p = 0.79 and N = 128,

$$R = \frac{P}{1 + 0.017P}$$

This prediction has been verified by a number of measurements.

One can also use the Butterfly machine to compare the self-scheduling model to a SIMD model. In the SIMD model, each of the P processors executes ARM until R is accepted. When the last value is finally produced, all of the processors then begin the execution of ARM to produce the next P numbers, and so on until all N numbers have been produced. On a MIMD machine, we refer to this technique as one of 'barrier synchronization'. Between every sequence of P accepted random numbers, a barrier is erected. The same algorithm is used as in the self-scheduling model, except for the fact that barrier synchronization occurs after every sequence of P numbers (i.e., after the completion of P tasks). On a SIMD machine, the DO-WHILE loop would continue execution until the last of the P numbers were generated. Thus, although the processors would be fully occupied, much of the processing would be wasted. In fact, with the above choices of parameters, the total execution time on the MIMD machine is about 2.5 times smaller than on the SIMD machine. This theoretical result has been verified by various Butterfly measurements and is shown in the GIST-log of Figure 2.

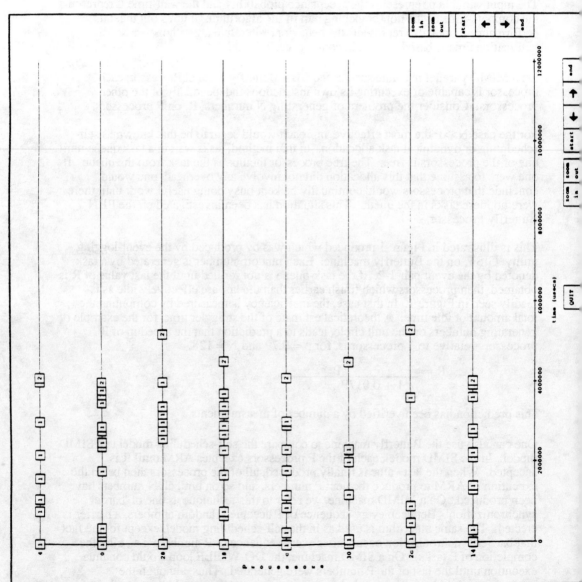

Event 1: Task Starts
Event 2: (Sometimes hidden by Task-1 boxes) Task Ends

P = 0.2, L = 64, Chunk-Size = 1

Full Self-Scheduling of the ARM Algorithm

Fig. 1 Dynamical allocation of tasks

Event 1 (Barrier): Task
Starts
Event 2:   Task Ends

P=0.2, L=64, Barrier
Synchronization

Emulation of 8-Processor
SIMD Machine

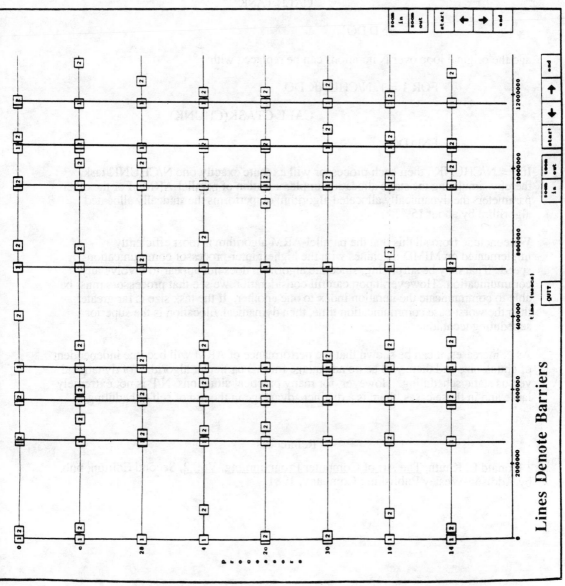

Lines Denote Barriers

Fig. 2 SIMD emulation

MIMD machines support a variety of models of task allocation apart from the self-scheduling model described above. When there are large communications costs between processors (distributed, as opposed to shared, memory) there is often a preference for static task allocation. In this scheduling system, tasks are pre-assigned to processors. Thus, in the ARM example, each of the P processors would be assigned to the tasks of computing N/P random numbers. We can simulate static allocation in a self-scheduling environment through the use of chunking, as follows. Suppose that N independent iterations of TASK are to be done. Then a new task, called CTASK (CHUNK) can be defined as

PROCEDURE CTASK(CHUNK)

{FOR 1 = 1,CHUNK DO

CALL TASK

END DO}

and the original loop over N iterations can be replaced with

FOR 1 = 1, N/CHUNK DO

CALL CTASK(CHUNK)

END DO

If P = N/CHUNK , then each processor will execute exactly one N/CHUNK task, thereby simulating the static allocation implementation of parallel-ARM. For the other parameters the dynamically allocated algorithm outperforms the statically allocated algorithm by about 15%.

We conclude from all this that the parallel-ARM algorithm is most efficiently implemented on MIMD machines with the highest inter-processor communication speed. That may be surprising, since the algorithm does not appear to involve any communication. However, upon careful consideration we see that processors must be able to communicate the iteration index to one another. If the task size is far greater than the worst-case communication time, then dymanical allocation is the superior scheduling technique.

As N increases, it can be shown that the performance of ARM will become independent of chunk-size and there will be no advantage for this particular algorithm of dynamical versus static scheduling. However, for many practical situations, N/P is not extremely large and in those cases, there is a distinct advantage in the use of self-scheduling.

References

1 Donald E. Knuth, The Art of Computer Programming, Vol. 2, Second Edition, pub. by Addison-Wesley Publishing Company, 1981.

Semi–Analytical Shape Functions in a Parallel Computing Environment

J.A. Puckett and R.J. Schmidt

Semi–analytical shape functions have been used successfully to model many physical systems in continuum mechanics. The orthogonality of appropriately selected shape functions allows a continuous system to be effectively transformed into a series of smaller discrete subsystems that can be solved independently. The uncoupling of the subsystems produces a highly parallel algorithm that is well suited to either a shared or a distributed memory environment. The semi–analytical shape functions are discussed with regard to the algorithmic implications. Examples are presented to illustrate the methodology.

The P–S Shape Functions in a Parallel Computing Environment
J.A. Puckett and R.J. Schmidt

A shape function which combines polynomial and sine series functions (P–S Shape Functions) is used to develop an efficient finite element algorithm for the solution of the elliptic equations that model many phenomena in continuum mechanics. This nontraditional approach results in an algorithm that is massively parallel, adaptive, and easily implemented using standard numerical techniques. The key to these favorable features is the orthogonality of the sine series which uncouples element matrices associated with the series terms. Because of the special structure of the system equation, condensation methods are used to form a series of matrix equations that are much smaller than the original system equation. These equations can be triangularized independently and thus the algorithm is highly parallel.

CHAPTER 25

Fast Laplace Solver by Boundary Integral based Domain Decomposition Methods*

E. Gallopoulos[†]
D. Lee[†]

Abstract. A new method called Boundary Integral-based Domain Decomposition is proposed for the fast solution of the Laplace equation on regular and irregular regions. The idea of the method is to adopt the domain decomposition approach by partitioning into subdomains and use an integral equation formulation to compute the interface values. The method provides a particularly efficient algorithm for parallel processing. Preliminary results from numerical experiments on an Alliant FX/8 vector multiprocessor are reported.

1 Introduction

We propose a new method for the fast solution of the Laplace equation on regular and irregular regions. It is called Boundary Integral based Domain Decomposition (BIDD), to encapsulate the ideas behind it: i) partitioning the region into subdomains, ii) computing the interface values by using an integral equation formulation and iii) solving a boundary value problem for each of the subdomains independently.

Most of domain decomposition (DD) techniques are based on the Schwarz alternating procedure (SAP) and reduce the general problem to several subproblems on the subdomains and a linear system for the interface values. The interface values are computed iteratively alternating among the adjacent subproblems. Computers with large grain parallelism are natural vehicles for demonstrating the advantages of DD techniques. However SAP requires continuous communication among subproblems. This may reduce significantly the efficiency of the parallel algorithm, especially as the number of subdomains grows.

Integral equation methods are used to solve harmonic problems in elasticity and electrical engineering. They can be computationally expensive however when solving at a large number of points.

The advantage of BIDD is that for homogeneous problems, it allows the complete decoupling of the problem, although the solution of a system of equations for the interfaces is still required. As the number of interface points is small in comparison with the total number of gridpoints, we avoid some of the cost of integral equation methods. It can also be an attractive alternative to other proposed methods for irregular regions [20]. Another advantage is that the maximum error will occur on the boundary thus allowing its easy estimation. The method has potential advantages when implemented on a parallel computer with a hierarchical memory [15]. The memory hierarchy,

*Research supported by the National Science Foundation under Grants No. US NSF DCR84-10110 and US NSF DCR85-09970, the US Department of Energy under Grant No. DOE DE-FG02-85ER25001, by the US Air Force under Contract AFSOR-85-0211, and an IBM donation.

† Center for Supercomputing Research and Development, University of Illinois at Urbana Champaign, Urbana, Illinois 61801.

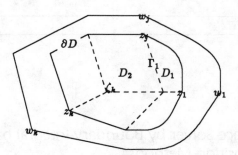

Figure 1: BIDD for an irregular region

in order of units of increasing capacity and memory access time may consist of registers, data caches, local memory and globally shared memory. The decomposition increases the chances for the data manipulated for each subdomain to fit in memory modules closer to the computational unit. The gains in performance using algorithms which are designed taking the memory hierarchy into account was demonstrated in [8].

2   Problem formulation

Given a bounded domain D and function g continuous on ∂D the Dirichlet problem is to find u satisfying:

$$\nabla^2 u(z) = 0 \text{ with } z \in D \text{ and } u(z) = g(z) \text{ with } z \in \partial D. \tag{1}$$

As formulated in Section 1, BIDD is an abstract solution method. Its specification requires i) a domain partitioning strategy, ii) a method to compute the interface values, and iii) the method(s) to compute the solution at the subdomains.

Although the domain shape may already impose constraints on the decomposition, this should be done with certain objectives in mind. For example the workload for each of the subdomains (depending on the number of gridpoints, the subdomain solver used, etc.) should be balanced and the subdomain dimensions should increase data locality. The computational cost of solving the integral equations also dictates that the number of subdomains (and resulting interface points) should not be too large.

The integral equations of potential theory furnish us a host of schemes for the calculation of interface values [11,22]. We choose the method of *fundamental solutions*, also known as *charge simulation method*, hereafter denoted as CSM. An approximation $\hat{u}$ to the solution u is sought as a finite linear combination of fundamental solutions $\{\phi_1, \ldots, \phi_N\}$ of $\nabla^2 u = 0$. Hence $\phi_j(z) = -\frac{1}{2\pi} \log |z - w_j|$ and

$$\hat{u}(z) = \sum_{j=1}^{N} \sigma_j \phi_j(z) \tag{2}$$

The singularities w_j lie in the exterior of $\bar{D} = D \cup \partial D$. Hence singular integrands are avoided. In physical terms, the method amounts to placing charges of strength σ_j at points w_j around the domain. Each of these charges generates a potential field. In CSM we want the charges to combine to a (single-layer) potential $\hat{u}$ which approximates u in D. The number and location of the charges will depend on the boundary values and the geometry. The situation is described in Figure 1. We see that the ability to continue the solution of 1 harmonically across the boundary and up to a closed curve passing through the w_j's plays an important role to the method's success. The σ_j are calculated so that $\hat{u}$ satisfies

$$\min_{z \in T} \|u(z) - \hat{u}(z)\|_p \tag{3}$$

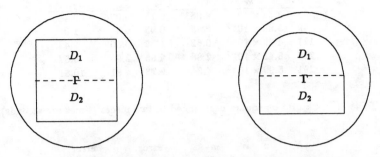

Figure 2: Rectangle and half racetrack with BIDD

where T is a discrete set of *observation points*[1] $T = \{z_1, \ldots, z_n\}$ where $T \subset \partial D$. In [16] the problems of charge placement, strength determination and best approximation with functions of this form for various domains and boundary conditions are discussed. The origins of the method are from [19]. [1,2,6,7,10,12,13,14,17,18,21,23] also contain valuable contributions. In [7] the link between boundary integral methods and CSM is drawn. In discrete form 3 becomes

$$\min \|u - G\sigma\|_p \qquad (4)$$

where $G \in \Re^{n \times N}$ is an influence matrix and $u_j = g(z_j)$. Depending on the relation between n and N, and the chosen norm $\| \cdot \|_p$, different schemes can be used for 4. For example if $n = N$ one can use simple collocation on the boundary, solving the square system involved, and since n is usually small a direct method is preferable. However $n = N$ may not be sufficient for certain shapes and boundary conditions ([5]). If $n > N$ a least squares or Chebyshev method can be used depending on whether $p = 2$ or $p = \infty$ respectively.

If the solution must be computed at m interface points ζ_j, form the influence matrix $H \in \Re^{m \times N}$ with general term $-\frac{1}{2\pi} \log |\zeta_k - w_j|$ and compute $\hat{u} = H\sigma$. G can be ill-conditioned [4,3] and care must be exercised. The effects of ill-conditioning may be less serious to the solution $\hat{u}$ since H has the same format as G and if $N = n$ then $\hat{u} = HG^{-1}g$. Finally to solve in each subdomain, use a suitable direct or iterative scheme.

By choosing the fundamental solutions ϕ_j differently, the method can be readily extended to handle different equations.

3   Computational experiments

We used an Alliant FX/8 vector multiprocessor with double precision arithmetic. We provide a more detailed analysis of these and other results in a forthcoming paper. We only deal with a square and a half racetrack shown in Figure 2. The w_j are equidistributed on an enclosing circle. The boundary conditions were $g_A(x, y) = e^x \sin y$ and $g_B(x, y) = \frac{1}{4}(x^2 + y^2)$. Unlike for g_A, the exact solution for g_B (torsion problem) is not known. In this case the estimated error is found by computing the solution with a high resolution on the boundary. To estimate the error for g_B we use another advantage of the method: using the maximum principle for harmonic functions, an estimate for the maximum error can be obtained from $\|g - H\sigma\|_\infty$, with H chosen to be the influence matrix from the points w_j to the points $z_{k+\frac{1}{2}} \in \partial D$ lying in the middle of the boundary segment with endpoints z_k and z_{k+1}. N charges were equidistributed at distance R from the origin and were computed by a block QR algorithm [9]. No attempt was made to optimize the individual algorithms or exploit the symmetries to save work in the formation of the influence matrices.

Table 1 corresponds to a $k \times k$ grid on a square, $n = N = 30$, $R = 2$. T_{CSM} is the time needed to compute the solution on the interface Γ. T_{D_i} is the time for solving each of the subdomains using FISHPAK [24]. The total time is given in T_1. The expected time for a two cluster machine is given in T_2. To this time however, the overhead in sharing data at the global memory level must be added. By optimizing the algorithm, this extra cost could be absorbed. The time on

---

[1] The term is from [7]

k	T_{CSM}	T_{D_i}	T_1	T_2	T_{fish}
39	3.35×10^{-2}	0.0631	0.15	0.0967	0.14
99	3.60×10^{-2}	0.45	0.93	0.49	1.03
127	4.29×10^{-2}	0.62	1.25	0.66	1.43
255	5.41×10^{-2}	2.88	5.65	2.94	6.57
291	5.99×10^{-2}	4.84	9.78	4.91	10.81

Table 1: $k \times k$ grid on square partitioned to two pieces (times are seconds).

N	n	m_1	m_2	error for g_A	error for g_B
20	60	28	32	$.30 \times 10^{-2}$	$.36 \times 10^{-2}$
40	40	20	20	$.55 \times 10^{-4}$	$.55 \times 10^{-3}$
60	60	28	32	$.87 \times 10^{-6}$	$.28 \times 10^{-3}$
80	80	40	40	$.16 \times 10^{-7}$	$.80 \times 10^{-3}$
80	100	48	52	$.13 \times 10^{-7}$	$.11 \times 10^{-3}$

Table 2: Error estimates on half racetrack.

a single cluster for the direct solution of the entire problem using [24] is given in column T_{fish}. We see that for large n the expected speedup is larger than 2. This is because the complexity of the subdomain solver is $O(mn \log_2 n)$ and halving n there is a reduction due to both n and $\log_2 n$. The side effect is that T_1 is also smaller than T_{fish}. Table 2 corresponds to a half racetrack and $R = 1.7$. The solution can proceed as before, so we only discuss the errors in using CSM to calculate the interface values. There were m_1 and m_2 points on the boundary circular and rectangular boundaries respectively. We were interested in the accuracy of the computed interface values. For g_B the error was estimated as earlier in this section.

4   Conclusions

The fundamental ideas behind BIDD have been described. The method seems to offer many advantages for solving homogeneous problems. Our experiments indicate that BIDD can be very competitive in a parallel processing environment and involves many interesting problems for consideration.

References

[1] S. Bergman and J. G. Herriot. Numerical solution of boundary value problems by the method of integral operators. *Numer. Math.*, 7:42–65, 1965.

[2] C. Brebbia. Boundary integral formulations. In C. Brebbia, editor, *Topics in Boundary Element Research: Basic Principles and Applications*, pages 1–12, Springer-Verlag, 1984.

[3] S. Christiansen. A comparison of various integral equations for treating the Dirichlet problem. In C. T. H. Baker and G. F. Miller, editors, *Treatment of Integral Equations by Numerical Methods*, pages 12–24, Academic Press, 1982.

[4] S. Christiansen. Condition number of matrices derived from two classes of integral equations. *Mat. Meth. Appl. Sci.*, 3:364–392, 1981.

[5] P. J. Davis and P. Rabinowitz. Advances in orthonormalizing computation. In F. L. Alt, editor, *Advances in Computers*, pages 55–133, Academic Press, 1961.

[6] S. C. Eisenstat. On the rate of convergence of the Bergman-Vekua method for the numerical solution of elliptic boundary value problems. *SIAM J. Numer. Anal.*, 11:654–681, June 1974.

[7] G. Fairweather and L. Johnston. The method of fundamental solutions for problems in potential theory. In C. T. H. Baker and G. F. Miller, editors, *Treatment of Integral Equations by Numerical Methods*, pages 359–349, Academic Press, 1982.

[8] K. Gallivan, W. Jalby, U. Meier, and A. Sameh. *The impact of hierarchical memory systems on linear algebra algorithm design*. Technical Report 625, Center for Supercomputing Research and Development, September 1987.

[9] W. J. Harrod. Programming with the BLAS. In L. H. Jamieson, D. Gannon, and R. J. Douglass, editors, *The Characteristics of Parallel Algorithms*, pages 253–276, The MIT Press, 1987.

[10] U. Heise. Numerical properties of integral equations in which the given boundary values and sought solutions are defined on different curves. *Computers and Structures*, 8:199–205, 1978.

[11] P. Henrici. *Applied and Computational Complex Analysis*. Volume 3, Wiley, 1986.

[12] P. Henrici. A survey of I. N. Vekua's theory of elliptic partial differential equations with analytic coefficients. *Z. Angew. Math. Phys.*, 8:169–203, 1957.

[13] J. L. Hess and A. M. O. Smith. Calculation of potential flow about arbitrary bodies. In D. Kuchemann, editor, *Progress in Aeronautical Sciences*, pages 1–138, Pergamon Press, 1967.

[14] M. A. Jaswon and G. T. Symm. *Integral Equation Methods in Potential Theory and Elastostatics*. Academic Press, 1977.

[15] D. J. Kuck, E. S. Davidson, Lawrie D. L., and A. H Sameh. Parallel supercomputing today and the Cedar approach. *Science*, 231:967–974, February 1986.

[16] R. Mathon and L. Johnston. The approximate solution of elliptic boundary-value problems by fundamental solutions. *SIAM J. Numer. Anal.*, 14:638–650, September 1977.

[17] A. Mayo. Fast high-order solution of Laplace's equation on irregular regions. *SIAM J. Sci. Stat. Comput.*, 6:144–157, January 1985.

[18] S. Murashima and H. Kuhara. An approximate method to solve two-dimensional Laplace's equation by means of superposition of Green's function on a Riemann surface. *J. Information Processing*, 3:127–139, 1980.

[19] E. R. Oliveira. Plane stress analysis by a general integral method. *J. Engng. Mech. Div. ASCE*, 94 (EM 1):79–101, 1968.

[20] W. Proskurowski. Capacitance matrix methods - a brief survey. In M. Schultz, editor, *Elliptic Problem Solvers*, pages 391–398, Academic Press, 1981.

[21] L. Reichel. On the determination of boundary collocation points for solving some problems for the Laplace operator. *J. Comput. Appl. Math.*, 11:175–196, October 1984.

[22] V. Rokhlin. Rapid solution of integral equations of classical potential theory. *J. Comp. Phys.*, 60:187–207, 1985.

[23] H. Singer, H. Steinbigler, and P. Weiss. A charge simulation method for the calculation of high voltage fields. *IEEE Trans. Power Apparatus and Systems*, PAS-93:1660–1668, September 1974.

[24] P. N. Swarztrauber and R. A. Sweet. Algorithm 541: efficient Fortran subprograms for the solution of separable elliptic partial differential equations. *ACM TOMS*, 5:352–364, September 1979.

CHAPTER 26

Parallel Multilevel Finite Element Method with Hierarchical Basis Functions

G. Brussino*†
R. Herbin*‡
Z. Christidis*
V. Sonnad*

Abstract. The 'p' version of the finite element method using a hierarchy of basis functions is employed in a multilevel computational scheme that is analogous to the multigrid iteration scheme. An advantage of using this approach over the traditional multigrid method is that only a single grid is used, and it is possible to preserve the speed of multigrid techniques while avoiding the very difficult problem of generating a sequence of nested grids on complex geometries. Another advantage is that the prolongation and restriction operators are easily defined, independent of the geometry of the domain, or the details of the differential operator under consideration. A sequential implementation using standard Gauss-Seidel smoothing is shown to give satisfactory rates of convergence. The parallel version uses a block Gauss-Seidel iteration scheme, with the blocks being chosen to reflect the order of basis functions. Implementation of this technique on a Loosely Coupled Array of Processors at IBM Kingston is described, and results are presented for the Poisson equation on a square with Dirichlet boundary conditions.

1. Introduction. The multigrid method is increasingly being accepted as one of the most effective computational techniques for the solution of problems arising from the discretization of partial differential equations. The traditional implementation of the multigrid method requires a sequence of nested grids on a given region; for complex 3-dimensional geometries, defining this sequence of nested grids can lead to difficulties. Furthermore, the definition of restriction and prolongation operators is not obvious for irregular grids that are often encountered in finite element discretizations. An approach that is proposed to handle this difficulty, is to use the p-version of the finite element method. While this approach

* IBM Corp., Dept 48B/MS 428, Neighborhood Rd., Kingston, NY 12401

† Present affiliation: Alliant Computer Systems Corp. Littleton, Massachusetts 01460.

‡ Present affiliation: Ecole Polytechnique Federale de Lausanne, GASOV/ASTRID, 1015 Lausanne, Switzerland.

146

was originally proposed for adaptive computations, (Ref. [1]), the structure of the method makes it unusually well suited to *multilevel* computations, (Ref. [3]). The basic idea of the p-version is to use higher order polynomials in regions where higher accuracy is desired, instead of reducing the mesh size, (h-version). Ill-conditioning of the matrix is avoided by using orthogonal polynomials of the Legendre type for the higher levels, and Lagrange polynomials at the lowest level. This approach has been very successfully used for adaptive computations in structural applications, and a large body of results is available establishing the effectiveness of this approach in discretizing partial differential operators.

2. Hierarchic basis functions in a multilevel iteration scheme.

The essential feature that makes the finite element method with hierarchic basis functions highly suitable in a multi-level iterative scheme is that the matrix formed in the p-version is hierarchic in nature; i.e., the matrix formed by a lower order polynomial is fully contained in the matrix formed by a higher order polynomial : the matrix is itself nested. When performing multilevel iterations, the nested matrices resulting from the p-version play a role similar to the nested grids employed in standard multigrid methods in the h-version. Traditional multigrid techniques use nested geometric grids to control errors at different frequencies, while the multilevel iteration scheme with hierarchic basis functions works in a more direct fashion with the frequencies.

The multilevel iteration scheme analogous to the standard multigrid V-cycle can be easily described (Ref. [3]): Starting with the matrix corresponding to the highest order polynomial, perform a chosen number of steps with a selected smoother and calculate the residual at this level. The restriction to the next lower level consists simply of truncating the residual vector to the size of the matrix at that level. By a simple change of variables, the solution vector at the previous level can be chosen as the starting vector for the smoothing process, without the need to define any intermediate vectors. The process of smoothing and restriction is carried down to the lowest level, where the problem is solved to the required accuracy. Prolongation from the lowest level to the upper levels is implicitly achieved by performing smoothing steps at each level using the available vectors as starting vectors, and this is carried out until the highest level is reached, at which point the downward step is started.

It is seen that the major operation in the multi-level iteration scheme (apart from the exact solution at the lowest level), is the smoothing operation. Since the Gauss-Seidel method has been widely used in traditional multigrid methods we decided to use this as a smoother to test the effectiveness of the multilevel iteration scheme described above.

The problem solved was:

$$\frac{\partial^2 u}{\partial x^2} + \frac{\partial^2 u}{\partial y^2} = f(x,y)$$

with

$$f(x,y) = 2N(2N+1)\left(x^{2N-1}(y^{2N+1} - y) + y^{2N-1}(x^{2N+1} - x)\right) \text{ with } N = 10.$$

The boundary conditions are $u(x,y) = 0$ on the perimeter of a unit square. The solution to the problem with the chosen boundary conditions and forcing function, has a sharp gradient near one of the edges, making it a suitable test problem. The results of using the Gauss-Seidel smoothing procedure in the multilevel iterative scheme described above are shown in Table 1. The first column gives number of elements used in the discretization of the square. All problems used a hierarchic level of 7; (a hierarchical level of 7 corresponds to Legendre polynomials of order 8). The conjugate gradient method was used to solve the problem at the lowest level. The numbers in the third column are the iterations required for convergence. The criterion used to determine convergence was that the norm of the residual was less than 10^{-6}. It is seen that for large problems, it takes about 5 iterations to reduce the error by a factor of 10 ; this is not as rapid as standard multigrid methods, which can reduce the error by a factor of 10 (or more) at every iteration. However in view of the many other advantages of the present technique, this is a satisfactory rate of convergence.

Elements	Unknowns	G.S. Iter.	Block G.S. Iter
1	48	4	3
25	1521	33	17
100	6241	39	22
400	25281	42	29
625	39601	42	29

Table 1: Convergence of the Multilevel Iteration Scheme.

3.   Parallel version of the multilevel iteration scheme.

The Gauss-Seidel smoothing technique used in the multilevel iterative scheme described above is inherently sequential in nature, and is extremely difficult to implement in parallel especially on a large grain machine. One approach to implementing the Gauss-Seidel scheme in parallel is to use red-black ordering with domain decomposition. This has been shown to be an effective scheme for parallel implementation of a traditional multigrid method, (Ref. [5]). A number of other approaches for parallel multigrid methods have been implemented on various architectures, (see Ref [6] for a survey and also the references in Ref. [5]). However, it is difficult to utilize these techniques on complex geometries, and it was decided to try an approach that would be consistent with the general applicability of the present method.

The approach adopted was to use a block Gauss-Seidel iteration scheme. The block sizes are chosen so that at every hierarchical level, the diagonal block of the corresponding matrix represents the coupling terms among the terms of that level only. This requires that a system of equations (consisting of the diagonal block), be solved at every Gauss-Seidel smoothing step. Instead of an exact solution on this system of equations, we use two iterations of a conjugate gradient procedure to obtain an approximate solution. The overall multilevel procedure is followed as before. The number of iterations required for convergence using the block iterative scheme on the same set of problems described before, is shown in in the fourth column of Table 1. We see that there is an improvement in the convergence rate of the overall scheme, (although the work per iteration is higher). This demonstrates that block Gauss-Seidel smoothing with approximate solution of the diagonal block equations, is a valid multilevel iteration scheme. An examination of the above process shows that the major computational steps consist of sparse matrix-vector multiplications, which can be readily implemented in parallel, (Ref, [4]). Furthermore, there is no reference to the geometry of the domain, and hence the approach is valid for arbitrarily complex geometries. We describe below the architecture of the lCAP (loosely Coupled Array of Processors), systems at IBM Kingston on which the above algorithm was implemented.

4.   The lCAP parallel computer system. There are at present two parallel processing systems in the laboratory at IBM Kingston. Both share the same fundamental architecture consisting of a host with attached processors, (Ref. [2]). We describe here the first of these systems, called lCAP-1, which is hosted by an IBM 3081 and attaches to 10 FPS-164 array processors, (AP's). In the initial stages of the development of the system, there was only limited ability to achieve direct slave to slave communication; for this reason the architecture was termed lCAP, (loosely Coupled Array of Processors). However, there are now multiple, independent data paths available for direct communications between processors, and the architecture can equally well be termed a ring, or a tightly coupled architecture. The ring configuration is achieved by connecting a number of AP's to local memories in a nearest neighbor configuration; the tightly coupled architecture uses a single global memory linking all processors; a double bus linking all AP's allows for broadcasting capability. These configurations allow experimentation with algorithms that can make effective use of one or more of these data paths.

5.   Results Preliminary numerical results for the Poisson equation on a square have been obtained on the lCAP system using the global memory to transfer data between processors. The results of this preliminary implementation are shown in Fig. 1. Speedup is defined as:

$$S(p) = \frac{\text{Execution time on one processor}}{\text{Execution time with } p \text{ processors}}$$

We see that although the algorithm is highly parallel, the speedups obtained are not as high as one would expect. The major reason for this is that the cost of data transfer and synchronization on the lCAP system are relatively high, so that it is essential that the amount of computation between data transfers be large. Also, the startup costs of data transfers are high, so that the data must be transferred in large chunks to achieve satisfactory data transfer rates. For medium size two-dimensional problems, the above conditions are not met, and the high price of data transfer reduces the speedup. Another reason is that in this preliminary implementation, the solution at the lowest level was performed sequentially on all processors, thus reducing the parallel content of the algorithm.

It must be pointed out however that these results pertain only to the part of the computation concerned with solving the system of equations; the overall finite element procedure involves the generation of the matrices, (usually by numerical integration). This phase is a completely parallel process, (no data exchange required between processors), and hence the overall parallel efficiency of the finite element solution of problems using hierarchical basis functions can be expected to be higher than shown here.

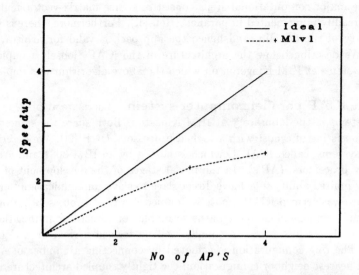

Figure 1: Speedup of the multilevel algorithm on the lCAP system; 13693 unknowns.

References

[1]   Babuska I., Szabo, B., and Katz, I.N., *The p-version of the finite element method*, SIAM J. Numer. Anal., 18, (1981), pp. 515-545.

[2]   Clementi, E., and Logan, D., *Parallel processing with a loosely coupled array of processors system*, IBM Kingston Technical Report, KGN-43, 1987.

[3]   Craig A.W. and Zienkiwicz, O.C., *A multigrid algorithm using a hierarchical finite element basis, in Multigrid Methods for Integral and Differential Equations*, edited by D. J. Paddon and Holstein, Clarendon Press, Oxford 1985, pp. 310-312.

[4]   Donati, R., Sonnad, V., *Coupling a Finite Element Frontal Solver Code to a Parallel Conjugate Gradient Solver*, IBM Kingston Technical Report, KGN-98, 1987.

[5]   Herbin, R., Gerbi, S., Sonnad, V., *Parallel Implementation of a Multigrid Method on the Experimental LCAP Supercomputer: Part II*, IBM Kingston Technical Report KGN-155, 1987.

[6]   Ortega, J. M., Voight, R.G., *Solution of Partial Differential Equations on Vector and Parallel Computers*, SIAM Review, 27, 2, (1985), pp. 149-240.

Invariant Imbedding and the Method of Lines for Multiprocessor Computers

Richard C. Allen, Jr., Lorraine S. Baca and David E. Womble

Recursive relations have been used to allow the solution of invariant imbedding equations with singularities. We demonstrate that these same relations can be used in an efficient implementation of invariant imbedding for multiprocessor computers and investigate the maximum attainable speedup.

The parallel implementation of imbedding can be used in conjunction with the method of lines to solve partial differential equations. We consider the problem of assigning lines to processors to minimize communication delays. We also consider the question of convergence of the resulting asynchronous algorithm. Timing and speedup data are presented for algorithms implemented on the NCUBE hypercube.

CHAPTER 27

The Parallel Waveform Relaxation Multigrid Method*

Stefan Vandewalle[†]
Dirk Roose[†]

Abstract. A great deal of attention has recently been given to the development of the highly parallel Waveform Relaxation method (WR) for solving very large systems of ordinary differential equations. Attempts to use WR to solve the equations arising from the numerical method of lines have not been very successful due to the slow convergence of the normally used Jacobi, Gauss-Seidel and SOR methods. In this paper we present a new algorithm that combines the parallel nature of WR with the fast convergence of multigrid to solve nonlinear, parabolic partial differential equations.

1. Introduction. In this paper we consider the following nonlinear, parabolic initial value problem, defined on an interval or rectangle Ω, for which we want to obtain a solution on the time interval $[0,T]$:

$$\frac{\partial u}{\partial t}(x,t) = K \nabla^2 u(x,t) + g(u,x,t) + f(x,t) \quad \text{with } x \in \Omega, \ t \in [0,T], \ K \in \mathbb{R} \qquad (1.1)$$

$$u(x,0) = u_{ic}(x), \ x \in \Omega$$

$$u(x,t) = u_{bc}(x,t), \ x \in \partial\Omega, \ t \in [0,T]$$

With the classical solution methods we either treat the space discretization or the time discretization as dominant. In the former case a sequence of elliptic boundary value problems is solved. In the latter case, also called *the numerical method of lines*, the problem is transformed into a large system of ordinary differential equations. Due to stiffness limitations this system has to be solved with implicit time-discretization methods. Both techniques lead to a system of nonlinear equations to be solved at each time step.

The use of global time steps has several disadvantages. The global time step has to be chosen small enough such that the solution is accurately represented in every part of the domain. This may lead to many *superfluous calculations* in regions where the solution is changing slowly. The introduction of time steps transforms equation (1.1) into a sequence

---

*This work was supported by the Belgian National Science Foundation (N.F.W.O).

[†] Katholieke Universiteit Leuven, Dept. of Computer Science,Celestijnenlaan 200A, 3030 Leuven, Belgium.

of problems, to be solved one after the other. In a parallel implementation *processor-synchronization* is forced before the calculation can proceed to the next time level. The initial computational task is reduced to a sequence of computationally much less expensive subtasks. These smaller tasks are more likely to have an *unfavorable communication to computation ratio*. It is indeed well known that parallel performance is easier to achieve for large size problems.

In this article we present an algorithm for solving nonlinear parabolic PDE's which addresses above problems.

2. Waveform Relaxation. Waveform Relaxation (WR) is a relatively new technique to solve large systems of ODE's. We discuss the Jacobi variant.
Consider the following system of N first order ODE's for $t \in [0,T]$

$$\frac{d}{dt}y_i = f_i(t,y_1,...,y_N) \quad \text{with} \quad y_i(0) = y_{i0} \quad \text{and} \quad i=1,...,N \tag{2.1}$$

The WR Jacobi algorithm to solve (2.1) can be formulated as follows:

initialization step:

   k := 0

   choose $y_i^{(0)}(t)$ for $t \in [0,T]$ and $i=1,...,N$

iteration steps:

  repeat

     for i=1,...,N solve $\dfrac{d}{dt} y_i^{(k+1)}=f_i(t,y_1^{(k)},...,y_{i-1}^{(k)},y_i^{(k+1)},y_{i+1}^{(k)},...,y_N^{(k)})$ (2.2)

     k := k+1

  until convergence.

In the initialization step an approximation $y_i^{(0)}(t)$ has to be provided for each unknown function $y_i(t)$. The obvious choice is to take each function constant and equal to the corresponding initial condition. In the iteration step each equation of type (2.2) is solved as an ODE in one variable. Thus instead of solving one ODE system in N unknowns, we repeatedly solve the N ODE's separately. This can be done in parallel.

A convergence analysis for linear systems is given by Miekkala and Nevanlinna in [4]. They study the Jacobi, Gauss-Seidel, SOR and SSOR schemes and calculate convergence rates. Nonlinear systems are studied in [5]. The authors basically prove uniform convergence on bounded time intervals under very relaxed conditions.

3. Application to parabolic PDE's. For notational convenience we restrict our attention to the one-dimensional problem (1.1) on the unit interval. The equation is discretized with central differences on an equidistant grid $x_i=ih$ with $i=0,...,N+1$, $h=1/(N+1)$ and $u_i(t) \simeq u(x_i,t)$.

$$\frac{d}{dt}u_i(t) = K \frac{u_{i+1}(t) - 2u_i(t) + u_{i-1}(t)}{h^2} + g(u_i,x_i,t) + f(x_i,t) \quad i=1,...,N \tag{3.1}$$

with $u_i(0) = u_{ic}(x_i)$ and $u_0(t) = u_{bc}(0,t)$, $u_{N+1}(t) = u_{bc}(1,t)$

Classical WR can now be applied to solve the ODE system (3.1). An analogue can be defined to each of the traditional iterative techniques for solving algebraic equations derived from the discretization of elliptic PDE's. However, the basic computation unit is no longer the calculation of an update to a single *grid point value*, but the solution of an ODE for a single *grid point function*.

The basic Jacobi WR scheme was implemented on a 64 node Intel iPSC hypercube. In Table 3.1 we give our results obtained for a 24 by 24, and a 8 by 8 discretization of the two-dimensional heat equation. For several processor configurations the efficiency,

$E_p = T_1/pT_p$, was calculated. Though many parameters influence the speed of the actual calculation (nature of the problem, computer representation of a function, choice of ODE integrator, error tolerances, ...) all our results indicate a very high efficiency.

	number of processors			
Table 3.1 : efficiency				
discretization	1	4	16	64
24 by 24	100%	99.7%	99.6%	99.3%
8 by 8	100%	99.1%	94.4%	92.0%

However, efficiency should not be confused with the numerical quality of an algorithm. The weak point in the algorithm is the initialization step in which the programmer has to provide an approximate solution on the whole time interval. The difference between the solution and the initial approximation may be substantial, resulting in a large number of iterations before convergence is obtained.

4. Multigrid Waveform Relaxation. To obtain higher convergence rates we combine WR with a multigrid iteration, using the Full Approximation Storage scheme (FAS) [1]. A similar technique for the Defect Correction Scheme was studied by Lubich and Ostermann [3]. In their article they analyze the convergence of the WR multigrid method applied to linear parabolic equations.

We briefly sketch the recursive algorithm for solving the nonlinear, elliptic problem $L(u)=f$. In the multigrid process a sequence of grids is used, with decreasing mesh size $h_0 > h_1 > ... > h_f$ and corresponding number of unknowns $N_0, N_1, ..., N_f$. The discretized problem on grid level k is denoted by $L^{(k)}(u^{(k)})=f^{(k)}$ with $u^{(k)}, f^{(k)} \in \mathbb{R}^{N_k}$ and $L^{(k)}(.)$ is the discretized, nonlinear operator. In the given procedure, r and $\bar{r}$ are restriction operators, which need not necessarily be equal, and p is the prolongation or interpolation operator. The constants ν_1, ν_2 and γ define the multigrid cycle.

procedure FAS(k)   (4.1)

 if k=0 **then** solve $L^{(0)}(u^{(0)}) = f^{(0)}$

 else

  perform ν_1 pre–smoothing steps on $L^{(k)}(u^{(k)}) = f^{(k)}$

  construct the level k−1 problem: $\bar{u}^{(k-1)} := \bar{r} u^{(k)}$

$$f^{(k-1)} := L^{(k-1)}(\bar{u}^{(k-1)}) - r (L^{(k)}(u^{(k)}) - f^{(k)}) \quad (4.2)$$

  solve the level k−1 problem: $u^{(k-1)} := \bar{u}^{(k-1)}$

      perform γ times FAS(k−1)

  correct the level k solution: $u^{(k)} := u^{(k)} + p(u^{(k-1)} - \bar{u}^{(k-1)})$

  perform ν_2 post–smoothing steps on $L^{(k)}(u^{(k)}) = f^{(k)}$

 end procedure.

Each step in algorithm (4.1) has a straightforward extension in the function space of WR. This will be clarified using the illustrative example (3.1).

● **problem definition.** The discretized nonlinear operator is

$$(L^{(k)}(u^{(k)}))_i := \frac{d}{dt} u_i^{(k)} - K \frac{u_{i+1}^{(k)} - 2u_i^{(k)} + u_{i-1}^{(k)}}{h_k^2} - g_i(u_i^{(k)}) \quad i=1,...,N_k \quad (4.3)$$

with $u^{(k)} = (u_1^{(k)}(t), ..., u_{N_k}^{(k)}(t))^T \in \mathbb{R}^{N_k}$ and $g_i(u_i^{(k)}) = g(u_i^{(k)}, x_i, t)$.

● **smoothing.** Pre- and post-smoothing is done by one or more Waveform Relaxation sweeps. Each ODE is solved separately and possibly in parallel, using whatever time steps needed to satisfy the tolerance requirement of the ODE integrator.

• **restriction and prolongation.** The coarse-to-fine transfer (prolongation) and fine-to-coarse transfer (restriction) can be done using identical formulae as in the elliptic multigrid method. However these formulae now operate on the *functions* associated with the grid points. E.g. the weighted restriction analogue equals

$$r \, u_i^{(k)}(t) := \frac{u_{i+1}^{(k)}(t) + 2u_i^{(k)}(t) + u_{i-1}^{(k)}(t)}{4} \qquad \text{for } t \in [0,T].$$

• **construction of the level k-1 problem.** To construct the right hand side of the level k-1 problem, (4.3) has to be substituted in (4.2). The resulting expression is similar to what is obtained in the elliptic case, except for the derivative terms $r\frac{d}{dt}u_i^{(k)}$ and $\frac{d}{dt}\bar{r}u_i^{(k)}$. As numerical differentiation is numerically unstable and computationally expensive, the derivative calculation should be avoided. Since both terms cancel when $\bar{r} = r$, only the special case of different restriction operators remains to be considered.
We rewrite the general level k-1 equation in the form

$$\frac{d}{dt}u_i^{(k-1)} - K\frac{u_{i+1}^{(k-1)} - 2u_i^{(k-1)} + u_{i-1}^{(k-1)}}{h_{k-1}^2} - g_i(u_i^{(k-1)}) = \frac{d}{dt}f_{i1} + f_{i2}. \tag{4.4}$$

with f_{i1} accumulating the functions which need to be differentiated and f_{i2} accumulating the functions which need not. After a change of dependent variable, $\bar{u}_i^{(k-1)} := u_i^{(k-1)} - f_{i1}$, (4.4) is rewritten in the usual form and can be solved accordingly.

5. Numerical results. Before discussing numerical results, some important implementation issues should be considered.

An effective computer representation for the grid point functions should be chosen according to some desirable features. The representation should be such that functions can be evaluated easily and efficiently. Linear as well as nonlinear arithmetic operations should be allowed on functions. The output of the ODE integrator should be compatible with the function representation. As memory is often limited on current distributed memory machines, the representation should be compact. We implemented each function by specifying its value on an equidistant sequence of time levels, combined with a cubic or higher order interpolation procedure.

Since the computationally most expensive part of the algorithm is the smoothing operation, the ODE integrator should be fast. Using variable step size and variable order techniques it should be able to solve nonlinear, stiff ODE's to a user specified accuracy.

The following two examples were calculated using LSODE [2] as the integrator, with absolute and relative error tolerances set to 10^{-6}.

• **example 1:** $\quad \dfrac{\partial u}{\partial t} = \dfrac{\partial^2 u}{\partial x^2}$ with $x \in [0,1]$ and $t \in [0,\frac{1}{4}]$

The one-dimensional heat equation with Dirichlet boundary conditions and known solution $u(x,t) = \sin(\pi x)\,e^{-\pi^2 t}$ was discretized on the unit interval with mesh size $h = 2^{-8}$. The problem was solved for $t \in [0,\frac{1}{4}]$, using a 0.01 time step to represent the grid point functions. In every iteration k the maximum difference between the known solution u and the approximation $u^{(k)}$ was calculated according to

$$\|u - u^{(k)}\| = \max_{t \in [0,T]} \max_i |u(x_i,t) - u_i^{(k)}(t)| \tag{5.1}$$

and the averaged reduction factor was calculated from

$$\bar{\rho}^{(k)} = [\, \|u - u^{(k)}\| \,/\, \|u - u^{(0)}\| \,]^{1/k}. \tag{5.2}$$

The results are listed in Table 5.1. A multigrid V-cycle ($\gamma=1$) was used, with and without post-smoothing ($\nu_2 = 1$ and $\nu_2 = 0$).

Table 5.1 : maximum error (5.1) and averaged reduction factor (5.2)							
k	0	1	2	3	4	5	$\bar{\rho}^{(5)}$
$\nu_2 = 0$	.92 10^0	.18 10^0	.40 10^{-1}	.96 10^{-2}	.25 10^{-2}	.65 10^{-3}	.234
$\nu_2 = 1$	.92 10^0	.47 10^{-1}	.22 10^{-2}	.12 10^{-3}	.56 10^{-5}	.33 10^{-6}	.052

In Table 5.2 we give the averaged reduction factor $\bar{\rho}(10)$ obtained for different values of the mesh size h. The results clearly show the boundedness of the convergence rate for the given multigrid cycle.

Table 5.2 : averaged reduction factor for V-cycle without post-smoothing								
h	2^{-2}	2^{-3}	2^{-4}	2^{-5}	2^{-6}	2^{-7}	2^{-8}	2^{-9}
$\bar{\rho}^{(5)}$	.11	.16	.22	.23	.24	.25	.25	.25

- example 2: $\quad \dfrac{\partial u}{\partial t} = \dfrac{\partial^2 u}{\partial x^2} + \dfrac{\partial^2 u}{\partial y^2} + e^{-u} + f \quad$ with $(x,y) \in [0,1] \times [0,1]$ and $t \in [0,1]$

As a second example we solved a two-dimensional nonlinear problem on the unit square, discretized on a 33 by 33 grid with Dirichlet boundary conditions and known solution $u(x,y,t) = e^{(x+y)t}$. The functions are represented by their values on equidistant time levels 0.05 apart. The results for V- and W-cycles ($\gamma=1$ and $\gamma=2$) are presented in Table 5.3.

Table 5.3 : maximum error (5.1) and averaged reduction factor (5.2)						
k	0	1	2	3	4	$\bar{\rho}^{(4)}$
V-cycle, $\nu_2 = 0$	.59 10^1	.23 10^1	.70 10^0	.22 10^0	.70 10^{-1}	.330
V-cycle, $\nu_2 = 1$	.59 10^1	.62 10^0	.61 10^{-1}	.62 10^{-2}	.63 10^{-3}	.102
W-cycle, $\nu_2 = 0$	.59 10^1	.17 10^1	.23 10^0	.60 10^{-1}	.13 10^{-1}	.217
W-cycle, $\nu_2 = 1$	.59 10^1	.39 10^0	.21 10^{-1}	.12 10^{-2}	.62 10^{-4}	.057

6. Conclusions. The algorithm has several interesting features. It takes advantage of independent time stepping. In contrast to the classical solution methods for parabolic PDE's, it does not introduce extra synchronization and has a good efficiency, even for small problems on large machines. This makes the Waveform Relaxation multigrid method a good candidate for implementation on massively parallel machines.

REFERENCES

[1] W. Hackbush, *Multi-Grid Methods and Applications* , Springer-Verlag, Berlin-Heidelberg, 1985.

[2] A.C. Hindmarsh, C.W. Gear, *LSODE and LSODI, two new initial value ordinary differential equation solvers* , ACM SIGNUM Newsletter, 15 (1980), pp. 10-11.

[3] Ch. Lubich and A. Ostermann, *Multi-Grid Dynamic Iteration for Parabolic Equations* , BIT, 27 (1987), pp. 216-234.

[4] U. Miekkala and O. Nevanlinna, *Convergence of Dynamic Iteration Methods for Initial Value Problems* , Siam J. Sci. Stat. Comput., Vol.8 No.4 (1987), pp. 459-482.

[5] J. White, F. Odeh, A.S. Sangiovanni-Vincentelli and A. Ruehli, *Waveform Relaxation: Theory and Practice* , Memorandum No. UCB/ERL M85/65 (1985), Electronics Research Laboratory, College of Engineering, University of California, Berkeley.

CHAPTER 28

Implementation of a Parallel Multigrid Method on a Loosely Coupled Array of Processors

R. Herbin*†
S. Gerbi*‡
V. Sonnad*

Abstract. A parallel multigrid method for the solution of elliptic partial differential equations has been implemented on the loosely Coupled Array of Processors system at IBM Kingston, an experimental MIMD machine with data communications via shared memory. We consider selected smoothers, restrictions and prolongations, using multicolour ordering of the grid points. Parallelization is obtained by an a-priori decomposition of the finest grid; each step of the multigrid method is then parallelized, up to the solution on the coarse grid, which is performed sequentially; data is communicated across the boundaries of the subregions thus defined. We present numerical results obtained for the parallel code, along with timing information utilizing up to 8 processors.

1. Introduction. Multigrid methods are among the fastest methods for the solution of large systems of equations arising from the discretization of partial differential equations. Our purpose here is to present results showing that this very fast sequential method can be efficiently implemented on a relatively coarse-grained parallel supercomputer such as the experimental lCAP system. The parallel implementation of multigrid methods has been studied by several authors, for different architectures (see [6] for a survey, and also the references in [4]).

Our approach here is to divide the domain of definition into as many subregions as the number of processors (up to 10 for the lCAP1 system), thus dividing all grids between processors. Data is communicated across the boundaries of each subregion from one processor to its neighbors via the shared bulk memory; no processor is idle at any given step of our parallel multigrid algorithm.

2. The Model Problem and Finite Difference Discretization. We have considered two model equations of the form:

$$\begin{cases} Lu = g \text{ in } \Omega = \ ]0, 1[\ \times \ ]0, 1[\\ u = 0 \text{ on } \partial\Omega, \end{cases} \tag{2.1}$$

with, $Lu = -\Delta u$, in the first case and in the second case :

* IBM Corporation, Mail Stop 428, Neighborhood Road, Kingston, New York, 12401.
  Current addresses:
† Ecole Polytechnique Fédérale de Lausanne, GASOV/ASTRID,
  Lausanne, 1015, Switzerland.
‡ Dept. of Computer Science, Yale University, Box 2158,
  Yale Station, New Haven, Connecticut, 06520.

$$Lu = \exp(x) \frac{\partial^2 u}{\partial x^2} + \exp(y) \frac{\partial^2 u}{\partial y^2}$$

The right hand side g, is taken so that the exact solution is : $u(x,y) = x\,(1-x)\,y\,(1-y)$ in both cases. Let h be a real number. We introduce L_h, the discrete operator obtained by approximating L by the second order Taylor's formula (5-point difference scheme) and the following discrete problem:

$$\begin{cases} L_h\,u_h = f_h \;\; \text{in} \;\; \Omega_h \\ u_h = 0 \;\; \text{on} \;\; \partial\Omega_h \; . \end{cases} \tag{2.2}$$

3. The Multigrid Algorithm. For the multigrid approach, one needs a sequence of grids (discretized domains) numbered from ℓ_0 (coarse grid) to ℓ (fine grid): Ω_{ℓ_0} ,......, Ω_{ℓ} , with respective meshsizes: h_{ℓ_0} , ..., h_{ℓ}, where $h_i = (1/2)^{i+1}$. So we simply write:

$$L_{\ell}\,u_{\ell} = f_{\ell} \;\; \text{in} \;\; \Omega_{\ell} \tag{3.1}$$

where L_{ℓ} is the discrete operator taking into account the boundary conditions.

The multigrid method can be described by a recursive program [3], [7], given below, where S_{ℓ}^{ν} denotes ν iterations of the smoother, which is a linear iterative method; r is the restriction from a grid to the next coarser one, d is the defect on this new grid (and thus the right handside for the next smoothing step), p is the prolongation from a grid to the next finer one.

procedure MGM (ℓ , u, f) : integer ℓ; array u, f;

comment : ℓ is the level of the fine grid , ℓ_0 level of the coarse grid ,
comment : $u = u_{\ell}^j$ is a given iterate , $u = u_{\ell}^{j+1}$ is the result.

```
        begin
            if ℓ = ℓ₀ then
                u = L⁻¹ℓ₀ f
            else
                begin integer j ; array v , d ;
                u = Sℓᵛ¹ (u, f)
                d = r (f − Lℓ u )
                v = 0
                for j = 1 to γ step 1 do  MGM (ℓ −1, v, d)
                u = u + pv
                u = Sℓᵛ² (u, f)
                end
            endif
        end
```

Several restriction and prolongation operators have been proposed, according to the type of equation and of discretization (see e.g. [3] , [7]). For simplicity, we consider the 7-point prolongation operators and the 5-point restriction operators.

4. Red-Black Ordering of Grid Points. The damped Jacobi method is generally considered to be a very easily parallelizable method, since within one iteration, each point is computed independently from the others. The smoothing properties of the Jacobi method, however, are not nearly as good as those of the SOR method, with a well chosen relaxation factor, [4].

We consider here a way of parallelizing SOR by a reordering of the nodes, [1]. The grid-points are partitioned by the classical red-black ordering. Using the lexicographical (row-wise) ordering, and assuming an odd number of nodes we split Ω_h, the set of grid points, into two sets of nodes : the red nodes with odd numbers, and the black nodes with even numbers. Let N be the total number of nodes $(N = (2^\ell - 1)^2$ for a grid of level ℓ as defined in Section 3.), and $M = E(N/2)$ be the number of black nodes. The 5-point discretization (2.2) of the Poisson equation leads to the following partitioned matrix form, :

$$\begin{bmatrix} \mathbf{D_r} & \mathbf{E} \\ \mathbf{E}^T & \mathbf{D_b} \end{bmatrix} \begin{bmatrix} U_r \\ U_b \end{bmatrix} = \begin{bmatrix} F_r \\ F_b \end{bmatrix} \tag{3.2}$$

where $U_r = (u_1, u_3, ..., u_N)^T$, $U_b = (u_2, u_4, ..., u_{N-1})^T$, $F_r = (f_1, f_3, ..., f_N)^T$,
$F_b = (f_2, f_4, ..., f_{N-1})^T$, $\mathbf{D_r} = \dfrac{4}{h^2} \mathbf{I_{M+1}}$, $\mathbf{D_b} = \dfrac{4}{h^2} \mathbf{I_M}$, and $\mathbf{E}$ is an $(M + 1) \times M$ matrix.

The Gauss-Seidel iteration on system (3.2) leads to :

$$\begin{cases} \mathbf{D_r} U_r^{k+1} = -\mathbf{E} U_b^k + F_r \\ \mathbf{D_b} U_b^{k+1} = -\mathbf{E}^T U_r^{k+1} + F_b \end{cases} \tag{3.3}$$

System (3.3), which defines the red-black relaxation, can be implemented in parallel, since the red and black unknowns are completely decoupled. Moreover, the SOR-RB method has better smoothing properties and for the chosen restriction has lower cost per iteration, than a scheme which does not use red-black ordering, [4]. If the partial differential equation to be solved has cross derivatives, then the unknowns at the red and black points are coupled; however the approach described here can be still be utilized by resorting to multi-colour schemes, which have been proposed for 9-point and 13-point finite difference schemes, [5]. These properties, and the fact that the red-black relaxation can be easily parallelized have led us to use the red-black relaxation as the smoother of our parallel multigrid method.

5. Parallel Implementation. The parallel system on which these results were obtained is termed ICAP1 (for loosely Coupled Array of Processors-1), and consists of an IBM 3081 host connected to 10 FPS-164 processors. Data can be communicated between processors either through a global memory or a bus. Data flow and synchronization between processors is controlled by inserting special directives within Fortran code; these directives are translated into system calls by a precompiler, (details about the ICAP system are given in [2]).

Parallelization is obtained throughout the multigrid algorithm by an apriori decomposition of the domain, with a one-to-one mapping of the subregions defined onto the processors. Note that for complex domains, this raises the dificult issue of load balancing.

Two arrays are needed in shared memory : one where the processors alternately read and write the unknowns at the boundaries, and another where the processors read and write the input and output data for the solution on the coarse grid. Each processor completes the entire multigrid algorithm only on the points assigned to it, except for the solution on the coarse grid, which is performed by all processors. Each processor exchanges data with its neighbours by reading from the shared memory the necessary information at its boundaries, and when needed writes into the shared memory the values computed at the boundaries. Synchronization is ensured between data exchanges with the necessary precompiler directives.

6. Numerical Results. Timing results have been obtained for the two model equations described in Section 2., and are presented in Figures 1 and 2. The convergence criterion used in all the tests was : $\| u_e - u_a \| \leq 10.0^{-6}$, where u_e is the exact solution, and u_a is the numerical solution. It was found in the course of the work, that because of the high communication costs, it was inefficient to perform the smoothing, and restriction operations down to the coarsest possible level, (in this instance, the coarsest level has only one point). Instead, it was more efficient to define a level containing 16x16 points as the coarsest grid, and solve for the problem at this level using a fast Poisson solver.

One approach that we have explored (for the Poisson equation only), is to use an iterative solver at the coarse grid, (with 10 steps of an SOR scheme with relaxation parameter $=1.73$). The results are shown in the Fig. 1 with the curves labelled NCS. While this does lead to reduced speeds in the sequential case, the overall parallel efficiency is higher. This is a promising approach and could be very effective with a more efficient solver such as the conjugate gradient method.

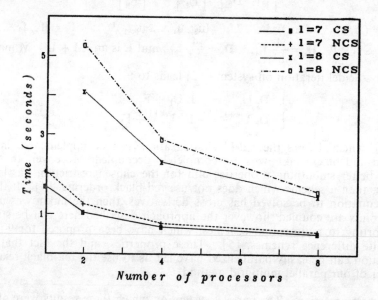

Fig. 1 : Execution time for the Poisson equation; $(v_1, v_2) = (1, 1)$. CS: Fast Poisson solver for the coarse grid solution; NCS : SOR at the coarse grid.

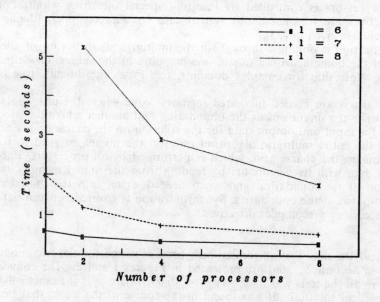

Fig. 2 : Execution time for the equation with exponential coefficients; $(v_1, v_2) = (2, 2)$.

In Fig. 3 we show speedup results (base 2), for the equation with exponential coefficients. The speedup with base 2 is the ratio of execution time using two processors to the execution time using p processors. The speedup has been measured with two processors as the base because of memory requirements; when the finest grid is of level 8 or higher, the code cannot be executed on one processor. We see that high speedups are obtained only for large problems; for smaller problems, the ratio of communication cost to computational cost increases, thus leading to reduced speedups.

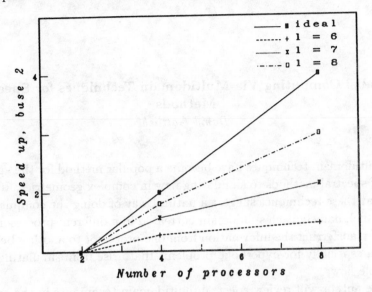

Fig. 3. : Speedup (base 2), for equation with exponential coefficients.

REFERENCES

[1] L.M. Adams, J. Ortega, *A multicolor SOR method for parallel computation*, IEEE., (1982), pp. 53-56.

[2] E. Clementi, D. Logan, *Parallel Processing with a Loosely Coupled Array of Processors System*, IBM Kingston Technical Report KGN-43, (1986).

[3.] W. Hackbusch, *Multigrid Methods and Applications*, Springer Series in Computational Mathematics 4, Springer-Verlag, (1985).

[4] R. Herbin, S. Gerbi, V. Sonnad, *Parallel Implementation of a Multigrid Method on the Experimental ICAP supercomputer: Part II*, IBM Kingston Technical Report, KGN-155, 1987.

[5] D.P. O'Leary, *Ordering Schemes for Parallel Processing of certain Mesh Problems*, SIAM J. Sci. Stat. Comput., 5, (1984), pp. 620-632.

[6] J.M. Ortega, R.G. Voight, *Solution of Partial Differential Equations on Vector and Parallel Computers*, SIAM Review., 27, 2, (1985), 149-240.

[7] H. Stuben, U. Trottenberg, *Multigrid Methods: Fundamental Algorithms, Model Problem Analysis and Applications*, in Multigrid Methods, Proceedings Koln-Porz, November 1981, (Hackbush and Trottenberg eds.), Lect. Notes in Math., 960, Springer, Berlin, (1982).

Parallel Computing Via Multidomain Techniques for Spectral Methods
David Gottlieb

Multidomain techniques have become a popular method for the application of spectral methods to simulating flow in complex geometries. It turns out that these techniques suggest a natural way of doing the computations in parallel, namely each subdomain is treated in a different processor. The way of transferring the information from one processor to another becomes crucial, especially for hyperbolic problems that arise from simulating compressible flows.

The author will review spectral multidomain techniques in the context of parallel computations. Ways of improving boundary conditions were explored that will speed up convergence. In particular, ways of composing matching conditions were tested for hyperbolic equations–in particular, the Euler equations of gas dynamics.

The plan is to use the Flex machine at ICASE to test the validity and efficiency of the multidomain technique. This research attempts to give an answer to the question whether it is preferable to parallelize the algorithm without physically subdividing the domain of computations, or whether it is more gainful to divide the domain and to treat the problem in different subdomain.

In both cases, the connection between boundary conditions and transfer of information between processors will be discussed and clarified.

Asymptotics and Domain Decomposition in Parallel Computing
Raymond C.Y. Chin

Mathematical models for scientific phenomena usually involve multiple scales, with the consequence of rapid temporal or spatial variations in the solution. While the details of these rapid variations may not be of interest, the numerical computations must somehow incorporate any effects on those aspects of the solution one is interested to determine.

Where the local behavior of the solution is known, analytic patches may be used, while in other regions special local grids will have to be employed in order to resolve the special features of the solution.

The sub–domains together with rules governing their interactions form a domain decomposition method. Domain decompositions derived from asymptotic analysis can take advantage of parallel vector architecture and represent, indeed, a truly "new" class of parallel algorithms.

In his talk, the author surveys the asymptotic methods employed to identify rapid varying regions and to analyze their local behaviors. He will discuss the incorporation of the local asymptotic solutions and the partition of sub–domains into a numerical scheme via a domain decomposition algorithm. An analysis of the domain decomposition algorithm within the framework of asymptotic preconditioning and deferred correction will be presented and the mapping of an asymptotics–induced domain decomposition method onto two parallel processing machines will be discussed.

CHAPTER 29

Parallel Processing of a Domain Decomposition Method*

Jeffrey S. Scroggs*

Abstract. A parallel algorithm for the efficient solution of a time dependent reaction convection diffusion equation with small parameter on the diffusion term will be presented. The method is based on a domain decomposition that is dictated by singular perturbation analysis. The analysis is used to determine regions where certain reduced equations may be solved in place of the full equation. Parallelism is evident at two levels. Domain decomposition provides parallelism at the highest level, and within each domain there is ample opportunity to exploit parallelism. A package called Schedule was used to exploit various levels of parallelism on the FX/8. Some run-time results will be presented.

1. Introduction. A parallel algorithm suitable for the efficient solution of singularly perturbed parabolic partial differential equations will be presented. The method involves a domain decomposition that is induced by a multiple scales asymptotic analysis of the problem. The equation solved is

$$u_t + A(x,t,u) \cdot \nabla u - \epsilon \Delta u - r(x)u = 0, \tag{1}$$

as $\epsilon > 0$ tends to zero. Here, ∇ is the gradient operator, and Δ is the Laplace operator. This equation arises in many applications, including Computational Fluid Dynamics, where it serves as a simplified model of the Navier Stokes (NS) equations. This equation possesses many qualities which make the NS equations so difficult to solve; namely, it may be used to model sharp discontinuities, such as shocks and boundary layers. We present results on problems with sharp discontinuities and boundary layer phenomena. We consider one spatial dimension here; however, the asymptotic results are extendible to the multi-dimensional case. As a prelude to solving a nonlinear problem, the analysis and computational method has been demonstrated on a

* This work supported in part by the National Science Foundation under Grant Nos. NSF DCR–85–09970 and NSF MIP–8410110, the U.S. Department of Energy under grant No. DOE DE–FG02–85ER25001, U.S. Air Force under Grant No. AFOSR–85–0211, the IBM Donation, and the Center for Supercomputing Research and Development. Some of the research was done while in residence at Lawrence Livermore National Laboratory, supported by the Applied Mathematical Sciences subprogram of the Office of Energy Research, U.S. Department of Energy under contract No. W–7405–Eng–48.

linear problem by Chin et. al. in [1].   We will focus on the nonlinear problem in this paper.

2. Asymptotic Analysis.
Asymptotic multiple scale analysis may be used to study problems with sharp discontinuities. In a neighborhood of a sharp discontinuity, one frequently would like to measure physical quantities using time and spatial scales very different than the ones used where the quantities are slowly varying. This is the essence of multiple scales. Asymptotic analysis provides us with analytic tools to identify and utilize the multiple scales in our problem. The asymptotics aid us in decomposing the problem and solution into constituent parts, leading to a solution method amenable to parallel processing. In addition, asymptotics also provide a means of 'preconditioning' the problem. In this setting, a preconditioned problem is more well conditioned for numerical computations, and hence may be solved by conventional techniques.

The solution method is formed from the synthesis of asymptotic and numerical analysis. Asymptotics will dictate a domain decomposition. The numerics, in turn, will provide a mechanism to recover information which might have been overlooked in the multiple scales analysis. We restrict this discussion to the nonlinear reaction convection diffusion equation,

$$u_t + uu_x - \epsilon u_{xx} - r(x)u = 0; \tag{2}$$

however, other forms of equation (1) may also be solved using these techniques [2,3]. When modeling shock behavior with this equation, the analysis decomposes the domain into two subdomains: an internal layer region where reaction, convection, and diffusion are all important physical mechanisms, and an outer region where we may drop diffusion effects. The problems associated with these subdomains are preconditioned versions of the original problem, and hence are easier to solve numerically. We will motivate and explain the subproblems; however, the complete theory is beyond the scope of this paper.

In the outer region, we solve the reduced equation, obtained by setting $\epsilon = 0$, via an iteration. Let the solution to the reduced equation be U. Then under suitable conditions [4],

$$|u - U| = O(\exp[-\epsilon^{-2}f^2(x,t)]) + O(\epsilon),$$

where $f(x,t)$ measures the distance from the shock. This result follows from a maximum principle argument similar to that of Howes in [5]. Thus, the reduced equation is an $O(\epsilon)$ approximation to the solution of equation (2), except in the internal layer subdomain. The reduced equation may be solved via an iteration of the form [4]

$$u_t^k + u^{k-1}u_x^k - r(x)u^k = 0. \tag{3}$$

This iterative process allows the capturing of behavior which might have been missed in the asymptotic analysis—an important feature of a method that depends so heavily on a priori analysis. Thus, we see that the asymptotics has reduced the complexity of the problem in the outer region by allowing us to drop diffusion effects, preconditioning our problem in the outer region.

The preconditioning for the equation in the internal layer subdomain is a transformation and scaling. The appropriate spatial scale for equation (2) in a neighborhood of a shock is $\chi = x/\epsilon$ (see Kevorkian and Cole [6]). The transformation is designed to follow the path of the shock, thus reducing convection. The scaling increases the diffusion coefficient, allowing the application of a standard numerical scheme which has negligible artificial viscosity. Thus, the asymptotics have preconditioned the problem in the internal layer subdomain.

We locate shocks using a technique quite different from the usual shock detection methods as it is based on a priori information [1]. Shock detection dictates the placement of the boundary of the internal layer subdomain. The asymptotics provide an $O(\epsilon)$ approximation for the size of this subdomain.

The asymptotic analysis has led to a natural decomposition into two classes of subdomains the internal layer region capturing regions of large gradients, and the outer region in which the solution is slowly varying. Placement of the subdomain boundaries is dictated by a synthesis of the monitoring of the size of the Jacobian of the shock following transformation with the error bounds obtained in the asymptotic analysis. In addition, we have used asymptotic analysis to precondition the problems in the subdomains, defining two simplified problems each of which is more well conditioned for numerical computations than the original problem. This analysis may be extended to solutions exhibiting boundary layer behavior [1].

By using asymptotics to dictate a domain decomposition, and to precondition the subproblems, we have demonstrated a new type of numerical method. This synthesis of asymptotics and numerics may be taken further, resulting in additional efficiency and accuracy For example, Chin and Krasney used asymptotics to dictate the use of a basis function for a finite element method [7]. We may incorporate this idea into our method by using basis function based on tanh. Thus asymptotics may be injected into our numerical method by using a hybrid finite element method to solve the internal layer problem. The goal of such blending of asymptotics and numerics is to incorporate as much a priori information into the method as possible, resulting in high accuracy and efficiency.

3. Parallelism.   We exploit several levels of parallelism within this method. Within this context, it is appropriate to discuss large–grain and small–grain parallelism. Large–grain parallelism arises from the domain decomposition, and is utilized by distributing the work in the subdomains across processors. Small–grain parallelism arises in the numerical methods used.

The small–grain parallelism arises in the particular numerical method used to solve in the subdomains. Each outer iteration requires the solution of equation (3) in the outer region, and equation (2) in the internal layer. The boundary conditions for the internal layer subdomain are provided by the solution in the outer region. In the outer region, equation (3) is solved by the method of characteristics. This requires the solution of many decoupled ordinary differential equations (ODEs). The solutions to these problems may be grouped into vectors, where each vector component is a solution to an ODE. Since the ODEs are decoupled, each vector may be operated on using vector instructions for all operations except determining the time step size. Thus, we treat the vector facilities as Single Instruction Multiple Data processors, and a parallel–vector machine, such as the Alliant, has two levels of parallelism in a virtual sense. Additional small grain parallelism is exploited within the linear algebra associated with a finite difference solution of equation (2) in the internal layer.

An important aspect of the development was the use of Schedule, a software package designed to be used in studying the feasibility of implementing explicit parallel algorithms in a transportable manner. This has allowed the expression of parallelism in a more abstract language, removing the burden of using specific machine dependent primatives, and allowing utilization of several levels of parallelism and dynamic allocation of user level processes on the Alliant FX/8. These were the main advantages of using Schedule since the Alliant system does not provide for explicit parallel programming, and does not readily support many layers of parallelism within a single program. Unfortunately, Schedule did not express the parallelism om the matrix solves efficiently; hence, it was not used for the timings.

Results and Conclusions. We will present two examples. The first will demonstrate avorable error behavior, and the second will show high resolution, robustness, and efficiency. hese numerical experiments were run on an Alliant FX/8.

This algorithm obtains higher accuracy as ϵ tends to zero. In general, as ϵ tends to zero, quations similar to equation (1) are more difficult to solve by standard techniques. For our first xample, consider equation (2) with $r = 0$, and initial and boundary conditions chosen so that anh(x/ϵ) is the analytical solution. For both $\epsilon = .001$, and $\epsilon = .0001$, the computed solution as within $O(\epsilon)$ of the exact. Thus, as the equation becomes more difficult for standard echniques, this algorithm becomes more accurate.

We demonstrate high resolution, robustness, and efficient use of state of the art omputational resources on our second example, which arises in modeling flow through a duct. Consider the equation,

$$u_t + uu_x - .001u_{xx} - 4\sin(2\pi x)u = 0,$$

here the spatial domain is $[0,1]$, and the boundary and initial conditions are $(0,t) = 1 = -u(1,t)$, with $u(x,0) = \cos(\pi x)$. The forcing term expresses the effects of the hape of a duct of width $\exp[2\cos(2\pi x)/\pi]$ [8]. The initial guess is obtained using finite ifferences on 100 spatial nodes, with a forward euler convection approximation, and a backward uler diffusion approximation. The stability constraints for this scheme required a large amount f artificial diffusion, resulting in an error of about $\epsilon^{1/2}$ in the location of the shock at time $= .3$. The method was able to adjust the location of the internal layer subdomain during the terative process. Thus, the method is not sensitive to the initial guess. The method oncentrates a large number points near the shock, obtaining high resolution.

We are able to obtain significant speedups, as demonstrated by Table 1. Improvements in he implementation could be made by exploiting more of the available parallelism, and by naking more effective use of the vector capabilities of the Alliant. For example we could have everal sweeps of the outer iteration going at once. In addition, we have not implemented the

Table 1.
Timings on an Alliant FX/8

CEs	RUN TIME (seconds)	SPEEDUP	EFFICIENCY
1	90.4	1.00	1.00
2	47.5	1.90	0.95
3	34.5	2.61	0.87
4	27.9	3.23	0.81
5	24.3	3.71	0.74
6	21.7	4.16	0.69
7	19.8	4.57	0.65
8	18.5	4.89	0.61

vector solves for the ODEs. Thus, while we have shown significant speedup, there is much untapped potential.

In summary, we have presented and demonstrated a numerical method for the solution of a nonlinear singularly perturbed reaction convection diffusion equation. The method was developed by synthesizing multiple scales asymptotics and numerical analysis. The viability of the method was demonstrated on some important canonical problems. The method is designed for equations similar to equation (1), which model shock and boundary layer behavior for high speed fluid flow. In the future, the method will be extended to systems of reaction convection diffusion equations and higher dimensions.

References

[1]   G. W. Hedstrom and A. Osterheld, *The Effect of Cell Reynolds Number on the Computation of a Boundary Layer*, **J Comp Phy**, 37 (1980), pp. 339–421.

[2]   R. C. Y. Chin, G. W. Hedstrom and F. A. Howes, *Considerations on Solving Problems with Multiple Scales*, in **Multiple Time Scales**, J. U. Brackbill, ed., Academic Press, Orlando, Florida, 1985, pp. 1–27.

[3]   R. C. Y. Chin, G. W. Hedstrom, F. A. Howes and J. R. McGraw, *Parallel Computation of Multiple-Scale Problems*, in **New Computing Environments: Parallel, Vector, and Systolic**, A. Wouk, ed., Philadelphia, 1986, pp. 136–153.

[4]   J. S. Scroggs, "The Solution of a Parabolic Partial Differential Equation via Domain Decomposition: The Synthesis of Asymptotic and Numerical Analysis", Center for Supercomputing Research and Development, Dept. of Comp. Sci., University of Illinois (to appear in 1988).

[5]   F. A. Howes, *Multi-Dimensional Reaction-Convection-Diffusion Equations*, in **Ordinary and Partial Differential Equations**, B. Eckmann A. Dold, ed., Springer–Verlag, New York, 1984.

[6]   J. Kevorkian and J. D. Cole, **Perturbation Methods in Applied Mathematics**, Springer–Verlag, New York, 1981.

[7]   R. C. Y. Chin and R. Krasny, *A Hybrid Asymptotic-Finite Element Method for Stiff Two-Point Boundary Value Problems*, **SISSC**, 4 (1983).

[8]   G. Rudinger, **Nonsteady Duct Flow**, Dover Publications, Inc., New York, 1969.

Domain Decomposition Methods for Three–dimensional Elliptic Problems

O. Axelsson

Decomposing a domain in $\Re^3$ by parallel vertical planes leads to a block structure of the stiffness matrix where it is natural to form the reduced Schur complement matrix corresponding to the dividing planes. Formation of the matrix is in general impossible so the Schur complement system must be solved by iteration. This requires only the computation of the action of the matrix but this action requires the solution of the same problem on the subdomains. We describe how an effective (optimal order) preconditioner can be constructed and how a recursive subdomain method can be used for the solution of the subdomain problems and of the preconditioner. Pre– and postsmoothing of the residuals on the fine mesh implies that the correction can be computed on the coarse mesh which can decrease the cost by a factor close to 1/8. We describe also how one can implement the method on a cluster of loosely coupled parallel processors where each cluster consists of more tightly coupled processors.

CHAPTER 30

Moving Point and Particle Methods for Convection Diffusion Equations

M. D. Rees[*]
K. W. Morton[†]

Abstract. We compare moving point and particle methods for the numerical solution of multi-dimensional convection-diffusion problems. Both methods possess very little or no numerical diffusion and yield algorithms very suitable for vector/parallel machines. They differ in this suitability, the type of information associated with each point and their conservation properties. Numerical results for a model problem are presented; the algorithms are being implemented on a Convex C1-XP2 vector processor.

1. Introduction. We consider the scalar equation

$$c_t + \nabla \cdot (a(x,t)c) = \nabla \cdot (b(x,t)\nabla c) \qquad (1.1a)$$

$$\nabla \cdot a = 0 \qquad (1.1b)$$

$$c(x,0) = c_0(x) \quad \text{given} \qquad (1.1c)$$

$$x \in \mathbb{R}^d, \ t \in [0,T_F] \quad , \qquad (1.1d)$$

where $a(x,t)$ (the velocity field) and $b(x,t)$ (the diffusion coefficient) are known functions of their arguments. Both particle and moving point methods employ a set of points which move with the local velocity field; using (1.1b), (1.1a) becomes

$$\frac{dc}{dt} = \nabla \cdot (b(x,t)\nabla c) \quad , \qquad (1.2)$$

where d/dt denotes the Lagrangian (convective) derivative. The key issue with both methods becomes the approximation of diffusion, as the set of points at which information is known is scattered. An important distinction between them is the type of information that is stored and the way in which it is used. The origins of such methods are found in

* Computing Laboratory, 8-11 Keble Road, Oxford, OX1 3QD. This author acknowledges the financial support of the Science and Engineering Research Council.
** Computing Laboratory, 8-11 Keble Road, Oxford, OX1 3QD.

Harlow [3]. More recently, moving point methods have been studied by Crowley [1], Farmer [2] and particle methods by Raviart [4,5].

The point positions are solutions of the ODEs

$$dx_j/dt = a(x_j(t),t) \tag{1.3a}$$

$$x_j(0) \text{ given} \quad . \tag{1.3b}$$

A sufficient distribution of points follows from:

<u>LEMMA</u> Denoting the components of $a(x,t)$ by $a_i(x,t)$, suppose

$$a_i \in L^\infty(0,T_F;W^{1,\infty}(\mathbb{R}^d)) \quad i=1,\ldots,d. \tag{1.4}$$

Then (i) the solution of (1.3a,b) exists and is unique and (ii) if $x(t)$, $y(t)$ are solutions of (1.3) then setting

$$C(t) = \exp(Lt) \quad , \tag{1.5a}$$

where L is a Lipschitz constant for $a(x,t)$, we have

$$C(t)^{-1}|x(0)-y(0)| \leq |x(t)-y(t)| \leq C(t)|x(0)-y(0)| \quad . \tag{1.5b}$$

I.e. the spacing of the points at subsequent times can be parameterised by the initial spacing (typically nodes of uniform mesh). Similar results hold for time-discretised methods (Euler, 4th order Runge–Kutta, et cetera). Equation (1.5) also suggests a way to insert points in regions of large gradients via an equidistributing principle.

Equations (1.3a,b) may be solved concurrently over the set of points as there is no coupling between them. This is an example of the 'natural parallelism' inherent in the algorithms.

2. The Particle Method.
The method we describe is due to Raviart. It differs from earlier particle methods (e.g. Harlow [3]) in that it is grid-free and each particle has an associated volume. Consider a box of fluid at time 0 with centre $x_i(0)$. As x_i moves, as in (1.3), due to velocity field, the volume, w_i, changes according to

$$\dot{w}_i = w_i(t)\nabla.a(x_i(t),t) \quad , \tag{2.1}$$

since the volume is proportional to the Jacobian, which itself satisfies (2.1). Integrating (1.1a) over the box and using the approximation

$$\int_{box} f(x)dx \simeq w_i f(x_i) \tag{2.2}$$

gives

$$w_i \dot{c}_i + c_i \dot{w}_i = w_i \nabla.(b\nabla c)_i \tag{2.3a}$$

i.e.

$$d(w_i c_i)/dt = w_i \nabla.(b\nabla c)_i \quad . \tag{2.3b}$$

The product $w_i c_i$ is termed the MASS of the PARTICLE i.

A smooth representation of the function c is obtained by replacing each point particle by a smooth 'particle' which has the same mass and has small support, $O(\varepsilon)$; the analogue in Harlow's PIC method [3] was 'cell-averages'. We introduce a function τ with the properties

$$\int_{\mathbb{R}^d} \tau(x)dx = 1 \tag{2.4a}$$

$$\int_{\mathbb{R}^d} x^j \tau(x)dx = 0 \quad 1 \leq |j| \leq k-1 \tag{2.4b}$$

$$\int_{\mathbb{R}^d} |x|^k \tau(x)dx < \infty \tag{2.4c}$$

e.g. $\tau(x) = \pi^{-d/2}\exp(-|x|^2)$ with k=2. Then we define

$$\tau_\epsilon(x) = \epsilon^{-d}\tau(x/\epsilon) \quad , \tag{2.5}$$

so that $\tau_\epsilon(x)$ also satisfies (2.4a,b,c). We obtain the representation

$$\Pi_\epsilon^h\, c(x,t) = \sum_{(j)} w_j(t)c_j(t)\tau_\epsilon(x-x_j(t)) \quad . \tag{2.6}$$

Raviart [5] describes an approximation for the diffusive term:

$$w_i\nabla\cdot(b\nabla c)(x_i) \simeq \sum_{(j)} w_iw_j(b_i+b_j)(c_j-c_i)\frac{(x_j-x_i)\cdot\nabla\tau_\epsilon(x_i-x_j)}{|x_j-x_i|^2} \quad , \tag{2.7}$$

representing pairwise interaction of particles.

Thus the particle method may be summarised by the system of ODEs:

$$d(w_ic_i)/dt = w_i\nabla\cdot(b\nabla c)_i \tag{2.8a}$$

$$(w_ic_i)(0) = w_i(0)c_i(0) \tag{2.8b}$$

$$dw_i/dt = w_i(\nabla\cdot a)_i \tag{2.9a}$$

$$w_i(0) = h^d \tag{2.9b}$$

$$dx_i/dt = a(x_i(t),t) \tag{2.10a}$$

$$x_i(0) = ih \quad . \tag{2.10b}$$

In the case assumed $(\nabla\cdot a \equiv 0)$, (2.9a,b) is trivial to solve. Similarly, if $b \equiv 0$, (2.8a,b) implies that the mass of particle i is constant. The basic error estimates are described in Raviart [4,5]. The convergence rate (in L_1, L_2, L_∞ norms) is found to be

$$0(\epsilon^k + h^m/\epsilon^{m+1}) \quad , \tag{2.11}$$

where k is as in (2.4) above and m is determined by the regularity of the velocity field. Thus the cut-off radius, ϵ, must be large compared to the interparticle spacing, h.

An important asset is the conservation property

$$\frac{d}{dt}\sum_i(w_ic_i)(t) = \text{boundary terms} \quad . \tag{2.12}$$

On the other hand, the implementation of boundary conditions for problems on non-periodic/infinite domains is awkward. In addition, the right-hand-side of (2.12) will contain terms which are distance ϵ from the boundary, and this may be unsatisfactory.

Note that equations (2.8,2.9,2.10) are highly suitable for solution in parallel; on the other hand, the diffusion approximation is very expensive to evaluate on a serial machine. Other cut-off functions include the generalised Gaussians

$$G_{2n}(x) \propto H_{2n+1}(x)\exp(-|x|^2)/x \quad , \tag{2.13}$$

where H_{2n+1} is the Hermite polynomial of degree 2n+1; it satisfies (2.4) with k=2n. Since these functions are non-positive (for some x), possible stability problems may arise if they are used.

3. The Moving Point Method. In this section, we describe a new moving point method for updating the values of c due to the effects of diffusion. Following the positional update, we have a set of scattered points. To update the c values, we triangulate the set using Delaunay triangles which gives a set of neighbours, $N(i)$, for each point i. Then we use mass-lumped finite elements or finite volumes on the triangles. In contrast to the particle method (which spreads the information over a large region), each point has a domain of influence extending over only the surrounding triangles; our 'smooth' representation of c is a piecewise linear interpolation of the point values over the triangles. The method is very similar to that found in Crowley [1].

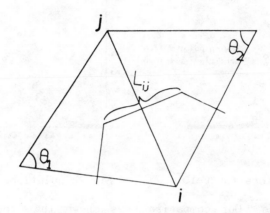

Fig. 1. A triangle pair describing the notation below.

From the above, for constant b, both f.e/f.v methods may be written as

$$\dot{w_i c_i} = b \sum_{j \in N(i)} (c_i - c_j) K_{ij} \tag{3.1a}$$

where

$$K_{ij} = - \frac{L_{ij}}{|x_i - x_j|} = - \frac{\sin(\theta_1 + \theta_2)}{2\sin(\theta_1)\sin(\theta_2)} \tag{3.1b}$$

and

$$w_i = \begin{cases} \text{volume of Voronoi cell} & \text{(finite volume)} \\ \text{(area surrounding triangles)/3} & \text{(finite element)} \end{cases} \tag{3.1c}$$

For Delaunay triangles, we have $\theta_1 + \theta_2 \leq \pi$; this ensures stability of the method via a maximum principle. In addition, we obtain the 'quasi-conservation' law

$$\sum_{(i)} \dot{w_i c_i} = \text{boundary terms} . \tag{3.2}$$

For a parabolic problem with no convection we obtain the optimal convergence rate $O(h^2)$ (see [6]); it is conjectured that this is generally retained for the convection-diffusion problem.

4. Numerical Results. We present results by both methods for

$$\varphi_t + \varphi_x = \upsilon \varphi_{xx} \tag{4.1a}$$

$$\varphi(x,0) = \begin{cases} \sin^2 \pi(4x-1) & \text{on } [1/4, 1/2] \\ 0 & \text{elsewhere} \end{cases} \tag{4.1b}$$

with 1-periodic boundary conditions and various υ with $T_F = 6.12$.

Figures 2a,b (below) demonstrate the expected superior accuracy of the moving point method over the particle method. Indeed, for this problem, the moving point method is exactly conservative.

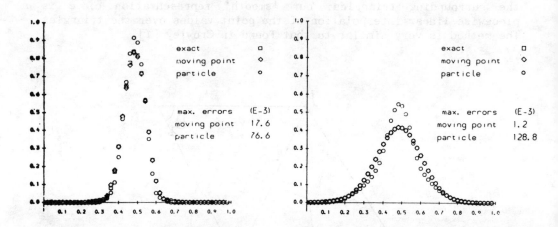

Fig.2a Solutions for $v=10^{-4}$.   Fig.2b Solutions for $v=10^{-3}$.

5. **Conclusions**   Our comparison has shown that the moving point method is more accurate than the particle method, although possessing only a quasi-conservation law. The particle method seems ideal for implementation on a shared memory machine, whereas the triangulation required in the moving point method inhibits the potential for parallelism – a possible remedy being some form of domain decomposition. Further numerical experiments are in progress, specifically in two dimensions, and preliminary results are very encouraging.

REFERENCES

[1] Crowley, W.P. Free-Lagrange Methods for Compressible Hydrodynamics in Two Space Dimensions. Lecture Notes in Physics (238): The Free-Lagrange Method, M.J. Fritts, W.P. Crowley, H. Trease, eds., Springer (1985).

[2] Farmer, C.L. A Moving Point Method for Arbitrary Peclet Number Multi-dimensional Convection-Diffusion Equations. IMAJNA (1985) v5 pp.465-480.

[3] Harlow, F.H. The Particle-in-Cell Computing Method in Fluid Dynamics. Methods in Computational Physics, v13, B. Alder, S. Fernbach, M. Rotenberg, eds., 1964, pp.319-343.

[4] Raviart, P.A. An Analysis of Particle Methods. C.I.M.E. Course on Numerical Methods in Fluid Dynamics, Como, July 1983.

[5] Raviart, P.A. Particle Numerical Models in Fluid Dynamics. Numerical Methods for Fluid Dynamics, K.W. Morton and M.J. Baines, eds., Oxford University Press, Oxford, 1985.

[6] Thomee, V. Galerkin Finite Element Methods for Parabolic Problems. Lecture Notes in Mathematics 1054. Springer (1984).

CHAPTER 31

Random-Walk Simulation of Diffusion-Reaction-Convection Systems*

D. J. Hebert[†]

Abstract

A simple stochastic model for populations of diffusing, interacting, and drifting particles is used to derive three simulation algorithms which are closely related to jump process models, random-walk and particle methods, and interacting particle systems. These algorithms are directly realized as simple vector or parallel programs. Model problems from biological and chemical self-organization studies provide examples for comparison of the resulting simulations and standard numerical solutions of diffusion-reaction systems of partial differential equations.

1. Introduction

The partial differential equations of diffusion-reaction-convection:

$$\frac{\partial u_k}{\partial t} = D_k \nabla^2 u_k + \nabla \cdot u_k \mathbf{v} + f_k(u)$$

are used extensively in the modelling of pattern formation in chemical and biological systems. The equations represent an idealization, approximation, or averaging of an underlying model of large populations of particles, cells, organisms, or attributes which are interacting locally while diffusing and drifting in space. The value of $u_k(x,t)$ is the density of the k-th population at the point x and at time t, D_k is the diffusion coeffient, v is the velocity of drift or convection due to environmental influences, and $f_k(u)$ is the reaction function which represents the local birth or death rate per unit volume per unit time. The PDE models have been studied extensively by mathematical analysis and by computer experimentation. The mathematical studies are usually limited to one space dimension, one or two populations, and simplified reaction functions.

In this talk we shall look at a simple branching random walk model of diffusion-reaction-convection which has a precise formulation in probability theory and a direct connection to the partial differential equations. The basic model supports three simulation algorithms suggested by mathematical studies of jump process models, random walk and particle methods, and interacting particle systems. The simulations require large numbers of simple independent local calculations and are suitable for easy implementation on vector and parallel computers. This has been demonstrated in some small

* This work was supported in part by the Pittsburgh Supercomputer Center.
† Department of Mathematics and Statistics, University of Pittsburgh, Pittsburgh, PA 15260.

FORTRAN programs for the CRAY X-MP. Pattern formation experiments are used to compare the performance of these algorithms and to compare to the numerical solution of the corresponding PDE systems.

Direct simulation of simple models has several advantages over numerical solution of partial differential equations. For computers which can do large numbers of simple calculations in parallel, the simple simulation models give small and fast programs. Inaccuracies may be largely confined to modelling errors rather than those of numerical approximations. Furthermore, for certain biological and ecological models in which local populations are small, the PDE approximations are extreme idealizations. Solutions of PDEs can not show the macroscopic effect of microscopic randomness. Finally, the simulation approach to modelling is not limited to models which lead to differential equations. Indeed, the connection with PDE theory may be lost by making small but physically meaningful alterations of the underlying model and resulting simulation algorithms.

2. The Model

Let us restrict attention to a two-population, two-space-dimension model which may be described as follows: The environment, a connected region in the plane bounded by piecewise smooth curves, is subdivided by a fine grid of square cells of width h. At each discrete time t, a large number $U_k(x,t)$ of individuals of population class k (k=1,2) is located in the cell x. The discrete density function u_k for the k-th population class is obtained by averaging U_k over neighboring cells and dividing by h^2. The nearest neighbor cells of x will be labelled as x_1, x_2, x_3, x_4 where $x_i = x + he_i$ and $x_{i+2} = x - he_i$ for i = 1,2 where e_i is the unit vector in the i-th coordinate direction. (Let $x_0 = x$.) Discrete time points will be separated by $\Delta t = \dfrac{h^2}{4}$. At each discrete time step an individual located at the point x may die or give birth with probability $|sf_k(u(x))|/U_k$. The living individual may remain at x with probability $p_0 = 1 - D_k$ or move to a neighboring point x_i with probability p_i where

$$p_i = \frac{D_k}{4} + \frac{\Delta t}{2h} v_i,$$

$$p_{i+2} = \frac{D_k}{4} - \frac{\Delta t}{2h} v_i,$$

for i = 1,2. To insure that p_i is a probability we must assume that $|v_i| \le \dfrac{2D_k}{h}$ which may be a severe limitation for convection dominated models.

Such a discrete model is made precise by constructing a discrete Markov process whose state space consists of mappings which assign individuals to their cell positions (see [4]). In this construction the expected values of U_k satisfy a discrete approximation to the system of partial differential equations.

Indeed, (dropping the subscript k) if the population at cell x at time t is U(x,t), let $U_i = U(x_i, t)$. The expected population at time $t + \Delta t$ is a sum of contributions due to populations at neighboring points and contributions of births and deaths:

$$E[U(x,t+\Delta t)] = p_0 + p_3(x_1)U_1 + p_4(x_2)U_2 + p_1(x_3)U_3 + p_2(x_4)U_4 + \Delta t f_k.$$

The definitions of p_i transform this equation immediately to

$$E[U(x,t+\Delta t)] = U_0 + \frac{D_k}{4}\sum_{i=1}^{4}(U_i - U_0) + \frac{\Delta t}{2h}\sum_{j=1}^{2}(v_j(x_{j+2})U_{j+2} - v_j(x_j)U_j) + \Delta t f_k$$

which can be rewritten (using the fact that $4\Delta t = h^2$) as the difference equation

$$\frac{E[U(x,t+\Delta t)] - U_0}{\Delta t} = \frac{D_k}{h^2}\Delta_h^2 U_0 + \frac{1}{2h}\sum_{j=1}^{2}\Delta_{2h} v_j U_j + f_k.$$

Theorem: There is a Markov process whose states are location maps, i.e. maps which assign individuals to cell locations, such that these formulae are made precise [4].

3. The Algorithms

Direct simulation of this basic model is possible but not really practical even on present super-computers. However, some useful models, whose simulations require much less memory and computation, may be derived directly from this one.

A jump process algorithm (JPA)

The first of these is called the jump process algorithm (JPA) since it is almost a direct simulation of the jump process model of chemical reaction-diffusion described by Nicolis and Prigogine [6] and Haken [2]. Convergence of the random population densities of the jump process in L^2 and uniformly in probability to the solutions to the PDEs was shown by Arnold and Theodosopulu [1] for an important class of reaction functions and in one space dimension.

In the basic branching random walk model the population U(x,t) for fixed x undergoes a jump in size at each discrete time step. The population $U(x,t+\Delta t)$ is a sum of contributions of neighboring points at time t. The jump process algorithm proceeds as follows at each discrete time point t:

An Array A contains U(x,t) for fine grid cells x; an array B of the same size is initialized to 0 for later accumulation of the values $U(x,t+\Delta t)$;

a random jump in size for U(x,t) is generated as a Poisson random variable with expectation $\Delta t f_k$;

a random splitting of U(x,t) into 5 parts is generated as a multinomial random variable with probabilities $p_0,...,p_4$;

the parts $U(x_i,t+\Delta t)$ accumulate on the array B;

B is copied to A and reset to 0.

Speedup is obtained in this algorithm as well as in the ones described below by computing an array of random numbers only once to be permuted at each time step rather than recalculated, and by averaging the population values over a courser grid for the computation of the velocities v_i and reaction functions f_k.

The density carrying particle algorithm (DCPA) and the density scattering algorithm (DSA):

The second derived model may be described as an algorithm (DCPA) which is similar to random walk and particle methods of numerical analysis. The derivation is based on a discrete version of the Feynman-Kac integral derived in [4] which says that the algorithm gives the same expected population densities as the basic branching random walk model. In fact, each of the stochastic algorithms presented here may be regarded as Monte Carlo estimations of this discrete Feynman-Kac integral.

The idea of DCPA is to assign a population U(x,0) to each particle in a collection initially located at the cell x. Each particle then performs a convected random walk on the fine grid of cells carrying the initial population size which is changed along the way according to course-grid values of the reaction function. The macroscopic population at time t is obtained by local averaging.

The third derived algorithm (DSA) is related to certain interacting particle systems discussed by Liggett [5]. Its implementation as a algorithm may be roughly described as follows: As in DCPA, initial populations U(x,0) are assigned to particles initially located at x. The particles do a random walk for a number of steps n_s. Then at each cell a new population is obtained as an average of the populations carried by particles presently in the cell. These populations are then changed by reaction and new particles carry the new populations again for n_s steps.

4. Simulations and Comparisons.

The three models described above have been implemented as FORTRAN programs which have been running on a wide range of computers including a micro, a 3B2/400, several VAX computers, and the CRAY X-MP/48. Vectorization for the CRAY was accomplished by using large inner loops with few calculations and no logic within the loops. The resulting programs are highly portable.

Parallel algorithms are equally easy to formulate and are expected to be more efficient than the vectorized versions. Many experiments have been performed to compare the algorithms among themselves and with numerical solutions to the same PDEs and to test the algorithms on pattern formation studies.

The test problems are chosen from among the simplified models of chemical reaction-diffusion, population diffusion, and morphogenesis in which the model of local reaction is some variation of the Brussellator or Oregonator chemical reaction models, the Fitzhugh-Nagumo threshhold model, the Voterra-Lotka conservative model, the Van der Pol limit cycle model, or the Gierer-Meinhardt activator-inhibitor model. Experiments with Dirichlet and Neumann boundary conditions on domains of varying shapes have have produced such patterns as rotating spirals, expanding target patterns, travelling line waves, stationary waves, and oscillating patterns.

The three stochastic algorithms give the same qualitative results for the same physical and grid parameters. For example, figure 1 shows line waves produced for a threshhold reaction function by

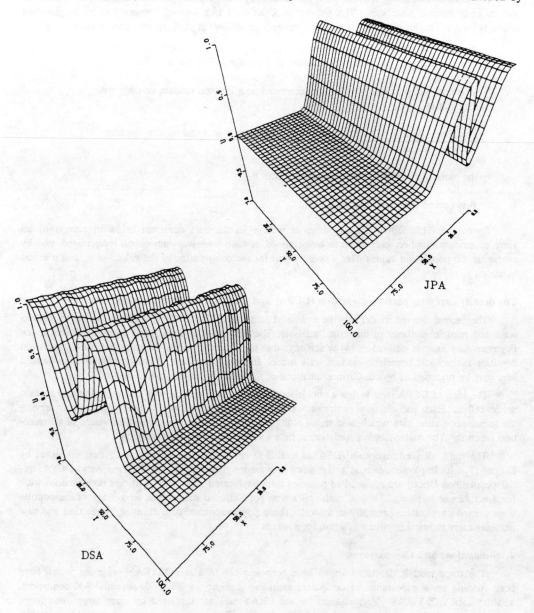

Figure 1. Line waves for a threshhold reaction function.

DSA and JPA with identical parameters. The noise in DSA maybe reduced by increasing the number of particles, while noise in JPA is apparently eliminated (at no cost) by increasing local population sizes in the underlying model. DCPA generally gives noisier results than DSA. Unusual boundary conditions and unusual domain shapes are easily incorporated. Figure 2 shows the contour map for an oscillating pattern on a domain with a hole produced by JPA.

Comparisons with the numerical package PDETWO [7], which is restricted to rectangular domains, show that the stochastic similuations also give results which are qualitatively similar to numerical solutions. Numerical solutions are much faster on many problems, but when large numbers of grid points are necessary, the stochastic simulations become competitive. For densities which do not approach a stable steady state the space-time scale of patterns seems to depend on spacial grid size both

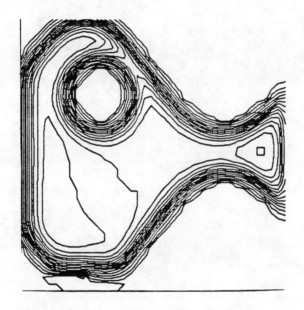

Figure 2. Oscillating density on a domain with a hole.

in numerical and stochastic experiments. For example, travelling line waves produced by PDETWO on a 40x40 grid and on a 100x100 grid with identical physical parameters show slighty wider waves with approximately half the wave speed on the finer grid. With JPA the finer grid PDETWO wave speed can be matched using a 40x40 course grid by adjusting the parameter n_s. In this case JPA is runs faster than PDETWO for similar results.

The questions of accuracy which these experiments raise are significant and are currently under study. The known convergence results (e.g. [1] and [3]) offer no help in understanding these grid dependencies.

On the CRAY X-MP the stochastic algorithms form a convenient experimental tool for the study of pattern formation in diffusion-reaction-convection systems. They are fast-running on a large number of models of interest, and the short programs are easily altered and compared. Quick results with correct qualitative features may also be obtained on smaller serial computers.

References:

1   Arnold, L. and Theodosopulu, M., *Deterministic Limit of the Stochastic Model of Chemical Reactions with Diffusion*, Adv. Appl. Prob. **12** (1980),367-379.

2   Haken, H., *Synergetics (3rd edition)*, Springer-Verlag, New York, (1981).

3   Hald, O. H., *Convergence of Random Methods for a Reaction-Diffusion Equation*, SIAM J. Sci. Stat. Comput. **2** (1981),85-94.

4   Hebert, D. J., *A Branching Random Walk Model for Diffusion-Reaction-Convection Systems*, Technical report ICMA-86-98, University of Pittsburgh, 1986.

5   Liggett, T. M., *Interacting Particle Systems*, Springer-Verlag, New York, (1986).

6   Nicolis, G. and Prigogine, I., *Self-Organization in Nonequilibrium Systems*, Wiley, New York, (1977).

7   Melgaard, D. K., and Sinovec, R. F., *Algorithm 565, PDETWO/PSETM/GEARB: Solution of systems of two-dimensional Nonlinear Partial Differential Equations*, ACM Trans. Math. Softw. **7** (1981),126-135.

Schwarz Splittings and Template Operators
Wei Pai Tang

Schwarz' alternating method (SAM) is an old mathematical technique dating from 1860. The method was commonly employed as a tool for theoretical analysis, but it is still not yet widely used for large scale scientific computations. The earlier experiences showed that SAM converged slowly. Our analyses show that some new methods based on SAM are competitive with other solution techniques. Some generalizations of SAM, namely Schwarz splittings (SS), are presented here. Several acceleration schemes are also discussed in this talk. In particular, when these techniques are applied to the solution of the model problem, an optimal complexity can be achieved. For many important applications, such as performing parallel computations in a non–shared memory environment, using composite grids and also applying fast solvers in an irregular region, SS's are found to be useful techniques.

In order to identify the types of problems for which these new techniques are most suitable, a new structure for the linear operators called Template operators has been developed. Some decay results for the elements of the inverses of sparse operators are given. These results provide a theoretical basis for determining when these SS techniques can be used successfully.

CHAPTER 32

Adaptive Mesh Refinement for Parallel Processors

Marsha J. Berger *

1. Introduction.

The goal of this work is to determine what kind of speedups are possible on a realistic application on today's parallel computers. The application I work on uses adaptive mesh refinement to solve the Euler equations for two-dimensional shock hydrodynamics. This adaptive mesh refinement strategy (henceforth AMR) was initially developed in [a]. More recently, it was been combined with the higher order Godunov methods of [b,c] to compute mach reflection for an oblique wedge. This combined code is 8000 lines of Fortran. It is used in several laboratories across the country. A typical run on a single processor of the Cray XMP takes 2 to 3 hours. The question is, what kind of speedups are possible using the other 3 processors?

One interesting question with adaptive mesh refinement, at least the kind of adaptivity AMR uses, concerns load balancing. The computational work in AMR changes dynamically in time, as well as being a function of space. Can the workload be efficiently divided so that all processors are kept busy, and no one waits idly for another CPU to finish? Another important question is the ease of parallel implementation. A major rewrite was out of the question, although certain subroutines had to be changed to be more amenable to multiprocessing. The parallel Fortran constructs available on different multiprocessors can make a huge practical difference. In particular, those of the Ultracomputer were much more convenient than those on the Cray.

In the next section, I very briefly describe the basic logic of AMR. I will only give the detail necessary for understanding my approach to parallelizing AMR. In section 3, I will review some alternative approaches to load balancing, including the one I use, called binary decomposition. Finally, section 4 discusses the parallelization of AMR, including how binary decomposition is applied to AMR, the parallel

---

* Courant Institute, New York University, 251 Mercer St., New York, NY 10012

constructs used, and the numerical results and timings on two different machines - the Cray XMP and the Ultracomputer at NYU.

Acknowledgements. Some of this work was done in collaboration with Phil Colella. I would like to acknowledge the computer time on the XMP given to me by Cray. I thank Philippe De Forcrand for getting me started using the XMP, and Anne Greenbaum for helping me on the Ultracomputer. I thank Phil Colella and Scott Baden for many helpful discussions on all aspects of these problems. This work was partially supported by grants from the AFOSR, DOE and DARPA.

2. The AMR Algorithm.

AMR uses nested grids with fine mesh spacing where more resolution is needed in the solution. This can be done recursively: several such nested levels of grid with increasingly finer mesh spacing may be used to achieve the desired resolution. These finer grids are simply *superimposed* on the underlying coarse grid in places where the resolution of the coarse grid is insufficient. By working with complete grids, instead of for example grid points, the data structures can be kept simple and regular. For a serial code, this makes vectorization possible, a property we would like to maintain in a parallel code as well.

The adaptive mesh refinement algorithm contains 4 basic components.

(1)   An automatic error estimator determines the regions of high error. This determines where refinement is needed.

(2)   An automatic grid generator creates fine grids patches covering the regions that need refinement.

(3)   Simple data structures store the information describing the nested grids and manage the solution storage arrays.

(4)   The integration strategy applies the same integrator to all rectangular grids.

The integration strategy needs to be described in more detail, since over 75% of the CPU time is spent integrating the grids. A key point of the algorithm is that when a grid is refined by a factor r in space, it is refined by the same factor r in time. This means that the mesh ratio $\dfrac{\Delta t}{\Delta x}$ is constant on all grids, so the same integration method is stable on all grids. This also means that the computational work is concentrated on the fine grids, where it should be. For every step on the coarse grid, r steps are taken on the fine grid, r^2 steps on the next finer level grids, etc.

Of course, the grids can not continue to be integrated independently of each other indefinitely. They interact in two ways. When the fine and coarse grids reach the same physical time, the fine grids update the underlying coarse grid points. This update step is a simple averaging procedure. If i,j are the indices of a coarse grid cell, and k,l are the indices of the lower left fine grid cell contained in the coarse grid cell, a conservative updating procedure is

$$u_{i,j}^{coarse} \leftarrow \frac{\displaystyle\sum_{n=1}^{r}\sum_{m=1}^{r} u_{k+m,l+n}^{fine}}{r^2}. \tag{1}$$

The second interaction between grids happens at grid interfaces. Fine grids need "boundary" conditions that come from linear interpolation in space and time from the surrounding coarse grid. In order for the combined difference scheme to be conservative, the coarse grid needs to be modified correspondingly so that the difference equations at the coarse points immediately adjacent to a fine grid *see* the fine grid. Since the grids are integrated independently, (and in fact, the coarse grid is updated first, before fine grid fluxes have even been calculated), this last requirement is implemented as a fix-up step after each fine grid step. This fixup pass takes the following form. First, a provisional coarse value is computed at the coarse points adjacent to fine grids,

$$\bar{u}_{i,j} = u_{i,j}^n + \frac{\Delta t}{\Delta x} D_+ F^{coarse} + \frac{\Delta t}{\Delta y} D_+ G^{coarse}. \tag{2}$$

This step provides values at the coarse grid points needed for the interpolation of the fine grid boundary values. When the fine grids are advanced, all the boundary fluxes are saved. The coarse grid is then modified so that the coarse grid difference scheme is fully conservative at the fine grid interface. For example, at a coarse grid point immediately to the left of a fine grid, the flux F across the right side of the cell must be modified to give

$$u_{i,j}^{n+1} = \bar{u}_{i,j} - \Delta t \left[F_{i+1/2,j}^{coarse} - \frac{\sum F_{1/2,k}^{fine}}{r} \right]. \tag{3}$$

The coarse grid keeps a list of coarse grid points that need adjustment. Eq. (3) can then be implemented as a scatter-gather operation on machines with such hardware. Although it is common to neglect boundary work in work estimates, in our initial implementation, boundary work took 40% of the CPU time and had to be rewritten. Complete details about these steps are found in [d].

Figure 1 shows a typical calculation with this method, also taken from [d]. This particular run took 1 1/2 hours on the XMP. Although this is a well studied test problem, the use of adaptivity led to greater resolution than was previously possible, and a new feature in the solution, a Kelvin Helmholtz instability along the slip line, can be seen. The coarsest grid used only 100 by 20 cells, and 149 coarse grid time steps were taken. Three levels of grid refinement were used in this run. The grids were refined by a factor of 4 in each direction at each level. Table 1 summarizes where the computational time was spent.

grid integration	76.4%*
interpolation	13.1%*
error estimation	3.4%
grid updates	2.8%*
I/O	2.7%
grid generation	.6%
space management	.6%

Table 1. Flowtrace information shows how the computational time is spent.

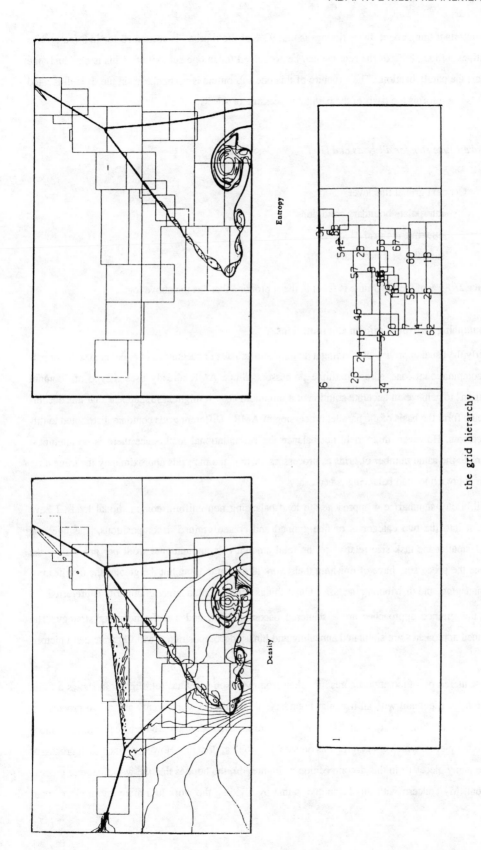

Figure 1. Mach reflection calculation.

An important thing about these timings is that 92% of the runtime is accounted for in 3 (categories of) subroutines. In fact, 89% of the runtime can be accounted for in one subroutine. This is obviously the place to start the parallelization. The structure of this one subroutine is particularly simple. It is illustrated in Figure 2, since it forms the basis of the parallel processing of AMR.

Take 1 time step for all grids at a level.

    Get next grid at this level
        interpolate boundary conditions
        integrate the grid
        save fluxes if necessary

Figure 2. 89% of the CPU time is spent in the subroutine that has the above logic.

3. Load Balancing for Non-Uniform Computations.

Clearly the usual approach of dividing a domain into squares or rectangles of equal area will not balance the computational load across multiple processors. Since AMR already uses a type of "domain decomposition" to achieve an accurate solution at a minimum cost, it might at first appear that this decomposition could form the basis of the parallel processing of AMR. Different grids could be distributed to different processors. However, this would not balance the computational load, since there is no attempt *a priori* to create the same number of grids as processors, or to make the grids approximately the same size. Some other approach to load balancing is necessary.

I will briefly summarize 4 approaches to load balancing non-uniform computational loads. I have divided them into the two categories of fine-grained and coarse grained decompositions. The finer the granularity (smaller the task size relative to the total amount of computational work per processor), the more chance the processors have of finishing their work at the same time, but the greater the overhead of creating, managing, and distributing the tasks. Other things being equal, a larger granularity is preferred.

The fine-grained approaches are a scattered decomposition, and a self-scheduling strategy. The coarse-grained approaches are simulated annealing and binary decomposition. I will illustrate them pictorially.

The scattered decomposition figures are taken from Morison and Otto [e]. Figure 3a shows a finite element mesh for a domain with an irregular boundary. To divide the work evenly among 16 processors, the domain is divided into many small pieces as if it were a regular domain. Some pieces will have more work than other pieces, which contain regions outside the problem domain. However, since each processor will receive many pieces (9 in this decomposition of Figure 3b), the hope is that the work load per processor evens out. My concern with this technique is that by dividing the work into many more pieces than

Figure 3. Scattered decomposition.

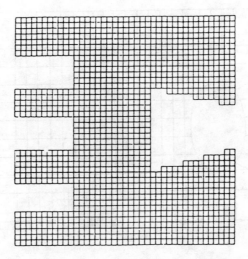

(a) computational mesh

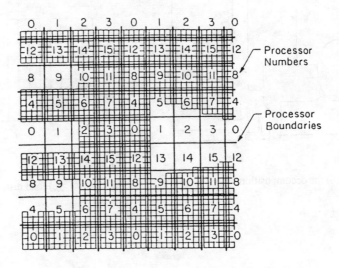

(b) the decomposition

Figure 4. Simulated annealing decomposition.

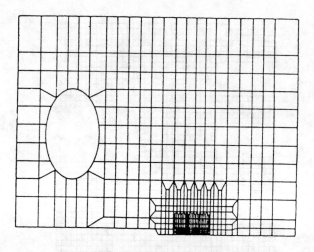

(a) computational mesh

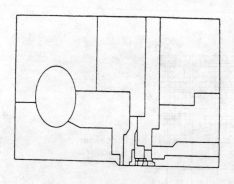

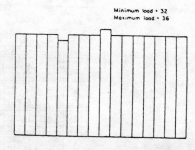

(b) the decomposition

(c) load distribution

processors, the inter-processor communication, typically proportional to the length of boundary of the sub-domains, has been greatly magnified.

Self-scheduling strategies are typically used on shared memory multiprocessors. In this approach, a queue of small tasks is maintained, say by one of the processors. When requested, the work is distributed to available processors. The "Fetch and Add" primitive on the Ultracomputer was designed to prevent such a queue from being a serial bottleneck. Here again, the task size cannot be too small or the work of main-taining the queue and distributing the tasks will swamp the original computational work. Self-scheduling has been used by Greenbaum in [f] to do sparse back-substitution, and by Phil Colella in parallelizing a front tracking code using SLIC. In this latter work, the small tasks were single rows of the computational domain. Depending on how much of the front lay in the row, the work per row varied by up to 50%.

The coarser-grained approaches to load balancing generally use as many domains in the decomposi-tions as there are processors. In the next approach, the load distribution problem is regarded as a discrete combinatorial optimization problem. The next step is to apply optimization techniques, such as the Monte Carlo method of simulated annealing. Figure 4 was taken from Flower, Otto and Salama [g]. Figure 4a again shows a finite element mesh for an irregular geometry. Figure 4b shows the domain decomposed using simulated annealing. Figure 4c shows the success of this approach in equidistributing the computa-tional work over the 16 processors. An advantage to this approach is that the objective function can con-tain many complicated terms, if desired, for a complete description of the computational and communica-tion complexity of the problem. Drawbacks, however, are that simulated annealing methods are very com-putational expensive, requiring hundreds of iterations to reach a solution. (The parallelization of the simu-lated annealing algorithm becomes a new computational problem by itself.) In addition, without additional work, the boundaries of each processors computational domain can be irregular, complicating the data structures needed to represent the decomposition.

The last method to be reviewed (and the one I use) is called binary decomposition. It is the most straightforward of the domain decompositions. The idea is to account for the workload in partitioning the domain, and to stick with simple shapes. In this approach, you have to have an *a priori* estimate of the computational work over the entire domain. The work can then be divided as equally in half as possible by a single vertical line, and the left and right halves given to two processors. If there are 4 processors, the decomposition continues recursively, using a horizontal line next, on each of the two halves. Figure 5, taken from [h], shows a sample binary decomposition for 16 processors. Figure 6, taken from [i], shows an application of binary decomposition to the solution of the incompressible Euler equations using a vortex method. In Baden's work, the measure of the computational work used was the number of vortices in each domain. If only vertical lines are used to partition the domain, you get strips. This has been used by Gropp in [j] in his earlier work on partitioning an adaptive mesh refinement program.

On the shared memory multiprocessors used in this work, the mapping of domains to processors is unimportant. However, some interesting theoretical properties of such mappings were studied in [h]. For example, consider the dual graph of the binary decomposition. (Regions sharing a common boundary segment have an edge connecting them in the dual graph). If this dual graph is mapped onto 4 nearest neighbor arrays, no fewer than 50% of the edges fall on edges of the machine graph, regardless of how skewed the partitioning is. This is important, since the edges that do not fall on machine graph edges incur higher communication costs. On an 8 nearest neighbor array, at least 79% of the edges fall on edges of the machine graph. Binary decompositions can also be mapped in a natural way onto a tree architecture. In this case, the communication cost is determined not by the longest boundary segment, but by the sum of such segments.

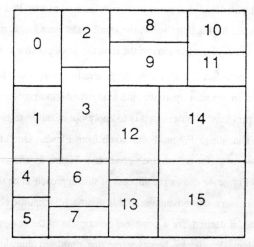

Figure 5. A binary decomposition for 16 processors.

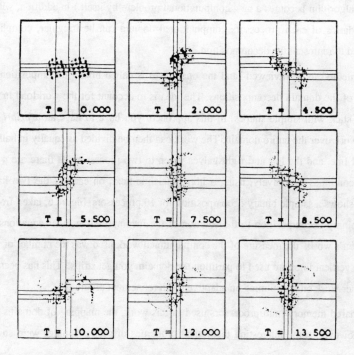

Figure 6. Binary decomposition applied to a vortex method.

4. The Parallelization of AMR.

Since the binary decomposition partitions the domain into rectangles, it seems well suited for application to AMR, which also partitions space into rectangles. Of course, with AMR the partition must be a function of time. Whenever the grid hierarchy is changed, the partition must be changed. The decomposition is implemented by partitioning the grids in the grid hierarchy at a given time for assignment to different processors. (Note that this is not the same as distributing the original grids in the hierarchy themselves).

When I started this work, my original idea was to partition the entire grid structure, based on all grids in the grid hierarchy. This is illustrated in Figure 7a. This is the largest granularity possible. If the coarse and fine grids in the same region of space are owned by the same processor, the updating step of the algorithm does not incur inter-processor communication. Although this is theoretically preferable, it is not practical. Such a decomposition does not balance the work at each grid level, only the total work of all levels. Since coarse grids are not load balanced, for example, there should be no synchronization after coarse grids are integrated, before moving on to fine grids. This means that before any of the r integration steps on the fine grid, the boundary values have to be interpolated from the coarse grids. The stencil we use in the higher order Godunov method requires 4 points to the side. Typically, the refinement ratio $r=4$ as well. This means 16 boundary values on each side of a grid must be set, or 32 in each direction. Although this is boundary work, this is not at all a negligible amount of extra work at each step.

The solution is to partition the grids by level. For example, if level 1 is just a simple rectangle, this gives the straightforward partitioning of figure 7b. The fine grids at level 2 are also partitioned by themselves. (Note that since most of the work is on the fine grids, the decomposition is very similar to the one in Figure 7a). As far as the updating step goes, for a refinement ratio of 4, one coarse point is updated by the average of 16 fine grid points. This update is performed every 4 fine grid time steps. Although at first thought this work is proportional to the area of the grids, in fact it is less time-consuming than the boundary work.

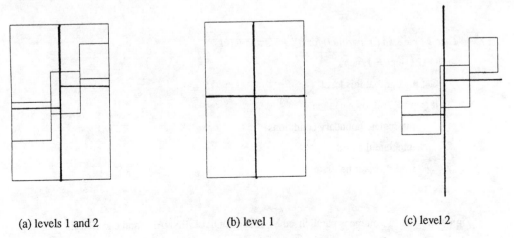

(a) levels 1 and 2   (b) level 1   (c) level 2

Figure 7. (a) Partitioning on the entire grid structure; (b) Partitioning by levels

Once the domain has been decomposed, grids in a given region are assigned a processor number. Only the owner processor may update a grid. Grids that span two such domains are subdivided, and each resulting grid is then assigned. Because of this division, there is some extra overhead associated with the decomposition, even if the program if run in serial mode. In my experiments, this serial degradation due to the extra grids is 3 - 5%.

The next step was the actual implementation using parallel Fortran constructs. AMR, like many if not most big scientific codes, contains chunks of computationally intensive, parallelizable code separated by long sections of serial code that don't take much time. The serial code does however account for most of the subroutines. The two machines I worked on, the XMP and the Ultracomputer, are quite different in their approach to parallel programs. Cray actually provides two mechanisms for parallelization. The main one is called macrotasking. With macrotasking, several CPU's are initialized for parallel threads of execution. Since the start-up time is high, the intent is to keep all the CPU's alive once they are initialized. However, it is difficult to "turn them off" for the serial sections. The other mechanism for parallel processing is called microtasking. This has evolved quite a lot, and the intent for its final use is still not clear. Initially, it provided do loop level parallelism on a self-scheduling basis. There is no intent to provide a pre-specified number of processors, nor give the user as much control over the CPU's in microtasking as in macrotasking. Although initially, subroutines could not be called within a microtasked loop, this restriction has been removed. However, microtasked loops cannot be nested.

In contrast, the Ultracomputer uses a "fork-join" model of parallelism. The user can explicitly initiate and terminate parallel section using a DOALL/ENDALL construct, for multiple CPU's executing the same code, or a PARBEGIN/PAREND construct for CPU's executing different sections of code. These constructs are limited to intra-procedural parallelism. However, unlike microtasking, they can be combined, on a self-scheduled basis, by nesting DOALL's.

This explicitly controllable parallel construct provides a particularly simple way to parallelize AMR. By adding 3 lines, the one subroutine accounting for 89% of the CPU time looks like this (cf. Figure 2).

Take 1 time step (in parallel) for all grids at a level.
    DOALL icpu = 1, ncpu
        Get next grid at this level
        if (icpu = grid_owner) then
            interpolate boundary conditions
            integrate the grid
            save fluxes if necessary
    ENDALL

Figure 8. This subroutine is parallelized by adding DOALL/ENDALL and one IF test.

If we only parallelize the one subroutine using 89% of the run time, perfect results would get a relative time of $89/4 + 11 = 33 1/4\%$, for a speedup of $S = 3.0$, and an efficiency $E = 75\%$. However, we get $S = 2.7$, which is an efficiency $E = 69\%$. This imbalance is due to 2 things:

(1)   The fine grids cannot be bisected exactly in half. A fine grid is constrained in AMR to start and end at a coarse grid point (so that every 4 fine grid points is a potential dividing place).

(2)   Perimeter (i.e., boundary interpolation) costs were not included in the work estimates.

Measurements show that the first problem causes 8-9% of the imbalance. The second problem, which is much more easily remediable, is responsible for only 2-3% of the imbalance.

For the next step, I parallelized the 5 subroutines that account for 98% of the CPU time. These next results use the Ultracomputer; the network and transmission lines to the XMP became too unreliable. Also, these timings omit the I/O time to output the results, which is not done in parallel. Table 2 shows timings of the 5 subroutines as well as the total CPU time, for a clearer idea of how successful the parallelization is. The "mod. code" column refers to the serial code but with the grids subdividied as they need to be for parallel execution.

	CPU time in 5 subr.	total CPU time
orig. code, 1 CPU	28097	28412
mod. code, 1 CPU	28918	29281
4 CPU's	8041	8403
speedup over mod. code	28918/8041 = 3.60	29281/8403 = 3.4
speedup over orig. code	28097/8041 = 3.49	28412/8403 = 3.38

Table 2. Timing results of parallelization of 5 subroutines.

The load imbalance here is again approximately 10%. The efficiency in parallelizing the 5 subroutines is thus 90% when compared to the modified code (where the grids are subdivided, causing extra boundary work), and 87% when compared against the original code. The overall efficiency is then 87% over the modified code, and 84.5% over the original code. On the Cray, there is an additional approximately 2% penalty for the modified code over the original code, due to the shorter vector lengths in the subdivided grids.

5. Conclusions.

We have taken a realistic application, and achieved a speedup of 3.4 using 4 processors. It involved only minor changes to the existing serial code, which admittedly was readily amenable to a domain decomposition approach to parallel processing. However, the ease of use of the parallel constructs make a huge difference in the amount of work, and debugging effort necessary to achieve this speedup. Unfortunately, the Cray XMP models available today do not actually allow all 4 CPU's to be used on a routine basis. In fact, the job priority is often lower if more than one CPU is requested. When supercomputer multiprocessors are more readily available, more experimentation can be done.

6. References

[a]   M. Berger and J. Oliger, "Adaptive Mesh Refinement for Hyperbolic Partial Differential equations", J. Comp. Phys. 53, March, 1984.

[b]   P. Colella, "Multi-dimensional Upwind Methods for Hyperbolic Conservation Laws", Lawrence Berkeley Laboratory Report LBL-17023, May, 1984. To appear in J. Comp. Phys.

[c]   P. Woodward and P. Colella, "The Piecewise Parabolic Method (PPM) for Gas-Dynamical Simulations", J. Comp. Phys. 54, April, 1984.

[d]   M.J. Berger and P. Colella, "Local Adaptive Mesh Refinement for Shock Hydrodynamics", Lawrence Livermore Report UCRL-97196, Sept. 1987. To appear in J. Comp. Phys.

[e]   R. Morison and S. Otto, "The Scattered Decomposition for Finite Elements", Caltech Report No. C^3P286, May 1985.

[f]   A. Greenbaum, "Solving Sparse Triangular Linear Systems Using FORTRAN with Parallel Extensions on the NYU Ultracomputer Prototype". New York University Ultracomputer Note #99, April, 1986.

[g]   J. Flower, S.W. Otto and M.C. Salama, "A Preprocessor for Irregular Finite Element Problems", Caltech Report No. C^3P292, June 1986.

[h]   M.J. Berger and S.H. Bokhari, "A Partitioning Strategy for Nonuniform Problems on Multiprocessors", IEEE Trans. Comp. C-36, May, 1987.

[i]   S.B. Baden, "Dynamic Load Balancing of a Vortex Calcuation Running on Multiprocessors", Lawrence Berkeley Laboratory Report LBL-22584, Dec. 1986.

[j]   W.D. Gropp, "Local Uniform Mesh Refinement on Loosely-Coupled Parallel Processors", Yale University Computer Science Dept. Report RR-352, Dec. 1984.

Shock Calculations on Multiprocessors
Björn Sjögreen

We present results from computing solutions to the compressible two dimensional Euler equations on a hypercube multiprocessor. We use modern high resolution finite difference schemes to get sharp shock profiles, mainly the Roe scheme and its generalizations to higher order accuracy. We discuss how to decompose the computational domain into different processors and we analyze whether we can increase the computational speed by not doing communication on every time level. Some results from solving the same problems on the Connection Machine will also be presented.

CHAPTER 33

Parallelization of Adaptive Grid Domain Mappings*

Calvin J. Ribbens[†]

Abstract. Adaptive grid domain mappings are used as a tool for grid adaptation in the numerical solution of partial differential equations. They allow the power of adaptive moving-grid methods to be applied to difficult problems, without introducing irregularities into the grid which may make good vector or parallel numerical methods harder to achieve. The three main computational steps associated with these mappings are: 1) constructing the mapping by tensor product least squares, 2) evaluating the mapping by tensor product spline evaluation, 3) inverting the mapping by a two-dimensional secant method. We describe parallel implementations of these three steps on the Sequent Balance 21000 multiprocessor. Issues affecting performance and speedup are discussed.

1. Introduction. We consider the parallelization of *adaptive grid domain mappings* (AGDMs), an approach to grid adaptation for the numerical solution of partial differential equations (PDEs). The goal of the AGDM framework, described more extensively in Ribbens [3] [4], is to allow the power of grid adaptation to be applied to difficult PDEs, while preserving the uniformity and regularity in algorithm and data structure which makes parallelism easier to exploit. The AGDM scheme is essentially a moving grid scheme. Consider a two dimensional domain R on which a PDE is posed. We define a mapping function $F(x, y)$ which maps points in R to points in a second domain S. The domain R is usually called the "problem" domain, and S the "solution" or "computational" domain (see Thompson [7], for example). The inverse mapping is denoted by $F^{-1}(s, t)$. These mappings are chosen so that a curvilinear adapted grid on R is mapped to a uniform rectangular grid on S. By using the $F(x, y)$ as a change of variables in the original PDE on R, we can define an equivalent problem on S. The irregularity introduced by grid adaptation is thus located in the transformed PDE, rather than in the grid and its associated data structures. The possibilities for exploiting parallelism seem better for methods based on regular grids such as the one placed on S than for irregular grids such

* This work was supported in part by NSF grant MS–8301589 and by AFOSR grant 84–0385.

† Department of Computer Science, Virginia Polytechnic Institute & State University, Blacksburg, VA 24061.

Table 1: Parallel performance in evaluating F^{-1}.

Processors	2	3	4	5	6	7	8	9
Speedup	1.99	2.95	3.95	4.84	5.79	6.61	7.53	8.31
Efficiency	.99	.98	.99	.97	.96	.94	.94	.92

as the one on R. The advantage of the AGDM approach seems particularly evident when compared with local grid refinement methods, although recently Berger [1] has had some success parallelizing such a scheme for the Cray XMP.

The question which this paper attempts to answer is, "do the adaptive grid domain mappings themselves parallelize well?". If the advantages for parallelism gained by applying an adaptive mapping are offset by a new bottleneck, namely the overhead of the mappings themselves, then the approach is not a good one. We consider the three main computational steps involved in the use of AGDMs: constructing the mapping F^{-1}, evaluating F^{-1}, and inverting F^{-1}. The three sections below deal with these steps in turn, discussing the methods and reporting on parallelization of each. We conclude with summary remarks and comments in Section 5.

The parallelization reported here is for a Sequent Balance 21000, a shared memory, common bus architecture. The machine we use has 10 processors. Programming is done in FORTRAN, plus the use of parallel programming directives and, to a lesser extent, direct calls to the parallel programming library. We use double precision for the experiments reported.

2. Evaluating the mapping. We represent the inverse mapping F^{-1} by two bicubic splines $x(s,t)$ and $y(s,t)$. In order to discretize the transformed PDE on S, F^{-1} and its derivatives must be evaluated many times, typically at each point in a rectangular grid. The sequential algorithm for evaluating F^{-1} at each point in a grid benefits from the computational savings inherent in the tensor product nature of the bicubic splines. For example, consider one coordinate of F^{-1}:

$$x(s,t) = \sum_{i=0}^{3} \sum_{j=0}^{3} \frac{c_{i,p,j,q}(s - \xi_p)^i (t - \eta_q)^j}{i! \, j!}, \tag{1}$$

where $\xi_p \le s < \xi_{p+1}$, $\eta_q \le t < \eta_{q+1}$, and $\{\xi_i\}$ and $\{\eta_j\}$ are the break points in the s and t directions, respectively. We can rewrite (1) as

$$x(s,t) = \sum_{i=0}^{3} \left(\sum_{j=0}^{3} \frac{c_{i,p,j,q}(t - \eta_q)^j}{j!} \right) \frac{(s - \xi_p)^i}{i!}. \tag{2}$$

The large parenthesized term in (2) is constant for each i, if t and p do not change. Thus, if we evaluate F^{-1} by rows of constant t, these terms (one for each i) need not be recomputed as long as s is in the same interval p.

The efficiency of the tensor product polynomial evaluation is easily preserved in a parallel version. The natural approach is to parallelize by rows—each processor getting a row of the points in S where F^{-1} is to be evaluated. This only requires the addition of the DOACROSS directive to our code, plus a very small amount of additional work (finding the correct polynomial piece q for a given t takes longer when each row is done in parallel). The example summarized in Table 1 is typical of the parallel performance of this step. The data is from a case with 100 grid lines and 5 spline pieces in each direction. Speedup and efficiency is nearly optimal. Speedup is defined as (sequential time)/(parallel time); efficiency is (speedup)/(number of processors). The slight degradation in performance as the number of processor increases may be due to increased contention for the shared coefficients, but with the number of processors available to us it is difficult to be sure.

Table 2: Parallel performance in inverting F^{-1} (Version 1).

Processors	2	3	4	5	6	7	8	9
Speedup	1.96	2.88	3.54	4.37	5.59	5.63	5.59	7.90
Efficiency	.98	.96	.89	.87	.93	.80	.70	.88

3. Inverting the mapping. In order to evaluate an approximate solution to the PDE at a point in the problem domain R, we must evaluate F. We do this by numerically inverting F^{-1}. In particular, to compute $F(x, y) = (s, t)$ we use a two-dimensional discrete newton's method to find a root of $F^{-1}(s, t) - (x, y)^T$. In the typical application, F^{-1} is inverted at each point in a grid on R. Thus, the sequential algorithm benefits from good initial guesses based on previous calculations.

There are two obvious levels at which to parallelize the inversion of F^{-1} on an $n \times n$ grid. One approach (Version 1) is to do the calculations by rows, resulting in n parallel processes. A second strategy (Version 2) is to invert F^{-1} at all n^2 points in parallel, so that each process is responsible for a single point, rather than an entire row. The degree of parallelism is clearly higher for Version 2, but it is easy to see that there is a performance penalty, since Version 2 will not be able to make use of previous work to generate good initial guesses. Version 1 does not suffer from this penalty, except for the first point on each row.

We implement these two versions. Modest changes to the sequential code are needed. Using only one processor, Version 2 is up to 50% slower than Version 1 for the problem sizes considered ($10 \leq n \leq 50$). The variation in the penalty is due to the type of mapping F^{-1}, not to the grid size n. The closer F^{-1} is to the identity map, the smaller the penalty in Version 2, since the initial guesses for this version assume an identity map.

Comparing the two versions when more processors are used, we are not surprised to find that Version 2 appears to be better than Version 1 eventually, as more processors are used. However, for a fixed number of processors, as the problem size n grows, Version 1 is eventually better. This suggests a serious problem in determining the single best way to parallelize the inversion of F^{-1}. For a given number of processors and a given mapping F^{-1}, there is a value of n beyond which Version 1 is the best strategy. Unfortunately, this cross-over point is generally different for each number of processors and each mapping F^{-1}.

A further aspect of parallelization which is illustrated by comparing the two versions is the importance of load balancing when the degree of parallelism is low. It is noted above that as n grows, Version 1 eventually wins. However, for certain relatively small values of n, Version 1 may happen to be best also. The reason has to do with the balancing of the computational load. Version 1 performs quite efficiently if the number of parallel processes happens to be a multiple of the number of processors. For example, the data in Table 2 are from an example with $n = 18$ rows of points. Notice that no appreciable gains in speedup are realized as we increase the number of processors from six to eight. The obvious reason is that with seven or eight processors, we still must make three passes through the data, during the third of which several of the processors are idle. The parallel efficiency when we use two, three, six or nine processors is quite good, as expected.

Finally, we report parallel efficiency for both versions on a typical example, with $n = 38$ grid points in each direction. Table 3 indicates that the efficiency for Version 1 is initially higher but steadily decreases. The parallel efficiency of Version 2 is lower than that of Version 1 initially, due to the penalty of poor initial guesses, but as processors are added, we see that Version 2 eventually wins because of its higher degree of parallelism.

Table 3: Parallel efficiency in inverting F^{-1}.

Processors	2	3	4	5	6	7	8	9
Version 1	1.00	.97	.96	.94	.90	.89	.90	.83
Version 2	.90	.90	.85	.86	.88	.89	.85	.88

4. Constructing the mapping. As described above, we represent each coordinate of F^{-1} by a bicubic spline. The coefficients of these piecewise polynomials are determined by least squares to best approximate a mapping from a uniform grid on the solution domain S to an adapted "mapping grid" on R. We have developed several efficient schemes for generating adapted grids (see [5] and [6]), but these are not discussed here. Note that the number of grid lines in each direction of the mapping grid need not be as high as the number in the discretization grid. The mapping grid need only be fine enough to capture the adaptivity desired.

The tensor product structure of the bicubic splines and the rectangular grid on S mean that the least squares problem can also be expressed in terms of tensor products. If the mapping grid is $n \times n$, and if we have $m - 3$ spline pieces in each direction, then instead of solving an $n^2 \times m^2$ overdetermined system, we can solve two $n \times m$ systems, one with n right sides and one with m right sides. Another advantage of the tensor product formulation is that the $n \times m$ matrices have much smaller bandwidth than the full $n^2 \times m^2$ matrix. We solve each of these small least squares problems by Cholesky factorization, since it is fast and the loss of accuracy due to the poor conditioning of the normal equations does not hurt the performance of the method. For details, see Ribbens [4].

Compared to the steps discussed in Sections 2 and 3, the construction of the mapping via tensor product least squares is a much more heterogeneous calculation. Several different and fairly expensive steps are identifiable. This makes parallelization more difficult because each of the computational steps must be parallelized if the overall efficiency is to be acceptable. Also, the best parallel strategy for an isolated computational kernel may not be the best choice when the kernel is part of a much more complicated calculation. We find this to be the case in several instances.

The three most expensive computational steps in the construction of F^{-1} are (along with their relative cost in a typical example): assembly of the normal equations and right hand sides (30%), forward and back substitutions (17%), conversion from B-spline representation to piecewise polynomial representation (25%). All three of these steps present difficulties. We again encounter tradeoffs between efficient sequential code and high levels of parallelism. Consider the first step more closely. The most expensive part of the assembly is computing the elements of an $m \times n$ matrix whose columns are the n right sides mentioned above. A straightforward parallelization of the sequential code (based on a program from de Boor [2]) yields no speed up at all, since locks are needed to protect a critical area in the innermost loop. Extensive changes to the code result in a version which, with one processor, is about twice as slow as the sequential version, but does achieve approximately 50% efficiency as the number of processors increases. The degree of parallelism for this version is $O(nm)$. There is yet another way to parallelize the assembly step which gives degree of parallelism $O(n)$. The $O(n)$ version is best for certain choices of n, m and the number of processors, but again, the point at which one version is better than the other is difficult to determine. Our best attempts so far have parallel efficiencies of around 40% using nine processors for the entire construction of F^{-1}. Some improvement could be made by parallelizing the remaining routines, and by taking advantage of the fact that $x(s,t)$ and $y(s,t)$ could be constructed in parallel.

5. Conclusions and comments. We make the following summary and concluding remarks:

- The evaluation of the bicubic splines parallelizes very well, and without significant code modification.
- The inversion of F^{-1} parallelizes fairly well also, although some re-writing of code is necessary, and a few difficulties do arise (see next two comments).
- When the degree of parallelism is low with respect to the number of processors, load balancing is extremely important.
- There is frequently a tradeoff between the most efficient code (in a sequential sense) and massive parallelism. Often, it is difficult to decide a priori when to switch from one approach to another. The number of processors, the problem size, and the numerical properties of a given problem are all seen to influence this decision. This suggests that it will be very difficult to build a system which automatically detects good parallel strategies given only a sequential program.
- Highly heterogeneous computations are very difficult to parallelize. This is true not simply because more work is required, but because the tradeoffs mentioned in the previous comment become even harder to identify, much less decide on. The best parallel strategy for a given computational kernel may differ depending on whether the kernel is considered alone or as part of a more complicated calculation.

REFERENCES

[1] M. J. BERGER, *Adaptive mesh refinement for parallel processors*, presented at Third SIAM Conference on Parallel Processing for Scientific Computing, Los Angeles, 1987, (these proceedings).

[2] C. DE BOOR, *A Practical Guide to Splines*, Springer-Verlag, New York, 1978.

[3] C. J. RIBBENS, *Domain mappings: a tool for the development of vector algorithms for numerical solutions of partial differential equations*, Ph.D. Thesis, Purdue University, August 1986.

[4] C. J. RIBBENS, *A computational framework for constructing adaptive grid domain mappings*, Purdue University, Computer Sciences Department Report CSD-TR 673, April 1987.

[5] C. J. RIBBENS, *A fast grid adaption scheme for elliptic partial differential equations*, Purdue University, Computer Sciences Department Report CSD-TR 678, April 1987.

[6] C. J. RIBBENS, *A priori grid adaption strategies for elliptic PDEs*, in Advances in Computer Methods for Partial Differential Equations VI, R. Vichnevetsky and R.S. Stepleman, (eds.), IMACS, New Brunswick, N.J., 1987, pp 102–107.

[7] J. F. THOMPSON, Z. U. A. WARSI AND C. W. MASTIN, *Numerical Grid Generation*, North Holland, New York, 1985.

Lattice Gas Methods for Solving Partial Differential Equations
Gary Doolen

Lattice gas methods are logical, discrete techniques for efficiently solving partial differential equations. These methods have been shown to solve 2D compressible Navier Stokes, reaction–diffusion, 2D MHD, and Burger's equations. They can exactly conserve mass, energy, and momentum; they are fast (about one billion cell updates per second on a CRAY XMP48); they are memory efficient (one CRAY word defines 10.6 cells). The codes are robust, efficient and short (approximately 50 FORTRAN lines); complicated boundaries are easily incorporated. They are fully parallel (N processors run N times faster). Estimates indicate that a $(512)^3$ processor computer could cost \$300,000. Small simulators (256 × 514 cells) have been built for around \$1000. Present lattice gas methods are constrained by Mach numbers that cannot exceed 1.4, density–dependent viscosity, velocity–dependent equations of state, and cell–averaging requirements.

The author will review recent developments, presenting details of 2D and 3D Navier Stokes solvers along with movies of complicated flows and concluding with a discussion of future challenges.

CHAPTER 34

A Gray-Code Scheme for Mesh Refinement on Hypercubes

William D. Gropp*
Ilse C.F. Ipsen†

Abstract. Adaptive methods for PDEs can be viewed as a graph problem. Parallel methods must distribute this graph efficiently among the processors. In doing this, the cost of communication between processors and the structure of the graph must be considered. We divide this problem into two phases: labeling of graph nodes and subsequent mapping of these labels onto processors. We describe a new form of Gray-code which we call an *interleaved* Gray-code that allows easy labeling of graph nodes even when the maximal level of refinement is unknown, allows easy determination of nearby nodes in the graph, is completely deterministic, and often (in a well-defined sense) distributes the graph efficiently across a hypercube. The theoretical results are supported by computational experiments on the Connection Machine.

1. Introduction

Parallel computing offers the possibility of greatly increased computing power. However, some problems are so large that even enormous parallel computers will not be able to handle them. Such problems include time-dependent partial differential equations (PDEs), which comprise regions with a fine-scale structure as well as regions with a coarse-scale structure. Such problems, solved on a uniform grid, may require on the order of 10^{13} floating operations per time step on a uniform three-dimensional mesh of a thousand points in each direction. The resulting resolution, however, often yields unnecessarily accurate solutions in the coarsely structured regions.

A tremendous amount of work may be saved by adapting the computations to the structure of the PDE at hand. The parallel implementation of such an adaptive method can be considered as the problem of managing a dynamic graph on a (static) network of processors. The properties desired of this graph are that its edges, as much as possible, map to physical processor interconnections, and that changes in the graph do not require

*Department of Computer Science, Yale University, New Haven, CT 06520. The work of this author was supported in part by the Office of Naval Research under contract N00014-85-K-0461 and by the Air Force Office of Scientific Research under contract AFOSR-86-0098.

†Department of Computer Science, Yale University, New Haven, CT 06520. The work of this author was supported in part by the Office of Naval Research under contract N00014-85-K-0461 and by the Army Research Office under contract DAAL03-86-K-0158.

a major re-arrangement regarding the assignment of nodes to the processors. Adaptive methods on parallel architectures have received little attention since the graph management problem is hard in general and does not parallelise very well. Our approach rests on the fact that problems suitable for adaptive refinement do *not* require random refinements but exhibit a certain *coherence*: the solution as well as the behavior of any discontinuities (such as shocks) are (nearly) continuous and thus the refinements are *localized*. This can be successfully exploited for operations on data structures as well as maintaining reasonably uniform balance of workload across the processors.

We have made the following choices for the solution of our problem. The initial domain on which a PDE is to be computed is discretised and can be represented as the selective recursive subdivision of a k-dimensional cartesian *grid of cells* (without loss of generality, the maximal number of cells per grid is assumed to be a power of two).

Due to its rich interconnection structure (as much as its commercial availability) the class of hypercube multiprocessors seems the one most suited to efficient handling of a graph management problem. All processors compute in parallel the function values on their cells; function values on or near a cell boundary are obtained by exchanging information with the processor that contains the cell sharing that boundary. In order to speed up computation a processor can distribute cells resulting from a subdivision to other processors; in this case there is communication between the processor containing the parent and those containing the children.

Our problem is now to assign cells to particular processors in such a way that processor utilization is high (all processors do worthwhile work most of the time), and that processors containing related cells are not too far apart so as to keep communication cost low. These two objectives are conflicting: if all cells are assigned to the same processor communication costs are zero but the load balance could not be worse. Alternatively, high communication costs and overheads may be incurred when trying to maintain a reasonable load balance. In particular, related cells (those that have to communicate with each other) should be in the same processor or in physically connected processors so that communication can proceed without any intermediate processors (which would have to spend time in forwarding information to target processors). The implicit assumption made here is that the of cost of communication between two processors is proportional to their distance, where distance denotes the smallest number of physical communication links that must be traversed to get from one processor to the other.

Because there are, in general, many more cells than processors and the number of grid levels and cells is unpredictable, we divide the process of assigning cells to processors into two stages:

1. Each cell is assigned a *unique* label.
2. Each label is mapped to a processor identifier (id).

The first stage preserves coherence: it is easy to find the labels of siblings, ancestors and descendants of a cell in the hierarchy of grids; once the label has been determined the corresponding processor can be found without much effort. The second stage makes it possible to ensure that related cells are allotted to physically close processors of the hypercube, and that the work load is distributed reasonably across all processors for our applications. Since the labelling strategy is static, a cell can determine the label of any other cell at any level without requiring external information.

Since the processors in a hypercube can be enumerated in such a way that the (binary) identifiers of physically connected processors differ only in one bit (Gray codes) [1], the labels and processor ids are represented as binary numbers. Therefore, the computation times of all our algorithms are measured in operations on bits, and their time complexities are always low-order polynomials in the lengths of the labels.

The usual method of labelling multi-dimensional grids proceeds by generating labels in such a way that the ith group of *contiguous* bits in the label represents the coordinate with respect to the ith dimension of the grid, and the coordinates associated with each dimension are generated as members of a Gray code sequence [1, 3, 4]. The obvious properties of such a labelling are that adjacent cells are easily mapped to physically interconnected hypercube processors and that each processor can systematically determine the label, and thus the processor, of an arbitrary cell, in particular a neighboring cell. However, it seems to be hard to achieve reasonably uniform processor utilization with a labelling based on a contiguous Gray code when the maximal level of refinement as well as the number of cells are not known in advance.

As a result, for the solution of PDEs on k-dimensional grids, we decided to employ a so-called *interleaved k-dimensional* Gray code that 'scatters' the bits associated with a coordinate: the (ik)th bit in a label represents the coordinate of dimension i in the grid. The length of a label increases with increasing level of refinement. Although interleaved Gray codes superficially resemble Quadcodes [2], the latter do not yield the small communication distances of interleaved Gray codes. In fact, for practical applications, interleaved Gray codes result in essentially constant communication times.

We first discuss the properties of one-dimensional Gray codes, used with locally refined grids. We then discuss the mapping of these grids to processors. Finally, we describe briefly the interleaved Gray codes for multi-dimensional problems.

2. Labels for Cells

In this section we summarize some results on Gray codes which are necessary for our results. Rather than describe a single Gray code for all cells or nodes in the refinement, we use a Gray code for each level of refinement. This allows us to develop properties within a single level, and then between levels of refinement. The labelling is done in such a way that labels of neighboring cells, and those of parent and children cells differ in only one bit and are easily determined. These ideas rely heavily on properties of Gray codes [1].

We establish the following results:

- Given an element $\mathcal{L}(i)$ from a binary reflected Gray code, the successor $\mathcal{L}(i+1)$ (right neighbor in the sequel) and predecessor (left neighbor in the sequel) can be determined by simple bit operations on $\mathcal{L}(i)$.

- Given a hierarchy of grids, each represented as a Gray code, the parents and children of a particular element $\mathcal{L}(i, l)$ of this hierarchy can be determined by simple bit operations on that element. Here, l is the *level* of the grid in the hierarchy, with level 0 being the coarsest or originating grid, containing a single element.

- Given a partial hierarchy of grids, where not every cell or element is refined, it is possible to find the neighbors of a cell within this partial hierarchy using only simple bit operations and the knowledge of the neighbors of the parent.

These results make it easy to find the neighbors of a cell in a one-dimensional, refined grid. By properly defining an *interleaved Gray code*, these properties can be extended to arbitrary dimension. Rather than use the "usual" tensor-product Gray code, we interleave the bits from each direction. The label of a child cell in d-dimensions is constructed by taking the label of the parent and appending d bits. The values of these bits are easily determined from the appropriate one-dimensional restrictions along the "coordinate" directions.

3. Mapping Labels of One-Dimensional Grids to Processors

Now we want to map the labels to processors in the hypercube such that the identifier (id) of a processor can be easily computed from the label of the cell, and cells whose labels differ in one bit are assigned to physically neighboring processors.

We take advantage of the fact that processors in a hypercube of dimension p can be enumerated according to a binary reflected Gray code sequence of length 2^p. Topological properties of hypercube graphs and their embeddings can be found in [1, 3, 4].

The assignment of labels to processor ids is easy if a grid of 2^d cells is mapped to a hypercube of dimension p where $p \geq d$: the cells in this case occupy processors in a subcube of dimension d, and the label of a cell is at the same time the id of a processor.

The more interesting case occurs when the number 2^d of cells exceeds the number of processors in the hypercube, and when grids are only partially refined. Let the processor to which a cell C is assigned be denoted by $P(C)$. Consider a hierarchy of $p + l$ grids where the finest grid contains 2^l more cells than there are processors, $1 \leq l \leq p$. Hence the label of each cell C has length $d = p + l$. Our strategy for mapping labels of length $p + l$, $l \leq p$, to p-dimensional hypercubes can then be defined as follows: a cell C with label $\mathcal{L}(C) = c_0 \cdots c_{p-1}.c_p \cdots c_{p+l-1}$, $1 \leq l \leq p$, is assigned to the processor with id

$$
\begin{aligned}
P(C) &= \{c_0 \oplus c_p\}\{c_1 \oplus c_{p+1}\} \cdots \{c_{l-1} \oplus c_{p+l-1}\}.c_l \cdots c_{p-1} \\
&= \{c_0 \cdots c_{p-1}\} \oplus \{c_p \cdots c_{p+l-1}0 \cdots 0\}
\end{aligned}
$$

where $\oplus$ is the XOR operator.

Remarks.

- One child always occupies the same processor as its parent and the other child is assigned to an adjacent processor.

- Cells whose labels differ in 1 bit are assigned to physically connected processors; thus, the processor of a cell contains the parent or it is physically connected to a processor that contains the parent and the sibling.

- The mapping is consistent: when $d < p$ and the number of processors in the hypercube exceeds the number of cells in the finest grid then $P(C) = \mathcal{L}(C)$ for any cell C.

- We consider here only the case where the number of refinements does not exceed the dimension of the hypercube ($l \leq p$). This is sufficient for massively parallel architectures like the Connection Machine.

One can always construct cases where our strategies will deliver nearly the worst processor utilization. However, for our applications such as problems with few regions of concentrated refinement, we can give upper bounds on the processor load imbalance, the physical distance to neighboring cells and the communication traffic. This is done for problems containing a moving refinement below. One could devise dynamic strategies that give preference to XORing those bits that are identical in all cells to be refined. We will not consider those strategies here as they are not predictable or static, and are likely to require a considerable amount of global information. Stratigies similiar to the one discussed here are applicable in higher dimensions, using the interleaved nature of the labeling.

4. Analysis of a Model Problem

To get a feel for how well this approach exploits coherence in the solution, we consider the problem of one region of refinement that moves across a domain. We chose the moving refinement problem not only because of its resemblance to many practical applications, but also because it allows us to examine different locations of intensive refinement within the domain and so to be able to take into account as many cases as possible.

The One-Dimensional Model

Consider a single region of refinement, moving across a one-dimensional region. Let this region be a set of contiguous cells belonging to the grid at level $p+l$. Since there is only one region of refinement, the region outside the refinement can comprise no two cells that are neighbors on the same level (that is, the cells become 'larger' with increasing distance from the refinement, and outside the refinement there can be no two cells of the same size). Given a p-dimensional hypercube, we assume at first that the refinement consists of 2^p cells belonging to the grid at level $l+p$ (as will be shown later, this assumption is not restrictive). There are two types of cells, *refinement cells* representing the refinement and *outside cells* representing the region outside the refinement; due to the Local Uniform Mesh Refinement, the outside cells are cousins (siblings of ancestors) of the refinement cells.

The following theorems can be proven about this situation, using our cell labeling and processor mapping:

- Communication between neighboring cells is of processor distance at most two.
- If the region of refinement contains at most $m2^p$ cells belonging to a grid at level $p+l$, $1 \leq l \leq p$, then each processor contains at most $m+1$ refinement cells and one outside cell. In other words, the maximum load per processor is $m+2$ cells. This is a sharp bound, and it is nearly optimal: in the best case, each processor contains $m+1$ cells.
- Under a certain communications strategy, there is at most one message on any processor-to-processor link.

The Two- and Three-Dimensional Models

Consider cells in a two-dimensional domain as a sequence or stack of one-dimensional domains (this is possible because of the interleaved Gray code: for each one-dimensional domain only the coordinates in one direction form a Gray code). Similarly, a three-dimensional domain is viewed as a stack of planes, where each plane consists of a sequence of one-dimensional domains.

In the multi-dimensional case one can distinguish one-dimensional domains in each coordinate. As the labels of the closest neighboring cells can differ in at most two bits in case of one dimension, they can differ in four bits in two dimensions and in six bits in three dimensions.

On a p-dimensional hypercube, each processor contains at most $2(\sqrt{m}+2)$ cells for a two-dimensional (box) refinement of $\sqrt{m}2^{p/2} \times \sqrt{m}2^{p/2}$ cells, and at most $3(m^{1/3}+2)$ cells for a three-dimensional (cubical) refinement of $m^{1/3}2^{p/3} \times m^{1/3}2^{p/3} \times m^{1/3}2^{p/3}$ cells. Thus, the workload per processor is linear in the number of dimensions and thus improves with increasing number of dimensions.

References

[1] Gilbert, E.N., *Gray Codes and Paths on the N-Cube*, The Bell System Technical Journal, (May 1958).

[2] Li, S.-X. and Loew, M.H., *The Quadcode and Its Arithmetic*, CACM, 30 (1987), pp. 621–6.

[3] Saad, Y. and Schultz, M.H., *Some Topological Properties of The Hypercube Multiprocessor*, Research Report 389, Dept Computer Science, Yale University, 1984.

[4] Wu, A.Y., *Embedding of Tree Networks into Hypercubes*, Jour. Par. Distr. Comp., 2 (1985), pp. 238–49.

CHAPTER 35

Automated Decomposition of Finite Element Meshes for Hypercube Computers

James G. Malone*

Abstract. A concurrent finite element formulation for linear and nonlinear transient analysis using an explicit time integration scheme has been developed for execution on hypercube multiprocessor computers. The formulation includes a new decomposition algorithm which automatically divides an arbitrary finite element mesh into regions and assigns each region to a processor on the hypercube. The decomposition algorithm is deterministic in nature and relies on a scheme which reduces the bandwidth of the matrix representation of the connectivities in the mesh. Numerical results obtained from a 32 processor Intel hypercube (the iPSC/d5 machine) are presented: speedup factors of greater than 31 have been obtained.

1. Introduction. In the finite element method, the spatial domain is discretized into a mesh of elements. An obvious way to take advantage of the concurrency in the problem is to decompose the mesh into a number of regions which are mapped to the processors of the hypercube. For efficient use of the hypercube the decomposition should result in a balanced computational load across the processors and interprocessor communication should also be minimized. This decomposition problem has been studied by a number of researchers, see Refs. [3], [4] and [6]. In this paper, an algorithm is presented which provides a completely automated way of obtaining almost optimal decompositions of arbitrary finite element meshes. A general concurrent finite element formulation for transient analysis using an explicit time integration scheme has been developed [5] in conjunction with the decomposition algorithm.

2. Concurrent Finite Element Formulation. In the following a knowledge of the finite element method [1] is assumed, a more detailed discussion of the concurrent formulation is given in [5]. One way to carry out a concurrent execution of the finite element method using p processors is to decompose the spatial domain into p regions and give each processor responsibility for one of those regions. The need for exchange of data among processors arises in the computation of nodal forces at a node located on a boundary segment

---

*Engineering Mechanics Dept., General Motors Research Laboratories, Warren, Michigan 48090.

common to two or more processors. Each processor can only sum contributions from those elements which are located within its own region. These partial sums are then transmitted to the processor having responsibility for that node and added to that processor's own partial sum to obtain the final value.

The unknown displacements $u^{t+\Delta t}$ at time $t + \Delta t$ are obtained by solving

$$(1/\Delta t^2) M u^{t+\Delta t} = R^t \tag{1}$$

where M is the mass matrix and R^t is the effective load vector at time t. A lumped mass formulation is used and therefore M is a diagonal matrix. At the solution step each processor proceeds to solve for the displacements $u^{t+\Delta t}$ at nodes which have been assigned to it. After the solution has been obtained, each processor transmits newly computed displacements at nodes located along boundary segments which are common to other regions to the processors which have responsibility for those regions.

3. Mesh Decomposition Algorithm. An arbitrary finite element mesh is fully defined by specifying the element types, coordinates of the nodes, and a list of the nodes associated with each element. Suppose there are n nodes in the mesh. Two nodes i and j are said to be connected if at least one non-zero entry will appear in the global stiffness matrix at the locations defined by the intersection of a row and column corresponding to displacement components at nodes i and j.

The connections in the mesh can be represented by an $n \times n$ symmetric matrix C defined by

$$C_{ij} \equiv 1 \tag{2}$$

if nodes i and j are connected, otherwise

$$C_{ij} \equiv 0 \tag{3}$$

For later use, we define here the half-bandwidth m of the matrix C by

$$m = \max_{C_{ij} \neq o} \{| i - j |\} \tag{4}$$

For simplicity in the exposition we assume that the number of processors p exactly divides the number of nodes n (modifications to the following are easily made when this is not the case). Let nodes

$$(i-1)n/p + 1, \ (i-1)n/p + 2, \ldots, in/p$$

be assigned to processor P_i for $i = 1, 2, ...p$. In the concurrent finite element analysis each processor P_k will have to exchange data with those processors P_l which possess an assigned node j which is connected to a node i (i. e. $C_{ij} \neq 0$) which has been assigned to processor P_k. Then, by definition of the half-bandwidth m (see Eq. (4)), processor P_k may need to communicate with up to $INT\,[m/(n/p)] + 1$ other processors P_l for $l > k$ and likewise with up to $INT\,[m/(n/p)] + 1$ processors P_l for $l < k$.

Therefore

$$Q = 2\{INT\,[m/(n/p)] + 1\} \tag{5}$$

is an upper bound on the number of processors with which P_k must exchange data. Interprocessor communication will be reduced by decreasing the upper bound Q, i.e., by ordering the nodes so that the bandwidth of the C matrix (and therefore m) is minimized.

The basic strategy behind the mesh decomposition algorithm is to renumber [2] the nodes so that the bandwidth of the C matrix is minimized. The elements are then assigned to the processors in a way that is consistent with this nodal numbering so that interprocessor communication will be close to the minimum possible. The method is as follows:

1. The nodes are reordered so that the bandwidth of the C matrix is minimized.
2. The elements are reordered [7] in ascending sequence of their lowest numbered nodes.
3. Suppose there are N elements of the same type (the case of a mesh with mixed element types requires some modification) and p processors. The elements are assigned in batches of approximately N/p elements to each processor in the order determined by the renumbering, i.e., elements numbered $(i-1)N/p+1, (i-1)N/p+2, \ldots, iN/p$ are assigned to the ith processor P_i .
4. Nodes located in the interior of a region are assigned to the processor having responsibility for that region. Boundary nodes common to two or more regions are assigned to that processor P_j whose label j is closest to the mean of the labels of the processors responsible for these regions.

4. Numerical Results and Discussion. The first set of numerical results is concerned with evaluating the performance of the concurrent finite element formulation by considering a problem in which the optimal decomposition of the mesh is easily obtained. The second set of numerical results will illustrate the effectiveness of the reduced bandwidth decomposition (RBD) algorithm for a problem with a non-uniform mesh over an irregular geometry.

The concurrent finite element formulation has been applied in an elastic plane stress transient analysis of a rectangular plate discretized by a mesh of triangular elements. For this problem, the decomposition results in communication between 'nearest neighbour' processors only. Execution times for meshes with 128, 256, 512 and 1,024 elements have been measured. Speedup versus number of processors in use is plotted for each mesh in Fig. 1. It can be seen that the speedup factor approaches the ideal value as the size of the mesh increases (i.e. as the computational load per processor increases).

The performance of the mesh decomposition algorithm is illustrated by application to a nonuniform mesh containing 997 elements over an irregular geometry, the triangular bracket shown in Fig 2. An elastic plane stress transient analysis has been carried out using three different decompositions of the mesh:

1. Assignment of the elements in a random manner to each processor.
2. Assignment of the elements to the processors in the order in which they are generated by the mesh generating program. This provides some degree (depending on the particular mesh generator) of assignment of sets of *contiguous* elements to a single processor.
3. Assignment of elements to processors by the reduced bandwidth decomposition (RBD) algorithm discussed in this paper.

For each of these decompositions approximately equal numbers of elements are assigned to each processor. A plot of speedup versus the number of processors is shown in Fig. 3 for each of the three decompositions. The improvement in speedup brought about by the reduced bandwidth decomposition algorithm is evident. A speedup factor of greater than 31 when executing on 32 processors has been obtained for a mesh with about 4,000 elements.

The RBD algorithm can be applied to 2D and 3D meshes, non-uniform meshes, and meshes over irregular geometries. Meshes containing different types of elements can also be decomposed [5]. A general concurrent finite element formulation for transient analysis has been developed [5] in conjunction with the decomposition algorithm.

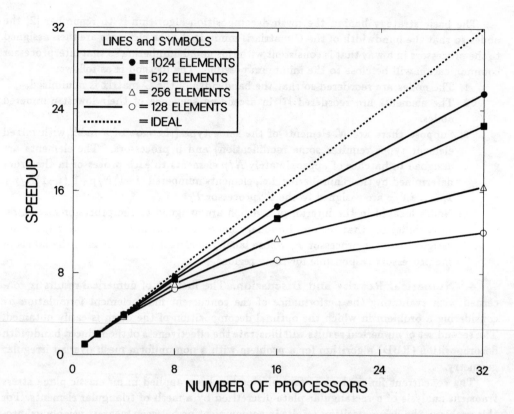

Fig. 1. Speedup factors for range of meshes.

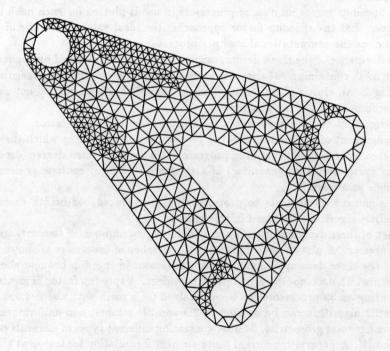

Fig. 2. Mesh for triangular bracket (997 elements, 586 nodes).

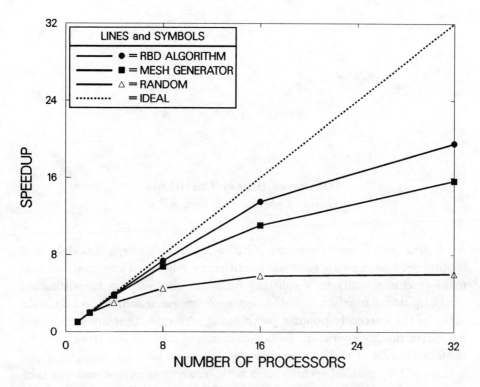

Fig. 3. Comparison of speedup for different decompositions.

However, other issues remain to be considered. During execution, local mesh refinement and changes in material properties such as those caused by yielding and unloading may result in unbalanced computational loads among the processors. A dynamić load balancing algorithm is required for these problems. Implicit time integration schemes give rise to banded systems of linear equations. The RBD algorithm can be used in conjunction with iterative equation solvers but direct equation solvers may require a different mesh decomposition strategy. Contact problems pose special dificulties in establishing *a priori* communication requirements. These issues will have to be considered if parallel processing is to be successfully applied to the full range of problems analyzed by the finite element method.

REFERENCES

[1] K. J. Bathe, *Finite Element Procedures in Engineering Analysis*, Prentice – Hall Inc, New Jersey, 1982.

[2] R. J. Collins, *Bandwidth reduction by automatic renumbering*, Intl. J. Num. Meth. Engrg., 6, 345-356, 1973.

[3] J. W. Flower, S. W. Otto, and M. C. Salama, *A preprocessor for irregular finite element problems*, CalTech/JPL Report C3P-292, July, 1986.

[4] G. A. Lyzenga, A. Raefsky, and B. H. Hager, *Finite elements and the method of conjugate gradients on a concurrent processor*, CalTech/JPL Report C3P-119, Dec. 1984.

[5] J. G. Malone, *Automated mesh decomposition and concurrent finite element analysis for hypercube multiprocessor computers*, Comp. Methods Appl. Mech. Eng., (to appear).

[6] R. Morison and S. Otto, *The scattered decomposition for finite elements*, CalTech/JPL Report C3P-285, May 1986, (to appear Parallel Computing).

[7] S. W. Sloan and M. F. Randolph, *Automatic element reordering for finite element analysis with frontal solution schemes*, Intl. J. Num. Meth. Engrg., 19, pp. 1153-1181, 1982.

Recursive Binary Partitions
George Cybenko and Tom Allen

A Recursive Binary Partition (RBP) is a partition of space or data that is obtained from a single recursive partitioning scheme. The generic scheme makes recursive calls to a splitting function. By varying the definition of the splitting function and the depth of the recursion, one can obtain many of the currently popular partitioning strategies that are being used for vortex methods, domain decomposition and multiobject tracking. The recursive nature of the scheme allows all RBP's to be easily computed and maintained on parallel architectures with binary recursive structures such as hypercubes, the Connection Machine, RP3 and the Butterfly.

CHAPTER 36

A Parallel Version of the Fast Multipole Method

Leslie Greengard*
William D. Gropp†

Abstract. This paper presents a parallel version of the Fast Multipole Method (FMM). The FMM is a recently developed scheme for the evaluation of the potential and force fields in systems of particles whose interactions are Coulombic or gravitational in nature. The sequential method requires $O(N)$ operations to obtain the fields due to N charges at N points, rather than the $O(N^2)$ operations required by the direct calculation. Here, we describe the modifications necessary for implementation of the method on parallel architectures and show that the expected time requirements grow as $\log N$ when using N processors. Numerical results are given for a shared memory machine (the Encore Multimax 320).

1. Introduction

Numerical methods for computing N-body interactions generally fall into two categories. Continuum methods are based on the fact that the potential satisfies the Poisson equation and use fast Poisson solvers to obtain the field [5]. They are hindered by the limited resolution of the imposed grid and the degradation of performance seen with highly non-homogeneous distributions of particles. Hierarchical methods [1, 2] are based on the fact that the field due a cluster of particles can be represented at a great distance by the net mass acting at the center of mass. Tree structures are used to partition space and group particles at various length scales, so that the center of mass approximation can be applied. The CPU time requirements of these methods generally grow as $N \log N$. They handle non-homogeneous distributions better than the continuum methods, but yield only approximate results.

The Fast Multipole Method (FMM) [4, 3, 7, 6] shares certain characteristics with the hierarchical solvers. Tree structures are imposed to partition space, and the strategy is

*Department of Computer Science, Yale University, New Haven, CT 06520. The work of this author was supported in part by the Office of Naval Research under contract number N00014-86-K-0310 and by a National Science Foundation Postdoctoral Fellowship.

† Department of Computer Science, Yale University, New Haven, CT 06520. The work of this author was supported in part by the Office of Naval Research under contract N00014-86-K-0310 and the National Science Foundation under contract number DCR 8521451.

similar, but analytic observations concerning multipole and Taylor expansions are used to produce results that are accurate to within round-off error. The CPU time requirements are of the order $O(N \log(1/\epsilon))$, where ϵ is the desired accuracy.

In this paper, we will describe a parallel version of the non-adaptive two-dimensional FMM and present numerical results for an implementation on a shared memory machine (the Encore Multimax 320). We note that Zhao [8] has independently developed a parallel implementation of a non-adaptive three-dimensional multipole method for the Connection Machine.

2. Mathematical Preliminaries

In this paper, we will consider as a model the N-body problem in the complex plane $\mathbb{C}$. That is, given the positions z_i and strengths q_i of N charged particles, we wish to compute the net potential ϕ and electric field $\mathbf{E}$ at each particle position from Coulomb's law. These are given by the expressions

$$\phi(z_i) = Re\left(\sum_{j \neq i} q_i \cdot \log(z_i - z_j)\right)$$

and

$$E(z_i) = (-Re(\phi'(z_i)), Im(\phi'(z_i))) \ ,$$

respectively.

Suppose now that m charges with strengths q_i and positions z_i are located within a disk of radius r centered at the origin (Fig. 1). Then, it is shown in [4], that for a point z with $|z| > r$, the potential $\phi(z)$ induced by the charges is given by a multipole expansion of the form

$$\phi(z) = Q \log(z) + \sum_{k=1}^{\infty} \frac{a_k}{z^k} \ , \tag{2.1}$$

where

$$Q = \sum_{i=1}^{m} q_i \qquad and \qquad a_k = \sum_{i=1}^{m} \frac{-q_i \cdot z_i}{k} \ .$$

The error in truncating the sum after s terms is

$$\left|\phi(z) - Q \log(z) - \sum_{k=1}^{s} \frac{a_k}{z^k}\right| \leq \left(\frac{A}{c-1}\right)\left(\frac{1}{c}\right)^s \ , \tag{2.2}$$

where

$$A = \sum_{i=1}^{m} |q_i| \qquad and \qquad c = \left|\frac{z}{r}\right| \ . \tag{2.3}$$

In order to obtain a relative precision of ϵ (with respect to the total charge), the number of terms required in the series representation of ϕ is approximately $-\log_c(\epsilon)$, independent of m, the number of source charges. The Fast Multipole Method is based on making explicit use of this result.

2.1. Translation operators

In the FMM scheme, it is necessary not only to form multipole expansions as in (2.1), but to carry out a sequence of analytic transformations of the expansion coefficients. These transformations are described in the next three lemmas. Details proof can be found in

[4]. The first, Lemma 2.1, provides a mechanism for shifting the center of a multipole expansion.

Lemma 2.1. (Translation of a Multipole Expansion) *Suppose that*

$$\phi(z) = a_0 \log(z - z_0) + \sum_{k=1}^{\infty} \frac{a_k}{(z - z_0)^k}$$

is a multipole expansion of the potential due to a set of m charges of strengths $q_1, q_2, \ldots, q_m$, all of which are located inside the circle D of radius R with center at z_0. Then for z outside the circle D_1 of radius $(R + |z_0|)$ and center at the origin,

$$\phi(z) = a_0 \log(z) + \sum_{l=1}^{\infty} \frac{b_l}{z^l}, \tag{2.4}$$

where

$$b_l = -\frac{a_0 z_0^l}{l} + \sum_{k=1}^{l} a_k z_0^{l-k} \binom{l - 1}{k - 1}, \tag{2.5}$$

with $\binom{l}{k}$ the binomial coefficients. Furthermore, for any $s \geq 1$,

$$\left| \phi(z) - a_0 \log(z) - \sum_{l=1}^{s} \frac{b_l}{z^l} \right| \leq \left(\frac{A}{c - 1} \right) \left(\frac{1}{c} \right)^s, \tag{2.6}$$

where A is defined in (2.3) and

$$c = \left| \frac{z}{|z_0| + R} \right|.$$

Lemma 2.2 describes the conversion of a multipole expansion into a local (Taylor) expansion inside a circular region of analyticity.

Lemma 2.2. (Conversion of a Multipole Expansion into a Local Expansion) *Suppose that m charges of strengths $q_1, q_2, \ldots, q_m$ are located inside the circle D_1 with radius R and center at z_0, and that $|z_0| > (c + 1)R$ with $c > 1$. Then the corresponding multipole expansion 2.1 converges inside the circle D_2 of radius R centered about the origin. Inside D_2, the potential due to the charges is described by a power series:*

$$\phi(z) = \sum_{l=0}^{\infty} b_l \cdot z^l, \tag{2.7}$$

where

$$b_0 = a_0 \log(-z_0) + \sum_{k=1}^{\infty} \frac{a_k}{z_0^k} (-1)^k, \tag{2.8}$$

and

$$b_l = -\frac{a_0}{l \cdot z_0^l} + \frac{1}{z_0^l} \sum_{k=1}^{\infty} \frac{a_k}{z_0^k} \binom{l + k - 1}{k - 1} (-1)^k, \qquad \text{for } l \geq 1. \tag{2.9}$$

Furthermore, for any $s \geq max \left(2, \frac{2c}{c-1} \right)$, an error bound for the truncated series is given by

$$\left| \phi(z) - \sum_{l=0}^{s} b_l \cdot z^l \right| < \frac{A(4e(s + c)(c + 1) + c^2)}{c(c - 1)} \left(\frac{1}{c} \right)^{s+1}, \tag{2.10}$$

where A is defined in (2.3) and e is the base of natural logarithms.

Lemma 2.3 provides a formula for shifting the center of a local expansion within a region of analyticity. This translation is exact, and no error bound is needed.

Lemma 2.3.   (Translation of a Local Expansion) *For any complex z_0, z and $\{a_k\}$, $k = 0, 1, 2, \ldots, n$,*

$$\sum_{k=0}^{n} a_k(z - z_0)^k = \sum_{l=0}^{n} b_l \cdot z^l \tag{2.11}$$

where

$$b_l = \sum_{k=l}^{n} a_k \binom{k}{l}(-z_0)^{k-l} \tag{2.12}$$

3. Informal Description of the FMM

In this section, we briefly outline the sequential FMM procedure. A more detailed discussion is available in [4, 6]. The algorithm uses a divide and conquer strategy to cluster particles at various levels of spatial discretization, and then uses multipole and Taylor expansions to evaluate the interactions between distant clusters. Once all distant interactions are accounted for by this expansion technique, the interactions between neighboring particles are computed by the direct application of the pairwise force law.

We now introduce the notation necessary for a description of the algorithm. Since we are considering the non-adaptive scheme, we assume that N charges are more or less homogeneously distributed within a square with sides of length one, and refer to this square as the computational box. We impose a hierarchy of meshes on the computational box which refine it into smaller and smaller regions. More specifically, mesh level 0 refers to the entire computational box, while mesh level $l + 1$ is obtained recursively from level l by subdividing each box into four equal parts. A tree structure is imposed on this hierarchy, so that if *ibox* is a box at level l, then the four boxes at level $l + 1$ obtained by its subdivision are considered its children. In general, the maximum number of refinements (the tree depth) is chosen to be on the order of $\log_4 N$, at which point there is on the order of 1 particle in each box at the finest level. For every box i at level l, we define the *nearest neighbors* to be the box itself and any other box at the same level with which it shares a boundary point. There are clearly at most 9 nearest neighbors.

Two boxes (at a given level) with sides of length D, are said to be *well-separated* if they are separated by a distance D. It is shown in [6] that, in using s-term expansions to account for the interactions between *well-separated* boxes, the error bounds (2.2),(2.6) and (2.10) apply with $c = (4 - \sqrt{2})/\sqrt{2} \approx 1.8$. For a given precision ϵ, we therefore choose $s = \lceil -\log_c(\epsilon) \rceil$.

Both multipole and local expansions are associated with each box. $\Phi_{l,i}$ is the s-term multipole expansion about the center of box i at level l which describes the far field potential due to the particles contained inside the box. $\Psi_{l,i}$ is the s-term local expansion about the center of box i at level l which describes the potential field due to all particles outside the box and its nearest neighbors. $\tilde{\Psi}_{l,i}$ is the s-term local expansion about the center of box i at level l which describes the potential field due to all particles outside i's *parent* box and the *parent* box's nearest neighbors. Finally, an *interaction list* is associated with each box i at level l. This is the set of boxes which are children of the nearest neighbors of i's parent and which are well-separated from box i.

The algorithm computes interactions between groups of particles at the coarsest possible mesh level. Two passes are executed.

N	T_{alg} $p=1$	T_{dir} $p=1$	T_{alg} $p=16$	T_{dir} $p=16$	Speedup Alg	Speedup Dir
625	14	54	1.2	3.45	11.7	15.7
1250	52	216	3.6	13.9	14.4	15.5
2500	68	872	4.6	54.9	14.7	15.9
5000	235	3490	15.5	220.8	15.2	15.9
10000	301	14020	19.7	910.4	15.3	15.4
20000	1008	56385	65.0	3560.4	15.5	15.8

Table 1: Table of times for algorithm (alg) and direct method (dir) on Encore Multimax 320. All times in seconds.

Initialization Choose a level of refinement $n \approx \lceil \log_4 N \rceil$, a precision ϵ, and set $s = \lceil -\log_c(\epsilon) \rceil$. Assign particles to boxes at finest mesh level.

Upward Pass

Step 1: Form multipole expansion $\Phi_{n,i}$ about the box center for each box i at finest mesh level. Uses equation (2.1).

Step 2: Recursively form multipole expansions about the centers of all boxes at all coarser mesh levels, each expansion representing the potential field due to all particles contained in the box. Uses Lemma 2.1.

In the downward pass, the local expansions $\Psi_{l,i}$ are formed for all boxes, beginning at the coarsest level. This process is somewhat more complex. Suppose, however, that at level $l - 1$, the local expansion $\Psi_{l-1,i}$ has been computed. Then Lemma 2.3 can be used to shift the expansion to each of the box's children. For each child box j at level l, what we have obtained is a local representation of the field due to all particles outside the parent's nearest neighbors, namely $\tilde{\Psi}_{l,i}$. The interaction list defined above is precisely the set of boxes whose contribution to the potential must be added to $\tilde{\Psi}_{l,i}$ to create $\Psi_{l,i}$. The initialization of this pass is simple. Since there are no well-separated boxes at level 0 or 1, we may set $\Psi_{0,i}$, $\tilde{\Psi}_{1,i}$, $\Psi_{1,i}$ and $\tilde{\Psi}_{2,i}$ to zero.

Downward Pass

Steps 3,4: Begin at level 2, and proceed to finer levels as follows: form $\Psi_{l,ibox}$ by using Lemma 2.2 to convert the multipole expansion $\Phi_{l,j}$ of each box j in the *interaction list* of box *ibox* to a local expansion about the center of box *ibox*, adding these local expansions together, and adding the result to $\tilde{\Psi}_{l,ibox}$. If finest level has been reached, process is complete. Otherwise form the expansion $\tilde{\Psi}_{l+1,j}$ for *ibox*'s children by using Lemma 2.3 to expand $\Psi_{l,ibox}$ about the children's box centers and continue procedure.

Step 5: Evaluate local expansions at particle positions to obtain the far-field potential and/or force.

Step 6: Compute potential (or force) due to particles in nearest neighbors boxes directly.

Step 7: For every particle, add direct and far-field terms together.

4. Description of the parallel algorithm

The algorithm described in the previous section has several opportunities for parallelism. The best of these opportunities are the completely parallel operations such as the computation of the initial moments and evaluation of the local expansions for each particle. The other parallel operations require some coordination between processors. Evaluation of the forces between particles in neighboring boxes (step 6) requires secure data access if Newton's third law is used; other than that, these operations are also perfectly parallel.

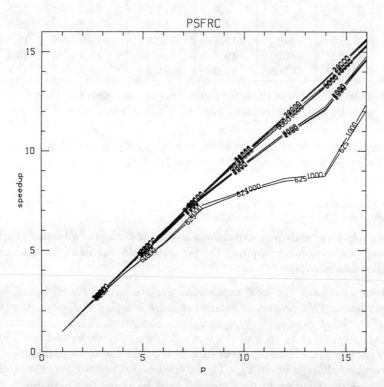

Figure 1: Speedup for the calculation of the far-field by the fast multipole method.

The *reductions* involved in the upward and downward passes can be done in parallel with some load imbalance.

The fact that the entire program is not completely parallel opens the question of how efficient the algorithm can be, particularly on a large number of processors. We will address this by estimating the computational complexity of the parallel algorithm.

We do not discuss the initialization of the algorithm since the initialization is essentially a parallel sort and is performed only at the beginning of a computation. Since a common use of the Fast Multipole Method is to compute the forces at each time step in a dynamical simulation, this initial sort may be amortized over all the time steps. Further, in a time dependent calculation, it is possible to exploit slowly varying changes in the potential to reduce the amount of computation; we will not consider this effect either.

For the actual implementation, we can consider each of the steps in the algorithm separately. We use the term "communication" to denote any coordination between processors. In a message passing system, this would be a message; in a shared memory system, this would be some critical section (e.g., a spinlock). We use N to denote the number of particles, n the number of levels, and p the number of processors.

Upward Pass

Step 1: Formation of expansions at finest level. There is no communication; the complexity is $4^n/p$.

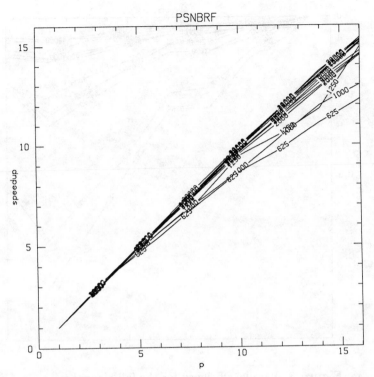

Figure 2: Speedup for the near-field by direct calculation.

Step 2: Merge upward. Communication is within box and with parent box. The complexity is

$$\sum_{i=0}^{n-1}\left\lceil\frac{4^i}{p}\right\rceil = \sum_{i=\log_4 p}^{n-1}\left\lceil\frac{4^i}{p}\right\rceil + \sum_{i=0}^{\log_4 p-1} 1$$

times the cost per box. This is one of the most important terms because it gives a limit on the available parallelism. The second sum shows the bottleneck in the reduction: when there are fewer than p boxes, some processors go idle. This bottleneck is an essential part of the algorithm. The ceilings in the first term are important for small n but become unimportant for $4^n \gg p$.

Downward pass

Steps 3,4: Convert the multipole expansions into local expansions and move down. The complexity is of the same form as above (but with a different constant).

Step 5: Evaluation of local expansions at the finest level. This is perfectly parallel and has complexity N/p.

Step 6: Compute the potential or force due to particles in neighboring boxes directly. This involves the neighboring boxes through direct interaction. There is no "reduction" overhead; however, there is some communication due to updating the field at a particle from several adjacent processors/particles. The complexity is roughly N/p. If Newton's law is not used, this is completely parallel (i.e., we can eliminate any possibility of memory contention at the cost of twice as much arithmetic).

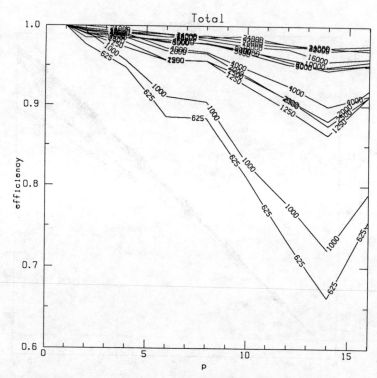

Figure 3: Parallel efficiency for the full algorithm.

Step 7: Add the components (direct and far-field) together. This is perfectly parallel; the complexity is N/p.

Summing the contributions from each step, the overall complexity is

$$\frac{aN}{p} + b\log_4 p + c(N,p)$$

where a and b are constants determined by the floating point speed and the requested precision, and c is a lower order term which includes things like the communication or synchronization overhead.

Note that with $\mathcal{O}(N)$ processors, the overall complexity is $\mathcal{O}(\log N)$. The parallel efficiency can be estimated as the ratio of the time on a single processors (without the overhead) to p times the time on p processors:

$$\text{efficiency} = \frac{aN}{p\left(\frac{aN}{p} + b\log_4 p + c(N,p)\right)}$$

$$= \frac{1}{1 + \frac{pb}{aN}\log_4 p + \frac{pc}{aN}}$$

$$\approx 1 - \frac{pb}{aN}\log_4 p - \frac{pc}{aN}.$$

4.1. Comments on the parallel implementation

Our parallel implementation is based on a version of the serial code described in [4]. The implementation is a "minimum distance" change, and does not attempt to rearrange the computation to be more parallelizable. In particular, it is possible to identify subclasses of boxes with which completely parallel operations may be performed; within these classes it can be proven that no data-access conflicts can occur. This can reduce the overhead in steps 2, 3, and 4 by reducing the number of memory locks (in a shared memory implementation) or the number of messages (in a message passing implementation).

5. Experimental Results

These results are for the non-adaptive algorithm, and do not include the work of the initial sorting of the particles. The results in Table 1 are for an Encore Multimax 320 with 18 processors. There are two important points to remember in interpreting these times. One is that they were taken on a time sharing system; even though no other users were present, various daemons will consume some resources. To reduce this effect, we used only 16 of the 18 processors in our experiments. The second is the effect of the choice of number of levels. While the complexity estimates predict time linear in the number of particles, in fact the actual times display a "ratchet" behavior as the number of levels increase. However, over a large enough range of number of particles, the behavior is linear.

Figures 1–3 show a breakdown of the results for the Encore Multimax 320. Figure 1 shows the speedup for the calculation of the far-field by the fast multipole method. Note that the results are clustered into four groups; these represent the number of levels (4–7). The speedup is lower when the number of processors is not a power of four; this a result of the poor load-balancing in the reduction stages (steps 2–4). Figure 2 shows the speedup for the calculation of the near field by direct calculation. The deviation from perfect speedup is due mainly to the overhead connected with secure access to data (critical sections). Figure 3 shows the overall efficiency. Note that for even small numbers of particles 75% efficiency is achieved and for 5000 or more particles, 95% efficiency is achieved.

A version of the three-dimensional multipole method has been implemented and studied by Zhao [8]. His results show the predicted $\log N$ growth as N ran from 64 to 16384. His timings, done on the Connection Machine, are somewhat slow. Different formulations of the algorithm presented here, in particular with respect to constant terms or terms in $-\log \epsilon$, should significantly reduce the timings.

6. Conclusions

Our results have shown that the fast multipole algorithm is very suitable for shared memory parallel computers. Both our experience and the results of Zhao indicated that it is suitable for message passing parallel computers as well. The overall complexity is (with $p = N$ processors) $\log N$; for fixed N it is $N/p + \log p$.

The non-adaptive algorithm described here has very regular memory access or communication patterns which can be exploited to reduce the parallel overhead. Many of these are intrinsic to the fast multipole method itself, and should be exploitable by the adaptive version.

References

[1] A. W. Appel, *An Efficient Program for Many-body Simulation*, Siam. J. Sci. Stat. Comput., 6(1985), pp. 85–103.

[2] J. Barnes and P. Hut, *A Hierarchical O(N log N) Force-Calculation Algorithm*, Nature, 324(1986), pp. 446–449.

[3] J. Carrier, L. Greengard, and V. Rokhlin, *A Fast Adaptive Multipole Algorithm for Particle Simulations*, Technical Report 496, Yale Computer Science Department, 1986.

[4] L. Greengard and V. Rokhlin, *A Fast Algorithm for Particle Simulations*, J. Comput. Phys., 73(1987), pp. 325–348.

[5] R. W. Hockney and J. W. Eastwood, *Computer Simulation Using Particles*, McGraw-Hill, New York, 1981.

[6] L. Greengard, *The Rapid Evaluation of Potential Fields in Particle Systems*, MIT Press, Cambridge, 1988.

[7] L. Greengard and V. Rokhlin, *Rapid Evaluation of Potential Fields in Three Dimensions*, Technical Report 515, Yale Computer Science Department, 1987.

[8] F. Zhao, *An O(N) Algorithm for Three-dimensional N-body Simulations*, Technical Report 995, Massachusetts Institute of Technology, 1987.

CHAPTER 37

Programming Abstractions for Run-Time Partitioning of Scientific Continuum Calculations Running on Multiprocessors*

Scott B. Baden†

Abstract. I will discuss a set of software abstractions for implementing various math-physics calculations on a team of processors. I tried out the abstractions on Anderson's Method of Local Corrections, a type of vortex method for computational fluid dynamics. I ran experiments on 32 processors of the Intel iPSC – a message-passing hypercube architecture – and on 4 processors of a Cray X-MP – a shared-memory vector architecture – and achieved good parallel speedups of 24 and 3.6, respectively. The abstractions should apply to diverse applications, including finite difference methods, and to diverse architectures without requiring that the application be reprogrammed extensively for each new architecture.

1. Introduction

A major application for multiprocessors is in obtaining solutions to partial differential equations arising out of various areas of science and engineering. A major outstanding difficulty in using them is how to construct robust software that can run efficiently on diverse systems without having to go through major changes in programming. This is particularly troublesome for calculations that apply computational effort non-uniformly over space according to time-dependent phenomena, and which must therefore be dynamically partitioned. I will discuss a set of abstractions that can hide many of the details entailed in dynamically partitioning and coordinating a computation among a team of processors and that can improve the robustness of software with respect to those activities.

The abstractions apply to the important class of calculations that spend most of their time in *spatially localized* computation in which two data points communicate far more information with respect to the computation done on them when they are close together than when they are far apart. Consider, for example, the particle-particle particle-mesh solution to the N-body problem. Such a calculation arises in problems in computational fluid dynamics, plasma physics, and particle physics; it entails following a set of particles that move under mutual interaction, congregating and dispersing unpredictably with time (see Figure 2). The particles move under the influence of a logarithmic potential, which computation divides into two parts: a local part that does roughly 90% of the computational work when the problem is large, and a relatively inexpensive global part whose data dependencies are not localized. The cost of computing the local part of the potential is a position- and time-dependent function of the local density of particles.

A simple way of applying a team of processors to spatially localized particle methods is to partition the domain into rectangular regions and assign the computation and data associated with each region to a

*This work was supported in part by the Applied Mathematical Sciences subprogram of the Office of Energy Research, U.S. Department of Energy, under contract DE-AC03-76SF00098; a California Fellowship in Microelectronics; Intel Scientific Computers; and Cray Research Inc.

†Mathematics Department, Lawrence Berkeley Laboratory, University of California, Berkeley, California 94720

processor. There are many ways to partition the domain, two of which are shown in Figure 1. A uniform partitioning, in which all partitions have the same area, is the most straightforward. Such a partitioning, however, utilizes only a small fraction of the total power of the processing team because work is distributed non-uniformly in space, as shown in Figure 1a. A better way is to partition adaptively into somewhat irregularly-sized regions that all complete in roughly the same time, as shown in Figure 1b. Such a

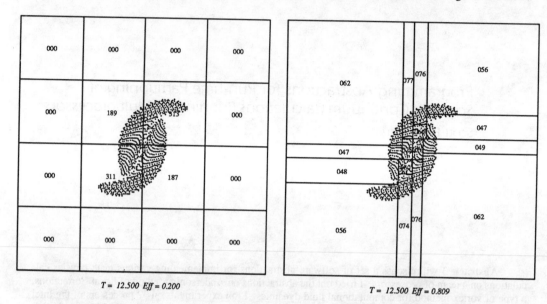

T = 12.500 *Eff* = 0.200   *T* = 12.500 *Eff* = 0.809

Figure 1. Partitioning of a particle calculation for vortex dynamics on 16 processors. A simple way to divide up the work is (a) to partition the domain uniformly into a regular pattern of box-like subproblems. This strategy, however, would underutilize the processors; only 4 of 16 would be given much work to do. The trouble is that the particles distribute themselves unevenly so that the completion time for a subproblem may not be proportional to its area. A better way (b) compensates for the uneven distribution of particles over the domain. This adaptive decomposition generates somewhat irregularly sized subproblems that all complete in roughly the same time, and it diminishes the running time of the computation by a factor of four. At the depicted time each processor's share of the workload is shown in the subdomain assigned to it, normalized to 1000 units of total work. A perfectly balanced workload would correspond to 062 units of work for each subproblem.

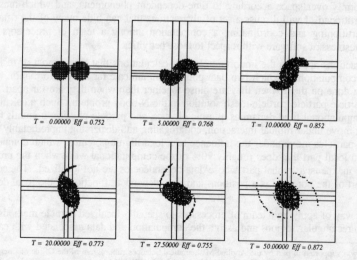

T = 0.00000 *Eff* = 0.752   *T* = 5.00000 *Eff* = 0.768   *T* = 10.00000 *Eff* = 0.852

T = 20.00000 *Eff* = 0.773   *T* = 27.50000 *Eff* = 0.755   *T* = 50.00000 *Eff* = 0.872

Figure 2. The distribution of particles changes with time, so the work must be periodically repartitioned. This series of snapshots was taken from the same calculation used to produce Figure 1.

strategy compensates for the uneven distribution of work, and can substantially accelerate the computation. Of course, the partitioning must be periodically recomputed, as shown in Figure 2, or otherwise the workloads would gradually drift out of balance as the particles redistribute themselves. Some processors would become overloaded, while others would sit idle waiting; the cost of the computation would steadily increase with time, as shown in Figure 3.

The decision to change the work assignments dynamically, rather than to assign work statically, can substantially complicate the user's software. The trouble is that the best way to handle the communication and the bookeeping that come as a side effect of shuffling work among the processors can depend on various overhead costs − such as memory latency or message startup time − that generally vary from system to system. Thus, the code required to effectively parallelize a calculation on a shared-memory multiprocessor differs substantially from that required to run on a message-passing architecture, and code can vary even among members of one family of architecture. To facilitate in the construction of robust software, I propose that the user program a generic multiprocessor whose partitioning and coordinating operations have the same semantics regardless of where implemented. I will discuss a particular generic multiprocessor called "genMP." GenMP can help desensitize substantial portions of the user's software from a change in various system parameters such as communication or memory latency, numbers of processors, processor interconnection structure, and the semantics of system library calls that handle various aspects of concurrency. GenMP isn't universal, however, and applies to localized computation only; the user will have to parallelize any non-localized computation himself − though separately from the local part − according to the particular architecture in use. The question of how best to parallelize non-localized computation is beyond the scope of this paper, and I consider only spatially localized computation.

2. The GenMP Abstractions

GenMP can be implemented by a layer of software on most any traditional multiprocessor system. It provides a set of run-time utilities that the user will invoke from his code. GenMP assumes a particular style of localized computation, a lattice model computation, in which the calculation maps onto a regular lattice of boxes, the work lattice, subdividing the domain. The computation updates the state of each box as a function of the previous state of only those bins within a given distance C, the local interaction distance. (In contrast, the data dependencies for the updates done in the global part of the computation are not constrained to be localized.) The cost of updating a bin generally depends on the state of the surrounding bins and can be reasonably estimated with an inexpensive auxiliary computation.

The lattice model of computation assumed by genMP is a reasonable one for a variety of computations in science and engineering, and for this reason I believe that it will prove useful for a diversity of applications such as:

- Finite difference calculations that use a fixed rectangular mesh trivially fit the model, as do methods such as Adaptive Mesh Refinement (AMR) [3] that employ dynamic grids.

- Localizable Particle methods such as Particle-Particle Particle-Mesh (PPPM) [11], and Rokhlin and Greengard's fast multipole method [10] are naturally organized around a lattice, and they spend the majority of their time computing direct interactions between nearby particles or doing other localized computation.

- Finite element methods may also be mapped onto a lattice [14], and divide naturally into localized and non-localized computation.

- Ray tracing for computer graphics may also be organized around a lattice, and has a localized communication structure (see Swensen and Dippé [9]).

In the interest of brevity I will consider the abstract problem of how to parallelize a lattice model computation using genMP. The interested reader should consult Baden [2] for the details regarding a

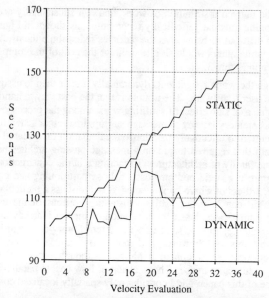

Figure 3. A comparison of static and dynamic load balancing on 32 processors of the Intel iPSC. If the workloads are partitioned only at the beginning of the calculation (static load balancing), the loads will drift gradually out of balance, and the time required to perform a velocity evaluation will steadily increase with time. In contrast, a dynamic load balancing strategy periodically rebalances the workloads and is able to maintain an almost steady running time. Here loads were rebalanced every fourth velocity evaluation.

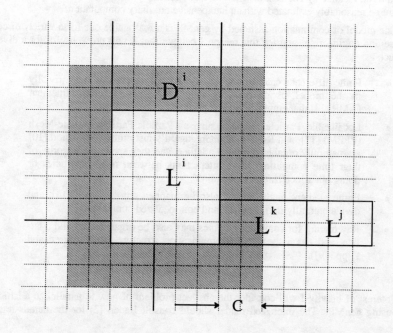

Figure 4. Task i is assigned L^i, a subregion of the work lattice, and an external interaction region D^i. D^i is a surrounding shell of bins and is C bins thick, where we have chosen $C = 2$. Since D^i and L^j do not intersect, subproblems L^i and L^j are locally independent. But D^i and L^k *do* intersect, and so subproblems L^i and L^k are locally dependent.

specific application. To parallelize a lattice model computation, we subdivide the work lattice into subregions, assign each such subproblem to a unique processor, and let each processor compute on its assigned subproblem in parallel with the others. Ignoring roundoff, results will be independent of the number of processors used. The computation begins with a distinguished task called the "boss." The boss reads in the input data and spawns P additional "worker tasks," where P is chosen by the user. These worker tasks participate in the numerical part of the computation. All execute the same program but each on a different set of data – a single subregion of the work lattice. Each worker executes out of a private address space and communicates with the others through a mechanism to be discussed. There is no shared memory.

Each worker maintains a private copy of its assigned part of the work lattice. It also maintains a copy of a surrounding collection of bins, called an "external interaction region," which augments the task's assigned sublattice (see Figure 4). The task uses this external interaction region to maintain copies of data belonging to other tasks that directly interact with its own; hence, the region is C bins thick, where C is local interaction distance previously discussed. As a consequence of using this distributed storage strategy, no task may access any bins beyond its external interaction region. Furthermore, a task may only indirectly access bins, owned by other tasks, that overlap its external interaction region. For example, when a task modifies a copy of a bin in the external interaction region, then the owner of the "original" won't know that the change was made. Similarly, when a task modifies an original any tasks that possess copies will be unaware of the changes. These changes must eventually be propagated, however, to ensure correctness; at certain points in the calculation each task must suspend computation and communicate with the other tasks in such a way that all bin-copies be consistent with the originals. To this end, all tasks periodically invoke a run-time utility called lBar. When a task encounters a call to lBar it communicates with all tasks overlapping its external interaction region and returns when it has finished communicating with all of them. This set of interacting tasks acts as a local barrier synchronization mechanism. Each task will generally encounter and leave the local barrier at a different time, according to the amount of work assigned to it. We refer to the barrier as being a local one because generally it involves only a local subset of tasks, rather than the entire set of tasks as in traditional (global) barrier synchronization; the name lBar stands for "Local BARrier synchronization." LBar is passed it two subroutines as arguments. These perform gather and scatter operations on the local user data structures. For details see [2].

To ensure they share the work evenly, the workers must periodically invoke a run-time utility called Partitioner. Partitioner readjusts each worker's assignment of bins according to a time-varying "work density mapping," supplied by the user. This mapping comes in the form of an array; each entry estimates the cost of updating one bin of the work lattice. All tasks leave Partitioner together and upon return each will be assigned a unique rectangular region of the work lattice. A task determines the set of indices for the bins assigned to it with the aid of querying functions provided as run-time utilities. As a result of calling Partitioner, some bins may change owners, and must therefore be transmitted to the correct task. A call to the lBar utility can handle the necessary exchanges of data. I chose to implement Partitioner with a recursive bisection algorithm similar to that used by Berger and Bokhari [4]. The user, however, is unaware of how Partitioner works, and any strategy that was fast and that rendered partitionings with a low surface area to volume ratio would suffice.

3. Computational Results

I evaluate genMP on the Intel Personal Scientific Computer (iPSC), manufactured by Intel Scientific Computers, and on the Cray X-MP, manufactured by Cray Research Inc.. I will show that the performance of either of these systems running genMP can scale reasonably well with the number of processors in use. A detailed description of the iPSC and the Cray X-MP is beyond the scope of this paper; see Baden for a summary of the relevant details, or the manufacturer's manuals [8, 12]. (The pamphlet by S. Chen et al. [5] on the Cray X-MP is a more accessible document than the manufacturer's Hardware Reference Manual.) Table 1 summarizes the relevant characteristics of the two machines.

	Cray X-MP	Intel iPSC
Communication Model	Shared Memory	Message-Passing
# Processors used [Max]	4 [4]	32 [128]
Megaflops/sec/cpu	100	0.035
Max Memory (megabytes)	128 total	0.5/cpu

Table 1. Design parameters for the Intel iPSC and the Cray X-MP. We used the iPSC model d5, with 32 processors, and the largest-model Cray, the X-MP/416, with 4 processors 16 megawords of main memory. The megaflop execution rates are typical *sustainable* rates for just one processor.

The application I used as a test problem was a vortex dynamics calculation chosen from fluid dynamics known. This calculation solves the *vorticity-stream function formulation* of Euler's equations for incompressible flow in two dimensions in an infinite domain:

$$\frac{D\omega}{Dt} = 0 \tag{3.1}$$

$$\omega = -\Delta\psi \tag{3.2}$$

$$\mathbf{u} = 0 \text{ at } \mathbf{x} = \infty, \tag{3.3}$$

where $\mathbf{u}(\mathbf{x}(t),t)$ is the velocity of the fluid at position $\mathbf{x}(t)$ at time t; ω is the vorticity, defined as the curl of $\mathbf{u}$; ψ is the stream function; $\frac{D}{Dt} = \frac{\partial}{\partial t} + \mathbf{u} \cdot \nabla$ is the material derivative and $\Delta = \frac{\partial^2}{\partial x^2} + \frac{\partial^2}{\partial y^2}$ is the two-dimensional Laplacian operator. (For an explanation of these equations consult Chorin and Marsden's introductory text on fluid mechanics [7]. Also see Chorin's original paper on the vortex blob method [6], or Leonard's survey of vortex methods [13].) The above equations were solved for an initial vorticity distribution that was constant inside two disks centered about the origin, and zero elsewhere. These disks are referred to as Finite Area Vortices. To discretize the above equations we place a collection of N marker particles, called vortices, on a regular mesh of points, and then compute the path of the vortices over a sequence of timesteps. The following system of ordinary differential equations describes the motion of the vortices:

$$\frac{d}{dt}\mathbf{x}_i(t) = \mathbf{u}_i(t), \; i = 1, \cdots, N \tag{3.4}$$

where $\mathbf{x}_i(t)$ is the position of the ith vortex at time t, $\mathbf{u}_i(t)$ the velocity, and ω_i is its strength, which is like a charge. A PPPM-type algorithm, Anderson's Method of Local Corrections [1], was used to compute the mutually-induced velocities on the RHS of (3.4). When the vortices number in the thousands this method typically spends less than 5% in a Poisson solver – global computation – and most of the remaining time in localized computation. The positions of the vortices were evolved by discretizing (3.4) in time with a second order Runge-Kutta time integration scheme. All software was written in FORTRAN 77. On the iPSC, the code was compiled with FTN286 and run under release 2.1 of the node operating system. On the Cray, the code was compiled with CFT (version 1.14), and run under COS (version 1.16). All arithmetic was done using 8-byte floating point numbers (double precision on the iPSC, single precision on the Cray). Experiments were run with various number of vortices N and processors P. Owing to the differences in processor speed, numbers of processors, and memory capacity, the values of N on the Cray were different from those used on the iPSC.

I use parallel efficiency as the figure-of-merit. Define η_P as the parallel efficiency on P processors:

$$\eta_P = \frac{T_1/P}{T_P}, \tag{3.5}$$

where T_P is the time to complete on P processors. T_1 is the time taken on a uniprocessor. For this special case of $P = 1$, various overheads that would be incurred on a multiprocessor, such as communication, are non-existent. By definition $\eta_1 = 1$. Table 2 gives the efficiency and speedup measurements obtained from the iPSC runs, and Table 3, the measurements obtained from the Cray. Efficiency was quite good on both machines. On the iPSC, η_P ranged from 90% with 4 processors to 74% with 32. The efficiency on the Cray was about 90% on 4 processors. Thus, if efficiency were somehow increased to 100%, that would speed up the iPSC computations by at most 35% $((1-\eta_P^{-1})\times100\%)$ and by 12% on the Cray.

Overall, genMP's overhead seems reasonable; it never exceeded 2.4% on the iPSC, and I estimate that it never exceeded 5% on the Cray (overheads on the Cray could not be measured directly). The computations vectorized on the Cray as well as they did in the uniprocessor version of the code, and ran at roughly 250 megaflops/sec. on 4 processors. Surprisingly, the iPSC's high message startup time – roughly 5 msec. – appeared to have very little impact on the running time of the calculation. GenMP incurred a low communication overhead during local barrier synchronization, for example, because it can transmit data in bulk rather than an element at a time, and this was facilitated by restricting the partitions to have simple shapes.

N	P	η_P (Efficiency)	S_P (Speedup)	%Lbar	%Part
386	4	90	3.6	0.3	1.2
796	8	85	6.8	0.8	1.2
1586	16	79	13	1.4	1.0
3180	32	74	24	1.6	0.8

Table 2. iPSC results, where the number of vortices N varies linearly with the number of processors P. The parallel efficiency η_P (reported as a percentage) and parallel speedup S_P decrease with P. By definition the $\eta_P = S_P / P$. Overhead costs are reported as %Lbar, the fraction of the total time spent in local barrier synchronization, and %Part, the fraction spent partitioning, including the cost of producing the work density mapping. All runs lasted 64 timesteps, two velocity evaluations were done per timestep, and loads were rebalanced every other timestep. Since the larger problems couldn't fit into the memory of a single processor, T_1 could not be measured directly, and the efficiency and speedup figures are pseudo-measurements.

N	P	S_P	η_P	η_P^{max}
12848	1	1.00	1.000	1.000
12848	2	1.95	0.973	0.994
12848	4	3.63	0.908	0.982
25702	1	1.00	1.000	1.000
25702	4	3.57	0.892	0.957

Table 3. Parallel efficiency and speedup for the X-MP runs. η_P^{max} is the maximum theoretical efficiency that could be achieved under ideal conditions, given we chose not to parallelize the global computation done by a Poisson solver. The runs with 12848 vortices ran for 400 timesteps, the larger runs for 240. Loads were balanced every timestep.

4. Summary

I have outlined a simple approach to parallelizing numerical software for multiprocessors that insulates the programmer from many of the machine-dependent and low-level details. Application-dependent code and system-dependent code need not become heavily intertwined; when code is transported to a new machine, the parts that would have to change to accommodate a different communication model are restricted mostly to code the programmer never sees. I tried out my ideas on a realistic application, and obtained good parallel speedups on architectures that represent two extremes in multiprocessor design philosophy.

My approach is to have the user program a generic multiprocessor, called "genMP," with abstractions for hiding the details of task decomposition and coordination activities from the user. GenMP employs domain partitioning to subdivide workly fairly among a team of processors, and local barrier synchronization to ensure correctness. The user must divide the data and computation for the local part of the problem into bins of a regular rectangular mesh, must supply work estimates for the computation in each bin, and must supply routines for converting these data to and from byte streams. GenMP will assign bins to tasks in order to even the workload and will allow each task to access and communicate the necessary boundary data. GenMP is intended for a diversity of calculations, previously identified, that fit a simple model of spatial locality. It is neither universal nor complete, however, and leaves some programming details up to the discretion of the user.

In order to explore its generality, I have begun to apply genMP to other kinds of applications; a boundary layer calculation that solves the incompressible Navier-Stokes equations in two dimensions (in collaboration with E. G. Puckett), an adaptive grid method for hyperbolic partial differential equations (in collaboration with M. J. Berger and P. Colella), and a three-dimensional vortex calculation (in collaboration with T. Buttke and P. Colella).

The research described here was part of my Ph. D. dissertation research [2] done in the Computer Science Division at the University of California at Berkeley. I gratefully acknowledge the encouragement and moral support of my thesis advisor, W. Kahan; Phillip Colella also helped to supervise the work. Many thanks go to Erling Wold for reading the final draft of this paper.

5. References

1.   C. R. Anderson, "A Method of Local Corrections for Computing the Velocity Field Due to a Distribution of Vortex Blobs," *J. Comput. Phys.* *62*(1986), pp. 111-123.

2.   S. B. Baden, "Run-Time Partitioning of Scientific Continuum Calculations Running On Multiprocessors," LBL-23625, Mathematics Department, University of California, Lawrence Berkeley Laboratory, June 1987. (Ph. D. Dissertation in the Computer Science Division at the University of California, Berkeley, Tech. Report # 87/366)).

3.   M. J. Berger and J. Oliger, "Adaptive Mesh Refinement for Hyperbolic Partial Differential Equations," *J. Comput. Phys. 53*,3 (March 1984), pp. 484-512.

4.   M. J. Berger and S. Bokhari, "A Partitioning Strategy for Non-Uniform Problems on Multiprocessors," *IEEE Trans. Comput. C-36*,5 (May 1987).

5.   S. S. Chen, C. C. Hsiung, J. L. Larson and E. R. Somdahl, "CRAY X-MP: A Multiprocessor Supercomputer," in *Vector and Parallel Processors: Architecture, Applications, and Performance Evaluation*, M. Ginsberg (editor), North Holland. To be published..

6.   A. J. Chorin, "Numerical Study of Slightly Viscous Flow," *J. Fluid Mech. 57*(1973), pp. 785-796.

7.   A. J. Chorin and J. E. Marsden, *A Mathematical Introduction to Fluid Mechanics*, Springer-Verlag, New York, 1979.

8.   *Cray X-MP Hardware Reference Manual*, Cray Research, Inc., 1986. Order number HR-0097.

9.   M. E. Dippé and J. A. Swensen, "An Adaptive Subdivision Algorithm and Parallel Architecture for Realistic Image Synthesis," *SIGGRAPH '84 Conference Proceedings*, Minneapolis, July 1984, pp. 149-158.

10.   L. Greengard and V. Rokhlin, "A Fast Algorithm for Particle Simulations," YALEU/DCS/RR-459, Yale Univ., Dept. of Computer Science, April 1986.

11.   R. W. Hockney and J. W. Eastwood, *Computer Simulation Using Particles*, McGraw-Hill, 1981.

12.   *iPSC User's Guide*, Intel Corporation, Beaverton, Oregon, October 1985. Order Number: 175455-003.

13.   A. Leonard, "Vortex Methods for Flow Simulation," *J. Comput. Phys. 37*(1980), pp. 289-335.

14.   B. Nour-Omid, A. Raefsky and G. Lyzenga, "Solving Finite Element Equations on Concurrent Computers," *Proc. ASME Symp. on Parallel Computations and Their Impact on Mechanics*, December 13-18, 1987.

Intrinsically Parallel Algorithms

Oliver A. McBryan

Most algorithms implemented on parallel computers have been optimal serial algorithms, slightly modified or parallelized. An exciting possibility is the search for intrinsically parallel algorithms. These are algorithms which do not have a sensible serial equivalent–any serial equivalent is so inefficient as to be of little use. We describe a multiscale algorithm for the solution of PDE systems that is designed specifically for massively parallel super–computers. Unlike conventional multigrid algorithms, the new algorithm utilizes the same number of processors at all times. Convergence rates are much faster than for standard multigrid methods–the solution error decreases by up to three digits per iteration.

We present results of the implementation of the algorithm on the 65,536 processor Connection Machine, and we compare the results with more standard algorithms implemented on that computer.

Iteration with Mesh Refinement for Systems of ODE's
Olavi Nevanlinna

We consider block Picard–Lindelöf iteration (or "wave form relaxation") for systems of initial value problems. A rigorous environment for the following fact is given: one can refine the mesh during the iteration in a reliable way so that the total amount of work is a small multiple of that for the decoupled system.

PART IV

Scientific Applications

CHAPTER 38

Parallel Vision Algorithms: An Approach*

Sharat Chandran*
Larry S. Davis*

Abstract. This paper considers the parallel implementation of the Hough Transform (a technique to detect collinear edge points) in a very general setting. We also consider two contrasting architectures, the Butterfly Parallel Processor [1], essentially a shared memory machine, and the NCUBE [2], a direct connection machine wherein processors are interconnected in the form of a hypercube, for the purpose of implementation.

Embedded in the Hough problem are questions of optimal processor allocation and parallel "peak" selection in image neighborhoods. We present fast practical algorithms (subject to inherent lower bounds) and discuss the relevant complexity issues.

1. Introduction. This paper discusses parallel algorithms and their implementation for the class of operations that involve computing two-dimensional histograms of image features. An important class of such operations are the so-called *Hough Transforms* [1], which can be used to detect straight lines or edges (or more generally, arbitrarily shaped point patterns) in an image. This problem is representative of computationally intensive intermediate level vision problems and is one that merits using a machine whose architecture is non Von Neumann.

This paper considers the problem in general. Our method applies, for instance, on two machines of very different characteristics. The first is the Butterfly Parallel Processor, a tightly-coupled, shared memory machine. Our programs were implemented on a 128 [3] node machine at the Center for Automation Research. The other machine is a sixty-four node NCUBE machine available at the University of Maryland's Center

---

* Preliminary versions of this paper were presented at the Fifteenth Workshop on Applied Imagery, Pattern Recognition, Concurrent Processing and Vision and at the Butterfly-Warp Users Group Meeting. The authors are with the Center for Automation Research, University of Maryland, College Park, MD 20742, USA.

[1] Butterfly is a trademark of Bolt Beranek and Newman, Inc.

[2] NCUBE is a trademark of NCUBE Corporation.

[3] Few nodes were used by the operating system, and some were simply not operational for hardware reasons; hence the number of processors on which the program ran was really 110.

for Advanced Research in Biotechnology. The processors in this machine have some local memory and are located at the corners of a hypercube.

The main issue in parallel processing is the efficient use of processors. Let P be the number of processors and let N be a measure of problem size. Since physical multiprocessors will be built with a finite number of processors, it may well be that the user pushes the limits of the machine by taking on a very large problem. (It is also not hard to visualize problems broken into modules and cases where the multiprocessor is essentially partitioned.) Here we have a case where $P \leq N$. It is also important to consider the case when $P \geq N$ since, as Arvind [2] points out, one wants processors to be *scalable* in such a manner that adding hardware resources results in increased performance; one wants to analyze the asymptotics of this behavior. Accordingly, one must independently analyze how the algorithm and the architecture behave as the number of processors and the problem size vary.

In all the cases, the goal is that the time to solve the problem sequentially be P times the time to solve the problem in parallel, but it is not clear whether this can be achieved for all input sizes. If we can achieve this, the parallel algorithm is *optimal*. Let $T_1(N)$ represent the time complexity of the fastest algorithm on a problem of size N using a single processor. Likewise, let $T_P(N)$ be the complexity of a parallel algorithm using P processors. Given any parallel algorithm A for a given problem along with a particular model of computation, then there is a processor efficiency function(PEF) $S(N)$ such that for $1 \leq P \leq S(N)$

$$T_P(N) = O(T_1(N)/P).$$

Given two algorithms A_1 and A_2, and the corresponding PEFs $S_1(N)$ and $S_2(N)$, we say that A_1 is more *processor efficient* than A_2 if there exists N_1 such that for all $N > N_1$, $S_1(N) > S_2(N)^4$. A subgoal in algorithm design thus is to maximize the PEF and one characterizes an algorithm A by saying that it is in the Class $K_{S(N)}$.

Alternatively, one can consider for any P, the *minimal* input size N for which the algorithm is optimal and define an analogous *data efficiency function* (DEF). Thus, given an optimal algorithm B, there exists a function $D(P)$ such that for all $N > D(P)$,

$$T_P(N) = O(T_1(N)/P)$$

The algorithm B is then said to be in the Class $J_{D(P)}$ and the subgoal is to minimize $D(P)$. Given two algorithms B_1 and B_2 and the corresponding DEFs $D_1(P)$ and $D_2(P)$, we say that B_1 is more *data efficient* than B_2 if $D_1(P) = o(D_2(P))$ [5]. Section 4.1.1 illustrates these concepts.

The remainder of this paper is organized as follows. Section 2 defines the specific type of Hough transform that we implemented and describes previous work done in this area. Section 3 describes the characteristics of the parallel machines we are considering. Section 4 presents the algorithms and discusses the relevant complexity issues pertaining to them as well as some useful implementation strategies. Section 5 contains some concluding remarks.

---

[4] In other words, $S_1(N) = \Omega(S_2(N))$.
[5] Given functions f and g of x, $f = o(g)$ if $\lim_{x \to \infty}(f/g) = 0$.

2. Definition of the Problem. A common problem in computer image processing is the detection of straight lines in digitized images. In the simplest case, the image contains a number of discrete feature points perhaps corresponding to the edge pixels obtained by a local operator (see below). The problem is to detect the presence of groups of collinear or almost collinear feature points. It is clear that the problem can be solved to any desired degree of accuracy by testing the lines formed by all pairs of points. However, the computation required for n points is approximately proportional to n^2, and may be prohibitive for large n.

Rosenfeld [3] describes a method originally proposed by Hough [4] and subsequently improved upon by Duda and Hart [1] to solve the problem. Essentially, a Hough transform is designed to detect collinear sets of edge pixels in an image by mapping these pixels into a parameter space (the Hough space) defined in such a way that collinear sets of pixels in the image give rise to peaks in the Hough space.

A brief discussion of the method follows. The edge pixels processed by the transform are obtained by applying, for example, a Sobel gradient operator to the image. Such operators compute digital approximations to the gray level gradient magnitude and direction at each (non-border) pixel of the image.

Now, given a set of collinear edge points $\{(x_1, y_1), (x_2, y_2), \ldots, (x_n, y_n)\}$, we know that they all must satisfy the *normal-angle* representation of a line:

$$\rho = x_i \times \cos \theta + y_i \times \sin \theta$$

The key observation is that points lying on the same straight line in the picture plane correspond to curves through a common point in the ρ, θ parameter plane. Thus, the problem of finding the set of lines in the image plane is reduced to that of finding common points of intersection of sinusoidal curves in the parameter plane.

The implementation of the Hough transform for detecting straight lines on a sequential machine involves a quantization of the parameter plane into a quadruled grid. The grid size is determined by the acceptable errors in the parameter values and the quantization is confined to a specific region of the parameter plane determined by the range of parameter values. A two-dimensional array (the accumulator array or the Hough array) is then used to represent the parameter plane grid, where each array entry corresponds to a grid cell. For each edge pixel, the algorithm increments the counts in all accumulator array entries that correspond to lines passing through that edge pixel. The process of incrementing accumulator array counts can be thought of as "voting" by the edge pixels for the parameter values of possible lines passing through these points. The time required to execute this algorithm is proportional to the size s of the grid, plus the number m of edge pixels times the number of votes v cast by each point i.e. O $(s + mv)$. The memory space is proportional to the size of the grid.

Once the Hough array is computed, we seek to find those array elements which have high values. (The number of these depends upon the constraints posed by the problem, but it is clear that we are going to seek a subset of the total available entries.) However, it is not just enough to pick the first few high values because, depending upon the quantization, the same edge pixel will contribute to a "thick" curve, i.e.

votes to entries that are adjacent. More importantly, because of the inherent non-linearity of the transformation, a "real" cell will be surrounded by "false" votes in its neighborhood. These problems are solved by searching for the local maximum in an $I \times I$ square neighborhood, perhaps after smoothing the Hough array.

The Hough transform has an extensive literature. In the sequential case, Hough originally proposed a slope-intercept parameterization. Duda suggested the normal-angle representation of a straight line in order to bound the values of the parameters. The technique has been extended by Kimme et al. [5] to detect other curves defined analytically, and by Ballard [6] to detect general curve shapes using the edge orientations at the image points. Yet another variation which avoids the use of transcendental functions is presented in [7]. Brown [8] discusses a memory efficient implementation of the Hough transform.

Continued interest has also been shown in the parallel version. Silberberg [9] describes a Hough algorithm on a mesh-connected computer. This has been subsequently improved upon by Rosenfeld et al. [10]. They consider multiprocessors executing instructions in a Single Instruction Multiple Data (SIMD) [11] fashion and where the amount of local memory available is critical. Merlin et al. [12] have a method for detecting a general curve in parallel and Ibrahim et al. [13] have discussed a simulation on the NON-VON architecture; they present two algorithms based on an SIMD and an MSIMD approach on a tree-like machine. Sher [14] has described an implementation of the general template matching problem on the WARP systolic multiprocessor and on a Butterfly. His implementation is very different from the one described here, as discussed later. Very recently, Olson et al[6] have performed many experiments for the Architecture Benchmarks [16]; we comment on them in a later section.

In the simplest case, the steps involved in detecting large collinear sets of edge or feature points in a binary picture may be summarized as follows: Let θ_{min} to θ_{max} be the range of angles that we are interested in and let *Accum* refer to the accumulator array. Then the simple algorithm is :

for each edge point (x, y)

    for $\theta = \theta_{min}, \theta_{max}, \Delta\theta,$
        begin
            $\rho = \lceil x \times \cos\theta + y \times \sin\theta \rceil$
            Accum $[\rho, \theta]$ = Accum $[\rho, \theta] + 1$
        end

Once the accumulator array is computed, we must next find the k highest local maxima. One can then construct another array which has as its entries all those elements which are the local maxima in all $I \times I$ neighborhoods (for some fixed value of I; in our experiments, we used $I = 3$.) . Next we could find the first k values (for instance by sorting the newly constructed array). In the sequential case, there is an obvious better algorithm which skips the intermediate step and directly finds the first k peaks. This is done using an array of k items (called final_list) which stores

---

[6] The final form of this appears in [15].

the angle, distance and Hough entry values. The corresponding simple algorithm that suggests itself is

for each (ρ, θ) pair entry

      if local_maxima(entry) then

      insert(angle, distance, value) in final_list.

It is not obvious if this algorithm will work better in the parallel case and we shall have more to say about this in Section 4.

3. Description of the Machines. In this section, we review those aspects of the parallel machines that are relevant to the problem at hand. For more information, see [17,18,19].

The NCUBE is a "message-passing" machine configured in the form of a 10-dimensional hypercube. Advantages of this topology have been discussed in [20,21] etc. Each of the 1024 identical custom built processors has "local" memory of 128K bytes and is connected to 10 other processors. The processor in the NCUBE is a general purpose 32 bit processor and includes high speed 32 and 64 bit IEEE floating point.

It is important to view the NCUBE in relation to a host Intel 286 minicomputer because code may be loaded onto the processors (interchangeably called "nodes") only from the host. Parallelism in the NCUBE is very natural after the host ships out (perhaps different) programs to the different processors. They start working on the data (provided by the host or some other node) and synchronization is achieved by sending messages. At the current stage [7] , the main mechanism for programming is a modified FORTRAN language. Parallelism will be reduced whenever the processors are idle, perhaps due to improper load balancing, and when messages are received, sent or forwarded by the processors.

The Butterfly is a tightly coupled "shared-memory" architecture machine with at most 256 MC68000 processors. Each processor sits on a card with a process node controller (PNC) and a megabyte of memory (which in some sense is local to this processor). Processors are interconnected in a "butterfly" [22] fashion, thus enabling processors to gain quick access to memory located physically on a different card. There is a single clock in the whole machine but asynchronous multiple instructions may be executed by the different processors on multiple data.

The standard approach to programming the Butterfly is the so-called Uniform System [23] wherein computational tasks that comprise an application are emphasized and the notion of processes is less relevant. The Uniform system views the conglomeration of memory as one huge shared memory seen in the address space of all identical processors and seeks to scatter application data to reduce memory contention. Parallelism is exploited by generating tasks using a "generator" (similar to the **map** function in LISP). This is typically visualized as a procedure which has as

---

[7] Though our algorithm suited the NCUBE very well, it was in its infancy of software development when the algorithm was implemented (early 1986) . For instance, the time spent in execution of the program viz. the duration between sending of the node program and receiving the information back to the host cannot be easily be measured. Secondly, it is possible to pick messages out of order in which they have been sent, i.e. by the type of message. Finally, input/output is possible only from the host and this restricts the pace of debugging programs. In view of these, we are not able to present performance measurements.

its arguments a description of the data and a worker function to analyze the data; the generator results in activating processors with different subsets of the data. The form of this subset is generally limited in the current release (version 2.2.3) and typically is a row of a matrix.

Parallelism in the Butterfly suffers because of memory contention, switch contention and overhead involved in generating the tasks. Unfortunately, in spite of the existence of a scheduler, it is not possible to generate more active processes than processors in the Uniform system. This has imposed a limitation on our experiments. Nor is it possible for processors to act "independently" since the same code is loaded on all the processors.

4. The Algorithms. We present our parallel algorithms in this Section. As mentioned earlier, one has to compute the Hough array and then extract peaks. On a systolic machine, such as the Warp, one may be interested in detecting the peaks even while the Hough array is being calculated, but on our machines there is no obvious advantage to this approach. We construct the Hough array and detect the peaks in a sequential manner but in both the steps we would like to allocate processors optimally.

4.1. Computing the Hough array. Let θ_{min} to θ_{max} determine the relevant range of angles. One may think of computing the Hough array in terms of mapping a three-dimensional $x \times y \times \theta$ space into a one-dimensional ρ space and then incrementing the appropriate (ρ, θ) cells by 1.

At this stage, it is not obvious how the processors should be allocated. For instance, we can identify the following options:

- Distribute processing by x: Assign one row of the binary image to each processor. It then computes ρ over the entire range of angles. (One could conceivably use a column instead of a row.)
- Distribute processing by region: Assign a square shaped region of the input binary image to a processor. It then computes ρ over the entire range of angles.
- Distribute processing by θ: Assign one angle to each processor. It then computes ρ for the entire binary image.
- Distribute processing by θ: As above, assign a processor to an angle. However, the processor analyzes only a part of the input image.

As an example, consider scaling the problem down to an input picture of size 1×2, an angle spread of 5° and an angular quantization of 1° (an assumption we will make throughout, for simplicity). Assuming that we have only two processors, we immediately see that it is preferable to distribute processing by θ. The only thing that we can hope for is that the input image is random. If we allocate one processor to each of the pixels, then one processor may end up doing no work at all since it may not be the case that both the pixels are edge pixels. On the other hand, distributing by θ implies that both processors are working at all times and will, in general, lead to better processor allocation. This corresponds to the case of "Distribute processing by θ" above and will be the pivotal idea upon which we shall base the allocation.

We present an allocation scheme for the general case (other schemes are also possible). Let $\Psi \equiv \theta_{max} - \theta_{min} + 1$ and let N be the number of rows in the input image. We will consider the following three possible cases:

(1) $$P < \Psi$$

(2) $$\Psi < P < \Psi N$$

(3) $$\Psi N < P$$

In the first case, the ith processor computes the Hough array for a set of θs; specifically, $\theta_{min} + i, \theta_{min} + i + P, \dots , \theta_{min} + i + \lceil \Psi(moduloP) \rceil$ [8] over the entire binary image.

In the second case, the unit of processing given to a processor is a row of the input image and an angle. For each edge point in the row, the processor computes (or looks up in a table) the appropriate value of ρ. Each processor can be given either overlapping sets of rows or "scattered" sets (our implementation).

Finally, in the last case, the unit of processing is now an angle and part of the input row. Thus, for instance, if $P = 2\Psi N$, then processors will be working on an angle and half a row of the input image.

4.1.1. Computational complexity for the Hough array computation. The purpose of this section is to illustrate the relation between the problem size and the efficiency of the parallel computation.

Once the input space is allocated to the different processors as described above, each processor independently computes a "version" of the Hough array based on the subset of the problem given to it. The Hough arrays are then collapsed into the complete Hough array using the standard [24] collapsing technique. We present the following simplified procedure, Algorithm Merge, to merge the Hough arrays on the hypercube (here, *mynodeid* is a unique address for a hypercube node):

for $index = 1$ to $\log_2 P$
    begin
    if mynodeid $\leq 2^{index}$ **then**
        neighbor $= mynodeid$ **xor** $2^{index-1}$
    if neighbor $< mynodeid$ **then**
        send (myhougharray, neighbor)
    else
        begin
        receive (hishougharray, neighbor)
        myhougharray $=$ myhougharray $+$ hishougharray
        end
    end

---

[8] A different partioning where each processor analyzes contiguous angles is possible.

Let H_size represent the size of the Hough array. For the purpose of analysis we assume that every pixel in the input $N \times N$ image is an edge pixel in the worst case and that H_size is asymptotically smaller than the input N. Then, in this model of computation,

$$(4) \qquad T_P(N) = O(N^2/P) + H\_size \log P$$

which we shall approximate as

$$(5) \qquad T_P(N) = O(N^2/P).$$

Recall that

$$(6) \qquad T_1(N) = O(N^2).$$

Munro and Patterson [25] have a lower bound result which states that

$$(7) \qquad T_P(N) \geq T_1(N)/P + \log_2(\min(P, T_1(N))) - 1.$$

If we substitute equation 6 into equation 7, we have

$$(8) \qquad T_P(N) \geq O(N^2/P) + \log_2(\min(P, N^2)) - 1$$

and thus asymptotically, this result is optimal within the constraints of our assumption. (See eqn. 5.)

As discussed in the introduction, it is useful to see for what values of input size this algorithm achieves optimal speedup. Essentially we want to determine when

$$T_P(N) = T_1(N)/P$$

viewing P and N in turn as the independent variable. We thus need to solve the equation

$$N^2/P + H\_size \log_2 P = O(N^2/P)$$

The solution to this is

$$N = \Omega(H\_size P \log P)^{1/2}.$$

The problem size should be at least $(P \log P)^{1/2}$ for the algorithm to be optimal or the DEF is $(P \log P)^{1/2}$.

Reversing the question we have, for optimality,

$$P = N^2/\log N$$

since the term

$$H\_size \log_2 P$$

evaluates to

$$2H\_size \log(N/\log N)$$

or

$$O(\log N)$$

In other words, the algorithm exploits at most $N^2/\log N$ processors to solve a problem of size N and this gives an idea of how good the algorithm is (the PEF is $N^2/\log N$). To summarize, the algorithm is in the Class $J_{(P\log P)^{1/2}}$ and Class $K_{N^2/\log N}$.

4.1.2. Implementation details. Recall from the description in Section 3 that on the Butterfly we do not have the ability to increase the number of processes beyond the actual number of physical processors, i.e. one cannot have virtual processors. This limits the set of experiments that we can perform since we cannot measure the speedup beyond 110 processors.

We remarked earlier that there are other data allocation schemes that may also achieve similar performance. Our scheme has the additional merit that on the Butterfly it is easy to implement. When the Uniform System generator produces tasks to be executed in parallel, it produces, conceptually, an *identification* number for each task and different processors execute tasks corresponding to different identification numbers. In our situation, when we allocated angles to each processor (corresponding to Equation 1 in Section 4.1), these numbers correspond to the different angles since now the unit of processing is an angle. As a result, the P processors evaluate the first P tasks in parallel and then process the next P tasks. (Each processor thereby does not work on the task corresponding to contiguous angles.) In the next situation (see Equation 2 in Section 4.1) we decided to allocate rows of the input image for a similar reason and thus each processor does not work on contiguous rows but rather on a scattered set of rows. It also helps that adjacent elements in a row of the matrix are also physically adjacent even though adjacent rows are scattered since *block* transfer of physically adjacent memory cells from global memory to the local memory of the processor is efficient on the Butterfly. Thus there is no contention when different processors are accessing the input image in parallel.

Once the individual Hough array versions are computed by the different processors, we need to merge them into one "correct" Hough array. Communication between processors on the shared memory Butterfly is achieved by directly accessing global memory, so in Algorithm Merge, we need not explicitly send or receive the individual Hough array versions to be merged. However, the "versions" that the different processors compute have to be allocated in global memory and there is an obvious duplication here; there are bound to be situations where this might be a limitation since, in particular, the current Uniform System forbids piecewise freeing of dynamic memory. Our implementation defined the "final" Hough array in global memory. The individual versions are never explicitly built but instead the copy in global memory is updated. It is well known that updating in global memory requires that the information be added atomically; so we use the "locking" mechanisms available in Chrysalis to ensure integrity. Thus, there is no need to allocate memory in the global address space for the individual versions.

Undoubtedly this approach reduces efficiency; however, one can manipulate the program variables to take advantage of the local and global memory aspects of the shared memory and reduce the contentions. Since programming on the Butterfly is done in C, we often need to follow a sequence of pointers in order to access the "locked" data structure. A useful strategy in such cases is to make local copies of the data thus restricting contention only at the final destination memory word.

For the Hough transform done on a window of an image for the Autonomous Land Vehicle, we observed linear speedup for a 32×32 input binary image with approximately twenty-five percent edge pixels when the angle spread was about twenty degrees. See Tables 1 and 2. In Table 1 we are able to obtain linear speedup. In Table 2, we see that as per our theoretical estimate, our problem size is too small to obtain good efficiency using processors as large as 110. (In these tables, we have shown the statistics obtained from the *Time Test* system facility in the Uniform System software. We use the notation from there; thus *#procs* is the number of processors; *Ticks* is the value of a counter which is a measure of the elapsed time of the program; *Time* is the actual time in seconds; and *Efficiency*, the quantity obtained by dividing the speedup by the number of processors.)

#procs	Ticks	Time	Efficiency
1	77090	4.64	1
2	36551	2.22	1
3	24257	1.46	1
4	17943	1.08	1
5	14454	0.87	1
6	11962	0.72	1
7	10301	0.62	1
8	8972	0.54	1
9	7975	0.48	1
10	7310	0.44	1
11	6646	0.40	1
12	5981	0.36	1
13	5649	0.34	1
14	5317	0.32	1
15	4984	0.30	1
16	4818	0.29	1

TABLE 1

Results obtained for a 32×32 window of an image on a sixteen processor Butterfly.

#procs	Ticks	Time	Efficiency
1	77090	4.73	1
2	39210	2.36	1
4	19605	1.18	.99
8	9969	0.60	.98
16	4984	0.30	.96
32	2658	0.16	.87
64	1994	0.12	.60
112	2160	0.13	.32

TABLE 2

Results obtained for a 32×32 window of an image on a larger machine.

We also ran a version of our program with an input array of 512×512 of binary (unoriented) edge detector outputs as per the Strategic Computing Vision Architecture Benchmarks [16]. Our results (with approximately twenty-five percent feature pixels) are indicated in Table 3. Table 4 has the figures reported for the Butterfly by researchers at the University of Rochester [15]. They have performed a number of experiments, fine tuning their algorithms. Our experiments compare with the unoptimized version of Table 4; there are many differences however; for instance, they use a gigantic 724×360 lookup table and their own indigenous Butterfly Image Processing Package. In Table 3 we see the effects of memory contention that we refered to above (when multiple processors are trying to update the global Hough accumulator).

The same idea may be used even in the processor allocation step. In addition, accesses to local memory cost less than accesses to global memory; so a "block" of data may be transferred to local memory and then used in computations. Again, when processors are allocated parts of the input matrix, allocation is done *alternatingly from the top and the bottom*. This results in less contention while accessing common data, for instance, overlapping rows.

4.2. Peak detection. More interesting theoretical ideas are involved in searching for the k highest peaks once the accumulator array is available and is in fact the subject of another report [26]. We propose different solutions here and consider their relative advantages.

Using a single processor, the problem cannot be solved faster than $O(N)$ because the kth largest value could be the Nth value (here N refers to the size of the Hough array and not the image size). Recall further that one can find the median of a list of numbers (or in general, the kth largest element) in linear time [27]. Once we do this, we can look at one element at a time, compare it to the kth largest element and thus find all the desired elements in linear time. Thus, although we search for the k highest peaks, our problem is in fact no more difficult than the median finding problem. Our goal is to solve the problem in constant time using N processors and, in general, to solve it in N/P time using P processors.

At first, one is inclined to sort the Hough bins using the accumulator value as the key. (Once sorting is done, it is trivial to pick the k highest peaks.) Using N processors, we cannot sort in less than $O(\log N)$ time [28]. As far as implementations go, though the above is fast, it is not practical because the constant hidden in front of the $O(\log N)$ description is very large. The value of k in most applications is low and the following algorithm is an efficient alternative (though asymptotically not faster than sorting).

Processor allocation is done by splitting (but not partitioning) the Hough array into P strips. Then each processor using a fast sorting algorithm finds the first k elements in its data. We then combine the individual k peaks in a tree-like fashion to produce the final desired k peaks.

This algorithm may be easily implemented both on the Butterfly and on the NCUBE. It is easy to observe that

$$T_P(N) = O(N/P \log(N/P) + k \log P),$$

the first term arising from the sorting part and the second from the merging. Asymptotically, this is not a good algorithm but for "reasonable" values of k and N, it is satisfactory.

#procs	Ticks	Time	Efficiency
1	120809922	7550.61	1
2	60208761	3763.04	1
4	30085180	1880.32	.99
8	15075985	942.24	.98
16	7558571	472.41	1
32	3782799	236.42	.87
64	1903878	118.99	.60
112	1097230	68.57	.32

TABLE 3

Results obtained for a 512 × 512 synthetic image.

#procs	(unoptimized) Time	(Optimized version) Time
1	??	139.66
2	??	69.32
4	??	35.28
8	151.75	17.38
16	77.99	8.76
32	52.30	4.57
64	30.60	2.5
112	18.72	1.49

TABLE 4

Results obtained by researchers at University of Rochester. Run times are for two different algorithms for 10000 feature points. Entries indicated by "??" have no reported values.

In fact, this is the version we have implemented. The results are summarized in Table 5. Note that the detection of peaks do take a significant amount of time even though for the problem of size 32 does not seem to be big enough for 112 processors.

We remark that for problems such as finding the maximum, a theoretical lower bound is $\Omega(\log N)$ *no matter what the number of processors is*. Since we can achieve this bound when $P = N$, the excess processors if P exceeds N cannot be fruitfully used. The allocation of processors for the case when N exceeds P is the natural one.

If we are indeed interested in asymptotics, we observe that we can use the linear median finding algorithm instead of the fast sorting algorithm in the initial stage (i.e. when each processor finds the k highest peaks in its data). The merging of these is done as before. In this case,

$$(9) \qquad T_P(N) = O(N/P + k \log P).$$

Once again, we are interested in when, as a function of N and P, this algorithm achieves a speedup of P over the sequential linear running time. If the problem size is at least $kP \log P$ optimal efficiency is achieved (the DEF is $kP \log P$) [9]. Once again, reversing the question, if we have at most $N/(k \log N)$ processors, then substituting into equation 9, we have

$$T_P(N) = O(k \log N + k \log(N/k \log N))$$

i.e.

$$T_P(N) = O(k \log N)$$

and thus

$$PT_P(N) = O(N)$$

the uniprocessor time. The problem is solved optimally and to summarize, the algorithm is in the Class $J_{(kP \log P)}$ and Class $K_{N/(k \log N)}$.

Using the above, we may interpret the results in Table 5. For the problem we are considering, we would expect the efficiency to tail off at 1000 processors. Similarly, we may ask what the theoretical problem size should be for $P = 128$. We notice that we are a factor of 30 off in the former case and a factor of 6 off in the latter; however, taking into account that we are using a big-oh notation and the errors inherent in asymptotic analysis, these may be reasonable. We conjecture that the non-optimality of this result stems from the multiple calls to the generator procedure on the Butterfly — this would be as reasonable a procedure as any to optimize!

#procs	Ticks	Time	Efficiency
1	46188	2.78	1
2	23094	1.39	1
4	11630	0.70	.98
8	5981	0.36	.96
16	3323	0.20	.85
32	2160	0.13	.67
64	1828	0.11	.30
112	1661	0.10	.24

TABLE 5

Finding 5 peaks from the Hough accumulator obtained for a 32 window with approximately 25% feature pixels.

5. Conclusions and future work. The Hough transform is a technique to detect line segments in an image by finding global consistencies in the data. Since it considers every feature point, the process is slow on a conventional Von Neumann machine. There is, however, a certain independence in the Hough array computation and whenever such independence exists in any problem, the use of machines whose architecture is not Von Neumann is justified. Matrix multiplication is an example of a problem wherein all the computations are independent and is an excellent test problem for a parallel machine. An appropriate partitioning of the data is required in such problems and in this paper, we have identified the "independent" component in

---

[9] [26] describes a better selection algorithm .

Section 4.1.1 and shown how to allocate processors on the NCUBE and the Butterfly. Once the Hough array is computed, one is usually interested in detecting peaks in order to find the best line segments. We have shown that this problem is no more difficult than the median finding problem and obtained an optimal parallel selection algorithm. Here the partitioning of data is less critical and the communication costs dominate the efficiency of the parallel computation. Nothwitstanding the theoretical optimality in our algorithms, the results we obtained suggests potential inefficacy of the architecture and the operating system.

An orthogonal issue that we have addressed in this paper is the relation between problem size and the number of processors. Given a machine of a fixed configuration, it is important to find the appropriate problem size to determine the efficacy of the architecture; in most problems, if the problem size is sufficiently large, one is bound to get a linear speedup on the finite number of processors that exist on the machine. We are interested in finding bounds on this parameter. Alternatively, when one has a fixed application, and the problem size is constrained by other issues, one is interested in determining a reasonable number of processors that one should use for the problem. It is clear that not all the processors may be useful if, for instance, the problem is too small. By introducing the idea of processor and data efficiency functions, we have illustrated a technique to obtain bounds on these numbers.

We have considered the parallel implementation of the Hough transform in which the sizes of the accumulator bins are equal. One problem with the Hough approach is that a feature point results in many votes that are of no consequence. One approach that was taken to reconcile the large memory requirements of such a schema (especially in higher order Hough spaces) in [29] was the use of *dynamic quantized spaces* and *dynamic quantized pyramids*. We are currently in the process of investigating the parallelization of this scheme.

Acknowledgements. Todd Kushner was involved in the writing of the C code for the Butterfly.

REFERENCES

[1] R. O. Duda and P. E. Hart. Use of the Hough transformation to detect lines and curves in pictures. *Communications of the ACM*, 15(1):11–15, 1972.

[2] Arvind and R. Iannucci. *Two Fundamental Issues in Multiprocessing*. Technical Report, Massachusetts Institute of Technology, 1986.

[3] A. Rosenfeld and A.C. Kak. *Picture Processing by Computer*. Volume 2, Academic Press, 1982.

[4] P. V. C. Hough. Method and means for recognizing complex patterns. U.S. Patent 3,069,654, 1962.

[5] C. Kimme, D. Ballard, and J. Sklansky. Finding circles by an array of accumulators. *Communications of the ACM*, 18(2):120–122, 1975.

[6] D. Ballard. Generalizing the Hough transform to detect arbitrary shapes. *Pattern Recognition*, 13(2):111–122, 1981.

[7] R. Wallace. A modified Hough transform for lines. In *Proceedings of the Conference on Computer Vision and Pattern Recognition*, pages 665–667, 1985.

[8] C. Brown. Peak finding with limited hierarchical memory. In *7th International Conference on Pattern Recognition*, pages 246–249, 1984.

[9] T. Silberberg. The Hough transform on the Geometric Aritithmetic Parallel Processor. In *Proceedings of the IEEE Workshop on Computer Architecture for Pattern Analysis*, pages 387–393, 1986.

[10] A. Rosenfeld, J. Ornelas, and Y. Hung. *Hough Transform Algorithms for Mesh-Connected SIMD Parallel Processors*. Technical Report CAR-TR-128, Center for Automation Research, 1986.

[11] M. Flynn. Some computer organizations and their effectiveness. *IEEE Transactions on Computers*, 21(9), 1972.

[12] P. Merlin and D. Farber. A parallel mechanism for detecting curves in pictures. *IEEE Transactions on Computers*, 24(1):96–98, 1975.

[13] H. Ibrahim, J. Kender, and D. Shaw. *The Analysis and Performance of Two Middle-Level Vision Tasks on a Fine-Grained SIMD Tree Machine*. Technical Report, Columbia University, 1985.

[14] D. Sher. *Template Matching in parallel*. Technical Report 156, University of Rochester, Rochester, New York, 1985.

[15] L. Bukys, C. Brown, and T. Olson. Low level image analysis on an mimd architecture. In *International conference on computer vision*, pages 468–475, 1987. See Benchmark results.

[16] R. Simpson, S. Squires, and A. Rosenfeld. Strategic computing vision architecure benchmarks. Private Communication, July 1986.

[17] *The Butterfly Parallel Processor Overview*. Bolt Beranek and Newman Inc, 1986.

[18] *Chrysalis Programmers Manual, Version 2.2.3*. Bolt Beranek Newman, 1986.

[19] *NCUBE Handbook, Version 0.6*. NCUBE Corporation, 1985.

[20] C. Seitz. The Cosmic Cube. *Communications of the ACM*, 28(1):22–33, 1985.

[21] M. Pease. The indirect binary n-cube microprocessor array. *IEEE Transactions on Computers*, 26(5):458–473, 1977.

[22] C. Kruskal and M. Snir. A unified theory of interconnection network structure. *Theoretical Computer Science*, 48:75–94, 1986.

[23] W. Crowther. *Using the Uniform System to program the Butterfly*. Bolt Beranek and Newman Inc., 1986.

[24] F. Preparata and J. Vuillemin. The cube-connected-cycle. *Communications of the ACM*, 24(5):300–309, 1981.

[25] I. Munro and M. Paterson. Optimal algorithms for parallel polynomial evaluation. In *12th Annual Symposium on Switching and Automata Theory*, 1971. Also see Journal of System Sciences,7, 1973.

[26] S. Chandran and A. Rosenfeld. *Order Statistics on the Hypercube*. Technical Report CAR-TR-228, Center for Automation Research, College Park, Maryland, 1986.

[27] A. Aho, J. Hopcroft, and J. Ullman. *The Design and Analysis of Computer Algorithms*. Addison Wesley, Reading, Massachusetts, 1974.

[28] M. Ajtai, J. Komlos, and E. Szemeredi. Sorting in $c \log n$ parallel steps. *Combinatorica*, 3:1–19, 1983.

[29] J. O'Rourke and K Sloan, Jr. Dynamic quantization: two adaptive data structures for multidimensional spaces. *IEEE Transcations on PAMI*, 6(3):266–279, 1984.

CHAPTER 39

Parallel Iterative Algorithms for Optimal Control*

Gerard G. L. Meyer[†]
Louis J. Podrazik[‡]

ABSTRACT. In this paper we present two parallel gradient based iterative algorithms to solve the linear quadratic regulator (LQR) optimal control problem with hard control bounds. In the first part of the paper, we introduce the algorithms in the context of the general class of problems to which they are applicable. The first algorithm is a parametrized gradient projection method and can be used to solve any convex programming problem. The second algorithm is a combination of the first algorithm with a constrained version of the Fletcher-Reeves conjugate gradient method and can be used to solve linear inequality constrained problems. We then use the two algorithms to solve the LQR optimal control problem with hard control bounds. In the second part of the paper, we embed our parallel step length and gradient projection procedures to produce two parallel algorithms which are suitable for real-time online implementation on a SIMD machine.

I. INTRODUCTION. Practical iterative methods to solve optimal control problems must exhibit fast convergence coupled with low computational overhead per iteration so that they may be implemented in a real-time online environment. In this paper we present two gradient based iterative interior methods for the parallel solution of the discrete LQR optimal control problem with hard control bounds. The algorithms are synthesized so that each iteration may be efficiently executed on a parallel computer. Unlike previous parallel approaches to the solution of optimal control problems which simply fold the computations to fit the number of processors [SCH81], [TRA80], our approach has been devised by explicitly considering the given computational environment. Consequently, one of the features of the parallel algorithms presented in this paper is that their structure, and hence their parallelism, is determined by the number of available processors, resulting in algorithms which are matched to the given parallel environment.

The first algorithm presented in this paper is a parametrized gradient projection method and contains a so-called ε-procedure for determining the active constraints in order to prevent the possibility of jamming and to ensure convergence; unlike approaches which generate a sequence of ε's, our approach uses a constant ε for all iterations. The second algorithm is a combination of the first algorithm with a constrained version of the Fletcher-Reeves conjugate gradient method and may be slightly modified to exhibit finite convergence on problems with quadratic cost.

We then use our algorithms to solve the LQR optimal control problem with hard control bounds which we rewrite as a bounded variable quadratic programming problem with special structure. A parallel implementation of our first algorithm is then presented; the parallel implementation of our

* Supported by the Air Force Office of Scientific Research under Contract AFOSR-85-0097.
†Electrical and Computer Engineering Department, The Johns Hopkins University, Baltimore, Maryland 21218.
‡ Bendix Environmental Systems Division, Allied-Signal Inc., Baltimore, Maryland 21284.

second algorithm follows directly and is not included here due to space limitations. Both implementations use the parallel step length and gradient procedures presented in [MEY87b] and blocked versions of the parallel recurrence solvers presented in [MEY87a]. We constrain the number of available processors, p, to lie in the range $1 \leq p \leq nN^{1/2}$, where n is the size of the system state vector, N is the number of stages in the control process and we assume $n \geq m$, where m is the size of the control.

II. THE CONVEX PROBLEM AND PRELIMINARIES. We consider the following problem:

Problem 1: Given $m +1$ maps $f^0(.), f^1(.), ..., f^m(.) : E^n \rightarrow E^1$ and a subset Ω of E^n defined by $\Omega = \{x \mid f^i(x) \leq 0$ for $i = 1, 2, ..., m\}$, find a point x^* in Ω such that for every x in Ω, $f^0(x^*) \leq f^0(x)$.

Definition 1: Let C be the class of all problems of the form of Problem 1 in which the maps $f^0(.), f^1(.), ..., f^m(.)$ are such that (i) $f^0(.)$ is continuously differentiable, convex and radially unbounded; that is, $f^0(.)$ is such that given any $x \in \Omega$, to every scalar α corresponds a scalar $\rho > 0$ such that $f^0(x) > \alpha$ whenever $\|x\| > \rho$, (ii) $f^i(.), i = 1, 2, ..., m$, are continuously differentiable, convex and define a nonempty constraint set Ω and (iii) the set Ω satisfies the Kuhn-Tucker constraint qualification at every solution x^*.

III. THE PARAMETRIZED GRADIENT PROJECTION ALGORITHM. Given a point $x \in \Omega$ and the parameter $\varepsilon \geq 0$, define the ε-active constraints index set $I(x,\varepsilon)$ as $I(x,\varepsilon) = \{i \mid f^i(x) \geq -\varepsilon\}$, given a subset I of the set $\{1, 2, ..., m\}$, define the subset $\Omega(I)$ of E^n as

$$\Omega(I) = \begin{cases} \{y \mid f^i(y) \leq 0 \text{ for all } i \in I\} & \text{if } I \neq \phi \\ E^n & \text{if } I = \phi, \end{cases}$$

and given a point $x \in \Omega$ and an index set I, let $w(x,I)$ be the projection of $x - \nabla f^0(x)^t$ onto the set $\Omega(I)$, that is, $w(x,I)$ satisfies $\|(x - \nabla f^0(x)^t) - w(x,I)\| = \min_y \{\|(x - \nabla f^0(x)^t) - y\| \mid y \in \Omega(I)\}$.

Lemma 1: If a point $x^* \in \Omega$ is optimal for a problem in C then $w(x^*,I(x^*,\varepsilon)) = x^*$ for all $\varepsilon \geq 0$. Conversely, if x^* and ε satisfy $w(x^*,I(x^*,\varepsilon)) = x^*$, then x^* is optimal.

We now give the parametrized gradient projection algorithm to solve a problem in the class C.

Algorithm 1: Let $x^1 \in \Omega$ and $\varepsilon \geq 0$ be given.

Step 0: Set $k = 1$.
Compute $\nabla f^0(x^1), I(x^1,\varepsilon) = \{i \mid f^i(x^1) \geq -\varepsilon\}$ and $p^1 = w(x^1,I(x^1,\varepsilon)) - x^1$.

Step 1: If $p^k = 0$ stop; else go to Step 2.

Step 2: Set $d^k = p^k$.

Step 3: Compute $\alpha^k > 0$ to minimize $f^0(x^k + \alpha^k d^k)$ subject to $(x^k + \alpha^k d^k) \in \Omega$.

Step 4: Let $x^{k+1} = x^k + \alpha^k d^k$.

Step 5: Compute $\nabla f^0(x^{k+1})$.

Step 6: Compute $I(x^{k+1},\varepsilon) = \{i \mid f^i(x^{k+1}) \geq -\varepsilon\}$.

Step 7: Compute $p^{k+1} = w(x^{k+1},I(x^{k+1},\varepsilon)) - x^{k+1}$.

Step 8: Set $k = k + 1$ and go to Step 1.

At each iteration, Algorithm 1 generates a search direction d^k which is computed by projecting the negative gradient onto the set $\Omega(I(x^k,\varepsilon))$. The parameter ε defines the "sufficiently-active" constraint region. If a point x^k is in that region, then $I(x^k,\varepsilon) \neq \phi$ and the corresponding search direction d^k is obtained by projecting $x^k - \nabla f^0(x^k)^t$ onto the set $\Omega(I(x^k,\varepsilon))$; otherwise, $d^k = -\nabla f^0(x^k)^t$.

Lemma 2: If $x \in \Omega$ is not a solution to a given problem in C and $\varepsilon > 0$ then there exists an $\varepsilon(x) > 0$ and $\delta(x) > 0$ such that $f^0(y + \alpha d) \leq f^0(y) - \delta(x)$ for all $y \in B(x,\varepsilon(x)) \cap \Omega$.

Theorem 1: If $\{x^k\}$ is a sequence constructed by Algorithm 1 to solve a problem in C then either $\{x^k\}$ is finite and the last point is optimal or $\{x^k\}$ is infinite and contains cluster points, each of which is optimal. Furthermore, when a problem in C is such that $f^0(.)$ is strictly convex, then that problem has a unique optimal solution x^*, and in that case $\{x^k\}$ converges to x^*.

In order to present the parallel implementation of Algorithm 1 used to solve the LQR problem, we first use it to solve a particular case of Problem 1 which lies in C, specifically the bounded variable quadratic programming problem given below.

Problem 2: Find $x \in E^n$ with components x_i, $i = 1, 2, ..., n$, that minimizes the performance index $f^0(x)$ $= 1/2\, x^t H x + b^t x + c$ subject to $x \in \Omega$, where H is an $n \times n$ symmetric positive definite matrix, b is an $n \times 1$ vector and Ω is the unit hypercube defined as $\Omega = \{x \mid \ |x_i| \leq 1, \text{ for } i = 1, 2, ..., n\ \}$.

We now give some properties of implementing Algorithm 1 to solve Problem 2.

Lemma 3: Given x^k and $\nabla f^0(x^k)$, let $p^k = w(x^k, I(x^k, \varepsilon)) - x^k$ for the case of Problem 2. Then each component p_i^k of p^k can be computed as

$$p_i^k = \begin{cases} +1 - x_i^k & \text{if} \quad i \in I(x^k, \varepsilon) \text{ and } x_i^k - \nabla f_i^0(x^k)^t \geq 1 \\ -1 - x_i^k & \text{if } i + n \in I(x^k, \varepsilon) \text{ and } x_i^k - \nabla f_i^0(x^k)^t \leq -1 \\ -\nabla f_i^0(x^k)^t & \text{otherwise,} \end{cases} \tag{3.1}$$

where we denote $\nabla f_i^0(x^k)$ to be the i-th component of $\nabla f^0(x^k)$.

Lemma 4: Given x^k and $\nabla f^0(x^k)$, let α^k minimize $f^0(x^k + \alpha^k d^k)$ subject to $(x^k + \alpha^k d^k) \in \Omega$ for the case of Problem 2. Then α^k can be computed as $\alpha^k = \min\{\alpha_u, \alpha_c\}$, where

$$\alpha_u = -<d^k, \nabla f^0(x^k)^t>/<d^k, H d^k>, \ \alpha_c = \min_{\alpha} \{\alpha \mid \alpha = \frac{sgn\,(d_i^k) - x_i^k}{d_i^k} \text{ for all } i \text{ such that } d_i^k \neq 0\}. \tag{3.2}$$

Remark 1: Since the cost function is quadratic and we are using the update $x^{k+1} = x^k + \alpha^k d^k$, the quantity $\nabla f^0(x^{k+1})^t$ can be updated using $\nabla f^0(x^{k+1})^t = \nabla f^0(x^k)^t + \alpha^k H d^k$.

We now give the following implementation of Algorithm 1 to solve Problem 2.

Algorithm 2: Let x^1 and $\varepsilon > 0$ be given.

Step 0: Set $k = 1$.
   Compute $\nabla f^0(x^1)^t = H x^1 + b$, $I(x^1, \varepsilon) = \{i \mid x_i^1 - 1 \geq -\varepsilon\} \cup \{i + n \mid -x_i^1 - 1 \geq -\varepsilon\}$ and p^1.

Step 1: If $\|p^k\|^2 \leq \varepsilon$, then stop; else go to Step 2.

Step 2: Set $d^k = p^k$.

Step 3: Compute $\alpha^k = \min \{\alpha_u, \alpha_c\}$, according to Eq. (3.2).

Step 4: Let $x^{k+1} = x^k + \alpha^k d^k$.

Step 5: Let $\nabla f^0(x^{k+1})^t = \nabla f^0(x^k)^t + \alpha^k H d^k$.

Step 6: Compute $I(x^{k+1}, \varepsilon) = \{i \mid x_i^{k+1} - 1 \geq -\varepsilon\} \cup \{i + n \mid -x_i^{k+1} - 1 \geq -\varepsilon\}$.

Step 7: Compute p^{k+1} using Eq. (3.1).

Step 8: Set $k = k + 1$ and go to Step 1.

IV. FLETCHER-REEVES MODIFICATION FOR LINEARLY CONSTRAINED PROBLEMS.
We now consider the following linearly constrained class of problems.

Definition 2: Let L be the class of all problems of the form of Problem 1 in which the maps $f^0(.), f^1(.), ..., f^m(.)$ are such that (i) $f^0(.)$ is continuously differentiable, convex and radially unbounded, (ii) $f^i(x) = <g^i, x> - h_i$, for $i = 1, 2, ..., m$ define a nonempty constraint set Ω and (iii) for any $x \in \Omega$, g^i for $i \in I(x, 0)$ are linearly independent.

Given $x^1 \in \Omega$, let $I(x^k, 0)$ be the index set defined by $I(x^k, 0) = \{i \mid <g^i, x> - h_i = 0\}$ and let G_k be the corresponding matrix whose rows are $(g^i)^t$ for all $i \in I(x^k, 0)$. The projection matrix P_k associated with the point x^k is $P_k = I - G_k^t (G_k G_k^t)^{-1} G_k$. Let $M(I(x^k, 0))$ be the linear manifold defined by

$$M(I(x^k,0)) = \begin{cases} \{x \mid <g^i, x> - h_i = 0 \text{ for all } i \in I(x^k, 0)\} & \text{if } I(x^k, 0) \neq \phi \\ E^n & \text{if } I(x^k, 0) = \phi. \end{cases}$$

We may then compute the projection of $\nabla f^0(x^k)$ onto the set $M(I(x^k, 0))$ as $\hat{p}^k = -P_k \nabla f^0(x^k)^t$.

Our approach to solving a problem in L is to approximate $f^0(.)$ as a quadratic and generate a new direction d^{k+1} which is conjugate with respect to the previous direction d^k whenever possible. We compute the initial direction $d^1 = p^1$ and subsequent directions d^k using $d^k = p^k + \beta^k d^{k-1}$, where $\beta^k = \|p^k\|^2 / \|p^{k-1}\|^2$. However, rather than projecting $x^k - \nabla f^0(x^k)^t$ onto the active constraint set

$M(I(x^k,0))$, we project onto the approximation set $\Omega(I(x^k,\varepsilon))$, that is, we compute $p^k = w(x^k,I(x^k,\varepsilon)) - x^k$. Consequently, we may produce a d^{k+1} which is conjugate to d^k whenever $P_{k+1} = P_k$ and $p^{k+1} = \hat{p}^{k+1}$; otherwise we restart the algorithm with $d^{k+1} = p^{k+1}$. Furthermore, if $f^0(.)$ is not quadratic, the algorithm should be restarted at least every $n - |I(x^{k+1},0)|$ steps as a spacer step to ensure global convergence.

We now give the parametrized gradient projection algorithm to solve a problem in the class L.

Algorithm 3: Let x^1 and $\varepsilon > 0$ be given.

Step 0: Set $k = \hat{k} = 1$ and $d^0 = 0$.
Compute $\nabla f^0(x^1)$, $\quad I(x^1,\varepsilon) = \{i \mid f^i(x^1) \geq -\varepsilon\}$, $\quad I(x^1,0) = \{i \mid f^i(x^1) = 0\}$ $\quad$ and $p^1 = w(x^1,I(x^1,\varepsilon))-x^1$.

Step 1: If $p^k = 0$ then stop; else go to Step 2.

Step 2: If $d^{k-1} = 0$ then let $d^k = p^k$ and go to Step 5; else go to Step 3.

Step 3: Let $\beta^k = \|p^k\|^2 / \|p^{k-1}\|^2$.

Step 4: Let $d^k = p^k + \beta^k d^{k-1}$.

Step 5: Compute $\alpha^k > 0$ to minimize $f^0(x^k + \alpha^k d^k)$ subject to $(x^k + \alpha^k d^k) \in \Omega$.

Step 6: Let $x^{k+1} = x^k + \alpha^k d^k$.

Step 7: Compute $\nabla f^0(x^{k+1})$.

Step 8: Compute $I(x^{k+1},\varepsilon) = \{i \mid f^i(x^{k+1}) \geq -\varepsilon\}$.

Step 9: Compute $p^{k+1} = w(x^{k+1},I(x^{k+1},\varepsilon)) - x^{k+1}$.

Step 10: Compute $I(x^{k+1},0) = \{i \mid f^i(x^{k+1}) = 0\}$.

Step 11: Compute $\hat{p}^{k+1} = -P_{k+1}\nabla f^0(x^{k+1})^t$.

Step 12: If $I(x^{k+1},0) \neq I(x^k,0)$ or $p^{k+1} \neq \hat{p}^{k+1}$ or $\hat{k} > n - |I(x^{k+1},0)|$ then set $d^k = \hat{k} = 0$.

Step 13: Set $k = k + 1$, $\hat{k} = \hat{k} + 1$ and go to Step 1.

Theorem 2: If $\{x^k\}$ is a sequence constructed by Algorithm 3 to solve a problem in L, then either $\{x^k\}$ is finite and the last point is optimal or $\{x^k\}$ is infinite and contains cluster points, each of which is optimal. Furthermore, when a problem in L is such that $f^0(.)$ is strictly convex, then that problem has a unique optimal solution x^*, and in that case $\{x^k\}$ converges to x^*.

V. DISCRETE LQR CONTROL PROBLEM WITH HARD CONTROL BOUNDS.
In this section we state the well-known discrete LQR optimal control problem with hard control bounds and then rewrite it as a bounded variable quadratic programming problem.

Problem 3: Given an m-input, discrete, time-varying linear system in which we are given the initial state, $z_0 \in E^n$, and $z_i = A_i z_{i-1} + B_i u_i$, $i = 1, 2, ..., N$, where for $i = 0, 1, ..., N$, $z_i \in E^n$ is the state of the system at time i and for $i = 1, 2, ..., N$, $u_i \in E^m$ is the control at time i with components $u_{i,j}, j = 1, 2, ..., m$, find the mN control vector $u = (u_1^t, u_2^t, ..., u_N^t)^t$ that minimizes the performance index

$$J(u) = \frac{1}{2} \sum_{i=1}^{N} \left(z_i^t Q_i z_i + u_i^t R_i u_i \right),$$

and satisfies $u \in \Omega = \{u \mid |u_{i,j}| \leq 1$, for $i = 1, 2, ..., N$, $j = 1, 2, ..., m\}$, where for $i = 1, 2, ..., N$, Q_i are $n \times n$ symmetric positive semi-definite matrices and R_i are $m \times m$ symmetric positive definite matrices.

Using the notation introduced in [MEY87b], Problem 3 may be rewritten as follows:

Problem 4: Find the control vector $u \in E^{mN}$ that minimizes the performance index $J(u) = 1/2 \, u^t H u + b^t u + c$ subject to $u \in \Omega$, where H is the $mN \times mN$ block symmetric positive definite matrix with block size $m \times m$ given by $H = (R + F_u^t F_z^t Q F_z^{-1} F_u)$, b is the $mN \times 1$ vector given by $b^t = z_0^t F_0^t F_z^t Q F_z^{-1} F_u$, and c is the scalar given by $c = 1/2 z_0^t F_0^t F_z^t Q F_z^{-1} F_0 z_0$, and the matrices R, Q, F_z, F_u and F_0 are defined as: Q is the $nN \times nN$ block diagonal matrix that consists of N $n \times n$ symmetric positive semi-definite blocks Q_i, R is the $mN \times mN$ block diagonal matrix that consists of N $m \times m$ symmetric positive definite blocks R_i, F_z is the $nN \times nN$ block lower bidiagonal matrix that consists of N^2 $n \times n$ blocks $\left(F_z \right)_{ij}$, F_u is the $nN \times mN$ block diagonal matrix that consists of N^2 $n \times m$ blocks $\left(F_u\right)_{ij}$ and F_0 is the $nN \times n$ block matrix that consists of N $n \times n$ blocks $\left(F_0\right)_i$ defined for all i and j in $[1,2,...,N]$ by

$$\left(F_z\right)_{ij} = \begin{cases} I & \text{if } i = j \\ -A_i & \text{if } i = j+1, \\ 0 & \text{otherwise,} \end{cases} \qquad \left(F_u\right)_{ij} = \begin{cases} B_i & \text{if } i = j \\ 0 & \text{otherwise,} \end{cases} \qquad \left(F_0\right)_i = \begin{cases} A_1 & \text{if } i = 1 \\ 0 & \text{otherwise.} \end{cases}$$

Since Problem 4 is of the form of Problem 2, Algorithms 2 and 3 can be used to solve Problem 4 and may be implemented using the approach for the evaluation of $g(u^1)$, α^k and $g(u^{k+1})$ given in [MEY87b] in which α^k and $g(u^{k+1})$ are computed by sharing common terms, where we use the notation $g(u^k) = (dJ/du)_{u=u^k}^t$.

VI. PARALLEL ALGORITHMS FOR OPTIMAL CONTROL PROBLEMS.

The model of SIMD parallel computation that we use consists of a global parallel memory, p parallel processors, and a control unit, where all processors perform the same operation at each time step.

We now embed the parallel procedures GRADIENT and DIRECTION given in [MEY87b] to obtain a parallel implementation of Algorithm 2 to solve Problem 4.

```
1. PROCEDURE PGPM(z_0,u^1)
2.   k := 1;
3.   g(u^1) := GRADIENT(z_0,u^1);
4.   p^1 := PROJECTION(u^1, g(u^1));
5.   WHILE ||p^k||^2 > ε_s DO
6.     d^k := p^k;
7.     π^k := DIRECTION(d^k);
8.     α^k := min {α_u, α_c};
9.     FORALL i ∈ {1,2,...,N} DO IN PARALLEL u_i^{k+1} := u_i^k + α^k d_i^k;
10.    FORALL i ∈ {1,2,...,N} DO IN PARALLEL g_i^{k+1} := g_i^k + α^k π_i^k;
11.    p^{k+1} := PROJECTION(u^{k+1}, g(u^{k+1}));
12.    k := k + 1;
14.  END WHILE
15. END PROCEDURE
```

A straightforward computation shows that the speedup S_p and efficiency E_p for procedure PGPM are bounded from below by $S_p \geq 0.6p$ and $E_p \geq 0.6$; similar results exist for the parallel implementation of Algorithm 3.

VII. CONCLUSIONS.

In this paper two efficient parallel algorithms have been presented to solve the discrete LQR optimal control problem with hard control bounds. The algorithms possess the desirable property that their structure, and hence parallelism, is determined by the number of available processors. Thus, unlike approaches in which the structure of the procedure changes with problem size, the procedures presented in this paper maintain the same computational and interprocessor communication requirements independently of the number of stages in the control problem.

REFERENCES

[MEY87a] Meyer, G.G. and L.Podrazik, L.J., A Parallel First-Order Linear Recurrence Solver, *Journal of Parallel and Distributed Computing*, Vol. 4, No. 2, April, 1987, pp. 117-132.

[MEY87b] Meyer, G.G. and L.Podrazik, L.J., Parallel Implementations of Gradient Based Iterative Algorithms for a Class of Discrete Optimal Control Problems, *1987 International Conference on Parallel Processing*, August, 1987, pp. 491-494.

[SCH81] Scheel, C. and McInnis, B., Parallel Processing of Optimal Control Problems by Dynamic Programming, *Information Sciences*, Vol. 25, No. 2, November 1981, pp. 85-114.

[TRA80] Travassos, R. and Kaufman, H., Parallel Algorithms for Solving Nonlinear Two-Point Boundary-Value Problems Which Arise in Optimal Control, *Journal of Optimization Theory and Applications*, Vol. 30, No. 1, January 1980, pp. 53-71.

CHAPTER 40

Ising Spin on a Shared Memory Machine: Computational Experience*

James L. Blue[†]
Francis Sullivan[†]

Abstract. Ising spin simulations occur in many areas of statistical physics, such as studies of phase transitions, alloy solidification, and domain growth. Because of the simplicity and utility of the Ising model, spin system simulations have attracted the attention of researchers since the earliest days of electronic computation. Computational intensity is due to the fact that, for a large system, several hundreds of thousands, or even millions of updates spin sites must be performed in order to approach a single equilibrium configuration, and often averages over many configurations are required.

Ising problems provide a very good example of how, from the point of view of algorithm design, two apparently disparate problem types can be attacked by almost the same techniques. For standard methods, the kernel process appears to be "iteration on regular grids." This observation suggests parallel algorithms similar in spirit to those used for solving elliptic pde's by finite difference approximations.

We have devised algorithms which have similarities with the techniques of red-black ordering and chaotic relaxation. Preliminary tests indicate that for a five processor shared memory machine one can expect a speedup of 3.5 to 4.5 times the execution time for a single processor.

1 Introduction

During the past decade there have been remarkable improvements in the amount of computing power available to computational scientists. These advances have been achieved partly because of the development of components with very high switching speeds, and partly because of the exploitation of new machine architectures. Further improvements in component speed are expected, but future radical change in computing power will probably come about because of advances in machine architecture. So far, vector pipeline machines have been the

* Research supported in part by the Department of Energy.
† Center for Applied Mathematics, National Bureau of Standards, Gaithersburg, Maryland 20899.

hardware used for the largest computation; in the next few years multiprocessing parallel machines will become more important.

The availability of supercomputers has encouraged scientists to attempt computations which would have been considered impractical only a few years ago. In some cases, success has been achieved and significant results have been obtained with the help of more powerful computers. In almost all cases, researchers have obtained some benefit from the use of fast machines. However, in many cases users have experienced inadequate speed-up in moving codes to vector machines. According to Amdahl's law, the unvectorized portion of a code eventually dominates in the sense that the cpu time spent in executing the unvectorized part of the program is not reduced. The greater the disparity between the vector speed and the scalar speed of the machine, the more striking the Amdahl's law effect becomes. Vectorizing compilers alleviate the problem somewhat, because they do a good job of vectorizing the parts of the code which are amenable to vectorization. However, no programming language processor can reorganize and restructure applications programs in a fundamental way; major rethinking of algorithms is often required to actually attain the speeds which supercomputers promise.

In the case of MIMD parallel computers, the problem is even more acute. One cannot simply give each of N processors an equal portion of a problem and achieve a speedup factor of N. In addition, a variety of parallel architectures are available, so that scientists are faced with questions more fundamental than those of optimizing the performance of a given class of problems on a given architecture. The application may in fact determine which type of machine to buy. Computer architects, on the other hand, now have more freedom of choice in trying to match machine design to problem type.

All of this highlights the need for techniques to measure and eventually to predict performance of parallel machines. Although computer performance has been measured for many years, the "science" of performance evaluation for novel architectures barely exists. Typical approaches so far have been code profiling and benchmarking. In the first case, software tools are used to determine in which part of a given program the execution time is spent. This information can be very useful, because it tells the programmer where to concentrate efforts in improving the code. The ideas of profiling and the benefits to be derived from it have been developed extensively in the work of Bentley [2]. There do remain questions on how to display and analyze the data generated by a code profiler. In addition, profiles can only lead to improvements in an already existing code; they do not point the way to new approaches.

Our approach to performance measurement is to develop a taxonomy of algorithmic paradigms based on how the data moves during a calculation, rather than on what the specific computational steps are. Thus, from this viewpoint, "Ising spin simulations," occur in two different places in the taxonomy as "applications," depending on which algorithm is to be used. As the taxonomy is constructed, benchmark codes typical of applications in each class of the taxonomy will be implemented on two different parallel computers in order to obtain detailed measures of performance.

The target machines for measurements are chosen as representative of the two main classes of MIMD parallel architectures - loosely coupled, message passing machines and tightly coupled, shared memory machines. The computers have been instrumented to allow detailed measurement and analysis of machine events, such as resource utilization and data motion. The measurements should lead to identification of the "kernel" process mechanisms in each benchmark, and also should result in benchmark implementation principles to optimize performance. Benchmarks will be distributed and promising methods will be scaled up to large calculations on mainframe parallel machines.

Our purpose in this paper is to describe the current taxonomy and to report on measurements of the Ising model as an instance of "iteration on regular grids." This is given in Section 3. In Section 2, we give some general information on the taxonomy.

2 Taxonomy

The taxonomy is based on connectivity and movement of data, rather than on problem areas. Thus, a specific application might appear in several classes of the taxonomy, depending on the algorithm used. Sorting, for example, appears in at least three different classes.

2.1 Iteration on Regular Grids

- For all grid points, i, find a new v_i from the old v_i and its old nearest neighbors v_{i+1}, v_{i-1}, v_{i+l}, v_{i-l}.

In this category are algorithms associated with finite difference approximations to partial differential equations. However there are other examples. In particular, the FFT algorithm can be written so that successive steps consist of a sequence of nearest neighbor updates, where the old values are multiplied by the appropriate complex roots of unity. Another example is a sorting algorithm, DIAMONDSORT [3], which performs well on vector machines. Because of the constraint of vector architecture, the method is designed to be "non-contingent" — the flow of control does not depend on the particular sequence to be sorted, only on its length. The kernel process is a sequence of vector merges which take the form of nearest neighbor updates.

2.2 Searching Ordered Structures

- Find all x such that;

$$m_i \leq x_i \leq M_i; \text{ for all } i = 1, \ldots, k$$

It has been known for a long time that sorting and searching procedures are pervasive in applications of computers to "non-scientific" areas. In fact, from the point of view of amount of computer time used, they are central to many scientific applications. In molecular dynamics codes, for example, sorting can be used as the basis of a method for constructing "neighbor tables" that greatly increase the efficiency of the code.

2.3   Event Scheduling

- Assign initial ranks.

- Do the rank-1 task.

- Re-order the ranks based on outcome.

Typical applications of algorithms in this class include adaptive quadrature, and one version of algorithms for Ising spin simulations. In the case of adaptive quadrature, the region of integration is subdivided into pieces, and an estimate of the relative error is made for each piece. The piece giving the largest error is the rank-1 task. To do the task we subdivide that piece and do a more accurate quadrature on the new subdivisions. This generates new tasks and ranks, etc. A similar idea can be used to speed up Ising spin simulations. Spin sites can be ranked according to the change in total energy which would result from reversing the spin at that site. The ranks are modified by the appropriate probabilities, but the principle is the same; a rank-1 task is performed (i.e. a spin is reversed) and the new ranks are assigned.

In both cases, sets (quadrature regions, spin sites) are continually being formed and ranked, and the item of interest is the current rank-1 set. In actual codes the set manipulations often take more time than the "calculations" (computing error estimates, evaluating energy changes). Data structures and algorithms which can be used for the set manipulations include heaps and more general order trees. The "kernel" process is "access the tree." Adapting such a kernel to a parallel machine raises interesting questions. On a loosely coupled architecture, for example, one probably should *not* try assigning sets to individual processors, because the membership of sets is constantly changing and so a lot of communication between processors would be required. On a tightly coupled machine, the usual worries about memory conflicts arise. In either case, dealing with the set accessing problem is the central issue.

2.4   Traversing Graphs

```
SEARCH(v)
    v=''old''
    FOR all w in L(v) DO
        IF w=''new'' THEN
            SEARCH(w)
        ENDIF
    ENDFOR
```

This is an outline of the recursive version of a depth-first-search algorithm. In its simplest form, it is tree traversal. Applications occur in many combinatorial problems, including searching for connected components of graphs, and triangulation.

3   Measurements on an Ising model

Ising spin simulations occur in many areas of statistical physics, such as studies of phase transitions, alloy solidification, and domain growth. Spin system simulations have attracted the attention of researchers since the earliest days of electronic computation [5]. They provide a very good example of how, from the point of view of algorithm design, two apparently disparate problem types can be attacked by almost the same techniques. For standard methods, the kernel process appears to be "iteration on regular grids." This observation suggests parallel algorithms similar in spirit to those used for solving elliptic pde's by finite difference approximations.

The Ising model is easily described. A spin variable $\sigma(i)$ is specified at the nodes of a uniform grid in two (or three) dimensions. At each grid site the spin can take on only the values +1 and -1. Spins are related to one another via the energy expression given by the Hamiltonian

$$E = -J \sum_{\{i,j\}} \sigma(i)\sigma(j)$$

where $\{i,j\}$ ranges over all nearest neighbor pairs of sites and J is the coupling constant for the problem under study. For an $\ell \times \ell$ two-dimensional problem, at site i we have the local energy expression

$$E(i) = -(J/2)[\sigma(i) \times (\sigma(i+1) + \sigma(i-1) + \sigma(i+\ell) + \sigma(i-\ell))].$$

In most cases periodic boundary conditions are imposed, so that $i \pm 1$ and $i \pm \ell$ are to be determined mod ℓ.

One wishes to compute various averages with respect to the probability $P(C)$, for a configuration of spins, C, to occur. A brute force evaluation of the averages is impractical, so that a method of "importance sampling" is used. This is acomplished by making a series of "moves" through configuration space, such that for each C, $P(C) \propto \exp(-E(C)/kT)$. Here $E(C)$ is the energy associated with configuration C, T is the temperature, and k is Boltzmann's constant. The classical algorithm for using Monte Carlo methods to sample configuration space is due to Metropolis, Rosenbluth, Rosenbluth, Teller and Teller [5], (called the $M(RT)^2$ algorithm for short). In the case of "spin-flip dynamics" a site i is chosen at random and the change in energy, $\Delta E(i)$, which would result in reversing the spin at that site is determined. Since only the site i and its four nearest neighbors are involved, it is easy to get an expression for $\Delta E(i)$. If $\Delta E(i) \leq 0$, then the move is "accepted" and the sign of $\sigma(i)$ is reversed. In case $\Delta E(i) > 0$, the move is accepted with probability $\exp(-\Delta E/kT)$. A similar algorithm can be devised for the case of "spin-exchange dynamics" where, rather than flips of single spins, swaps of neighboring pairs of spins are attempted.

It can be shown that the $M(RT)^2$ algorithm satisfies detailed balance and therefore defines a Markov process which samples the correct (Boltzmann) distribution of configuration of spins. For a large system, several hundreds of thousands, or even millions of updates of each site must be performed in order to approach a single equilibrium configuration, and often averages over many configurations are required.

3.1   Ising Spin Exchange on a Shared Memory Machine

Associating processors with spin sites in a one-one manner is not practical for a shared memory machine. Instead of mapping sites to processors, we use the fact that the expressions for the energy associated with a site are similar in form to the central difference approximation used to solve the Poisson equation,

$$\nabla^2 u = -\rho.$$

In the Poisson case, a grid site and its four nearest neighbors are related through the finite difference expression

$$-u(i - \ell) - u(i - 1) + 4u(i) - u(i + 1) - u(i + \ell) = (\Delta x)(\Delta y) * \rho(i).$$

A common strategy used in implementing iterative methods for solving the Poisson equation on a *vector* machine is to use the red-black ordering depicted below:

$$
\begin{array}{cccccc}
r & b & r & b & r & b \\
b & r & b & r & b & r \\
r & b & r & b & r & b \\
b & r & b & r & b & r \\
\end{array}
$$

Since no two red sites are nearest neighbors, all red sites can be updated "simultaneously." These values can then be used to update the black sites, and so-forth, alternating on each iteration between red and black sites.

The same idea can be applied to modify $M(RT)^2$. At each step, one of the two colors is chosen at random and then spin flips are attempted at all sites of that color [6]. The ordering of which sites to do first is immaterial.

For spin exchanges, eight different sites are involved in each move. For example to interchange sites k_1 and k_2 we have all of the depicted bonds to consider.

$$
\begin{array}{ccccccc}
 & & i_3 & & j_3 & & \\
 & & \Updownarrow & & \Updownarrow & & \\
i_2 & \Longleftrightarrow & k_1 & \Longleftrightarrow & k_2 & \Longleftrightarrow & j_2 \\
 & & \Updownarrow & & \Updownarrow & & \\
 & & i_1 & & j_1 & & \\
\end{array}
$$

In spin-exchange simulations, at least five colors seem to be required. (Sixteen colors have been used for large-scale calculations on the Cyber 205, resulting in an extremely fast *vector* code [1].)

The five-coloring is as follows:

$$
\begin{array}{ccccccc}
1 & 2 & 3 & 4 & 5 & 1 & 2 \\
3 & 4 & 5 & 1 & 2 & 3 & 4 \\
5 & 1 & 2 & 3 & 4 & 5 & 1 \\
2 & 3 & 4 & 5 & 1 & 2 & 3 \\
4 & 5 & 1 & 2 & 3 & 4 & 5 \\
1 & 2 & 3 & 4 & 5 & 1 & 2 \\
\end{array}
$$

All pairs of type 2 $\Longleftrightarrow$ 3, for example, can be considered for exchange without interfering with each other.

Version 1: "Synchronized"

Randomly choose one of the five colors, and randomly choose a direction from among { up, down, left, right }. All pairs of the chosen type can be worked on simultaneously by all the processors, dividing up the sites among the processors in any desirable way. Evaluating all pairs of a given type is called a "sweep" of that color. Sites of a given color can be swept in sequential order without introducing false dynamics into the simulation. The processors must be synchronized at the start of a color, when the pair type is chosen, to avoid memory conflicts; then no "locking" needs to be done. Preliminary tests indicate that for a five processor machine one can expect a speedup factor of 3.5 to 4.5 over the execution time for a single processor.

Version 2: "Independent"

If processes are *not* synchronized, each processor can work independently on a pair type. A memory "lock" must be used whenever spins are exchanged, because the the physics dictates that the balance between "+1" and "-1" spins must be maintained. It is possible that, because of memory conflicts, "old" spin information from one color will be used in some evaluations of $\Delta E(i)$ for another color. Numerical experiments and measurements seem to indicate that independent asynchronous evaluation causes no difficulties and is more efficient than synchronizing on the particular shared memory computer tested.

Instead of locking on every exchange, it is possible for each processor to save a list of needed exchanges and wait until the end of each sweep to exchange. Only one lock per sweep per processor is needed, but the overhead is higher.

For the independent case, a *proof* that the correct distribution is sampled will depend on the computer architecture, because the resolution of memory conflicts determines the age of the data used for computing $\Delta E(i)$. The method of chaotic relaxation uses similar algorithms which have been shown to be correct [4].

3.2   Measurements

Measurements were taken on a six-processor shared memory computer on a small simulation (52 by 52 sites). The variables used for measuring performance of the Ising models were timing measurements and counts of the number of spin exchanges per sweep. Definitions of the measurements are in Table 1. Measurements were recorded at three different temperature settings of the model: one high temperature setting ($\beta = 0.2$), and two low temperature settings ($\beta = 0.7$ and 0.9). Here $\beta = 4J/kT$; the critical temperature is at $\beta \approx 0.44$. (Note: In Versions 2 and 3, five times as many pairs are processed per processor sweep as in Version 1)

1. Average Sweep Time

   The average sweep time per processor (Table 2a) was significantly less for Version 2, independent processors, than for Version 1, synchronized processors, at all three temperatures. After normalization to equivalent numbers of exchange tests, the percentage reduction in sweep time per processor ranged from 13.6% at $\beta = 0.2$ to 26.6% at $\beta = 0.9$. The variant

Table 1

Definition of Measurement Variables

VARIABLE	DEFINITION
Sweep time per processor	Time from end of previous sweep to end of current sweep
Synchronization time	Time from first processor arriving at synchronization to last processor arriving at synchronization (Version 1 only)
Number of spin exchanges	Number of spin exchanges per processor sweep
Exchange locking time	Time to lock, do spin exchange, and unlock spin array (Version 2 only)

Table 2a

Average Sweep Time Per Processor

(μ sec)

β Setting	Version 1 (Synchronized)	Version 2 (Independent)	Version 2a (Indep.+wait)
0.2	1821*	7870	8185
0.7	1691*	6683	8012
0.9	1793*	6581	7965

* Represents one fifth as many exchange tests as the other two versions — sites of one color are divided among the processors.

of Version 2 which waits until the end of the sweep gave only a five to eleven percent reduction in sweep time over Version 1.

2. Synchronization Time

Synchronization time (Table 2b), the waiting time at the end of each sweep for all the processors to finish the sweep, averaged 9 to 13 percent of the total cycle time (10 sweeps). The time for synchronization appeared to be larger, both in absolute magnitude and relative to total sweep time, for the lower temperature tests. This is probably because the low temperature model is still "settling down" and most of the spin exchanges are done by the first processors to be started, resulting in larger differences in sweep time.

Table 2b
Synchronization Time
(Version 1 only)

β Setting	Wait Time (μ sec)	Percentage of Sweep Time
0.2	1646	8.83
0.7	1804	10.54
0.9	2329	12.91

3. Number of Spin Exchanges

Counts of the number of spin exchanges per sweep (Table 2c) reflect the sensitivity of the model to the different temperature settings. The average number of spin exchanges was significantly different for the low and high temperature settings. However, increases in the number of spin exchanges appeared to affect average sweep times only very slightly. As usual for Ising simulations, most of the processing time during a sweep is used for the tests and the generation of random numbers rather than for the exchanges.

Table 2c
Average Number of Spin Exchanges Per Sweep
(Total For All Processors)

β Setting	Version 1 (Synch.)
0.2	141.8
0.7	26.7
0.9	24.5

4. Exchange Locking Time

Timings of the spin exchange locks (Table 2d) consist of a random "snapshot" of the locks which occur during the operation of the model.

It seem clear that there were very few memory conflicts per sweep, but we did not record the total. We observed from the timings that the time to "fork" a process, i.e. to start up a new processor, is fairly large compared to the time to process a sweep. There is an initial start-up time (which for our model appeared to be on the order of ten full sweeps of the sites) before the full parallelism is effective.

Table 2d
Average Spin Exchange Lock Time
(μ sec)

β Setting	Version 2 (Indep.)
0.2	87.7
0.7	81.7
0.9	82.1

References

[1] J. G. Amar, F. E. Sullivan, and R. D. Mountain, *A Monte Carlo Study of Growth in the Two-Dimensional Spin-Exchange Kinetic Ising Model*, Phys. Rev. B, to appear.

[2] J. L. Bentley, *Writing Efficient Programs*, Prentice-Hall, Englewood Cliffs, N. J., 1982.

[3] H. Brock, B. Brooks, and F. Sullivan, *Diamond: A Sorting Method for Vector Machines.*, BIT 21 (1981), pp. 142-152.

[4] B. D. Lubachevsky and D. Mitra, *A chaotic asynchronous algorithm for computing the fixed point of a nonnegative matrix of unit spectral radius*, J. ACM 33 (1986), pp. 130-150.

[5] N. Metropolis, A. W. Rosenbluth, M. N. Rosenbluth, A. H. Teller, and E. Teller, *Equation of State Calculations by Fast Computing Machines*, J Chem. Phys. 21 (1953), pp. 1087-1092.

[6] G. O. Williams and M. H. Kalos, *A New Multispin Coding Algorithm for Monte Carlo Simulation of the Ising Model*, J. Statistical Physics 37 (1984), pp. 283-299.

CHAPTER 41

Large Scale FE Parallel Nonlinear Computations Using a Homotopy Method

Charbel Farhat[*]
Luis Crivelli[†]

Abstract Here we revisit Finite Element algorithms based on homotopy equations, for implementation on shared memory and local memory multiprocessors, which represent both extremes of today's high performance architectures. To achieve this goal, two nonnumerical algorithms for automatic domain decomposition are developed first. Then, a computational strategy that ties these with a numerical nonlinear algorithm based on homotopy equations is presented and discussed. Its implementation on parallel architectures features two levels of parallelism, namely, concurrency and vectorization. It requires little storage and minimizes synchronization and/or interprocessor communication. Numerical experiments conducted on Alliant FX/8 validate the computational strategy and assess its performance.

1. Introduction Discrete equilibrium equations arising from finite element nonlinear formulations may be written in the general compact form

$$\mathbf{r}(\mathbf{u},\mathbf{p},\theta) = 0 \qquad (1)$$

where $\mathbf{u}$ denotes the unknown vector of generalized displacements (rotations, temperatures, etc.) at the nodes of the discretized geometrical domain, $\mathbf{p}$ denotes a set of control parameters, θ is a functional of past history of the generalized deformation gradients, and $\mathbf{r}$ denotes the residual vector of out-of-balance generalized forces (moments, fluxes, etc.). Equation (1) covers all geometrical nonlinearities, material nonlinearities and several types of boundary condition nonlinearities.

The Newton-Raphson method and its numerous variants, collectively known as *Newton—like methods*, are the most popular class of methods for the solution of (1) on conventional computers. Despite their fast local convergence, these methods are known to suffer from serious computational disadvantages. The stiffness matrix $\mathbf{K} = \dfrac{\partial \mathbf{r}}{\partial \mathbf{u}}$ has to be recomputed and re-factorized if a direct solver is used. As a result, the solution cost for large-scale three dimensional problems remains high, even for today's single processor supercomputers.

---

[*] Assistant Professor, Department of Aerospace Engineering Science and Center For Space Structures & Controls, University of Colorado at Boulder, Campus Box 429, Boulder, Colorado 80309-0429

[†] Becario Externo del Consejo Nacional de Investigaciones Cientificas y Tecnias de la Republica Argentina and Department of Mechanical Engineering, University of Colorado at Boulder, Campus Box 427, Boulder, Colorado 80309-0427

Opportunities are now emerging to take advantage of parallel processing in nonlinear computations and thereby decrease the cost of a finite element analysis. To illustrate the potential of these new computing environments, an implementation of the conventional Newton-Raphson method on local memory multiprocessors was presented in [1], in conjunction with a general finite element distributed architecture. A direct equation solver was used to solve the incremental system arising at each iteration, so that storage requirements limited the size of the problems to be solved, while interprocessor communication costs required larger problems in order for the ratio work/communication to be optimized. Moreover, it is well known that conventional *Newton–like methods* may behave poorly near critical and bifurcation points and often fail to handle path-dependent problems such as plastic flow, where the stiffness matrix may oscillate wildly as the solution changes by small amounts.

Here, we are concerned with a robust nonlinear parallel computational strategy for large-scale two and three dimensional engineering problems. The basic algorithm is iterative and does not require a large amount of storage. First we briefly describe the selected numerical algorithm and point out its inherent two-level parallelism. Then, we discuss some issues of parallel implementation and propose two domain decomposition techniques. Performance assessment carried out on the Alliant FX/8 concludes this paper.

2. Relaxation Equation

The following vector algebraic homotopy equation associated with (1) was proposed by Felippa [2] in a different context than parallelism:

$$\mathbf{h}(t) = \mathbf{r}(t) - e^{-\alpha t}\, \mathbf{r}^{(0)} + e^{-\beta t}\, \mathbf{D}\, (\mathbf{u} - \mathbf{u}^{(0)}) = 0 \tag{2}$$

where α and β are complex conjugate with nonnegative real part and unit modulus and $\mathbf{D}$ is a nonnegative diagonal matrix. Basically, the idea is to drive at each iteration the residual vector $\mathbf{r}$ to zero as fast as possible. Here we modify (2) as following. After differentiating (2) twice, we set $\mathbf{D}$ to zero to obtain

$$\mathbf{K}\, \ddot{\mathbf{u}} + (\dot{\mathbf{K}} + 2\gamma\mathbf{K})\, \dot{\mathbf{u}} + \mathbf{r} = 0 \tag{3}$$

where $\gamma = (\alpha+\beta)/2$ and the dot denotes the derivative with respect to time. Clearly $\dot{\mathbf{K}}$ introduces only a second order effect, since it vanishes in the linear case. Hence, we set it to zero. Moreover, our interest is not in the exact solution of time dependent systems of equations, but in reaching $\mathbf{r} = 0$ as fast as possible, without overshooting in the iteration process. Hence, we replace (3) with

$$\overline{\mathbf{K}}\, \ddot{\mathbf{u}} + 2\gamma\overline{\mathbf{K}}\, \dot{\mathbf{u}} + \mathbf{r} = 0 \tag{4}$$

where $\overline{\mathbf{K}}$ is a diagonal matrix related to $\mathbf{K}$. Equation (4) can be also written as

$$\overline{\mathbf{K}}\, \ddot{\mathbf{u}} + 2\gamma\overline{\mathbf{K}}\, \dot{\mathbf{u}} + \mathbf{f}^e(\mathbf{p}) - \mathbf{f}^I(\mathbf{u}) = 0 \tag{5}$$

where $\mathbf{f}^e$ and $\mathbf{f}^I$ denote respectively the external and internal generalized forces. Equation (5) is discretized in pseudo-time with centered finite differences:

$$(1+\gamma h)\overline{\mathbf{K}}\, \mathbf{u}^{n+1} - 2\overline{\mathbf{K}}\, \mathbf{u}^n + (1-\gamma h)\overline{\mathbf{K}}\, \mathbf{u}^{n-1} - h^2(\mathbf{f}^e(\mathbf{p}) - \mathbf{f}^I(\mathbf{u}^n)) = 0 \tag{5}$$

The choices of $\overline{\mathbf{K}}$ and γ are dictated by convergence rate optimization. The pseudo-time step size h is obtained via the modification of an adaptive scheme due to Underwood [3] that maintains the stability of the integration procedure. The diagonal form of $\overline{\mathbf{K}}$ is necessary to preserve the explicit form of the central difference integrator.

3. Inherent Parallelism

It is worthwhile to make two important remarks at this point. First, note the vector form of (5), which clearly demonstrates the suitability of the algorithm for vector processors. By eliminating the need for matrices, one saves not only storage but also the overhead associated with interprocessor communication that is usually required by algorithms manipulating two dimensional arrays. Second, the explicit nature of the central difference method allows computations on different vector subcomponents to be performed in parallel without any processor synchronization and/or interprocessor communication, except for the evaluation of $\mathbf{f}^I(\mathbf{u})$ which requires special attention. The internal generalized force vector is obtained by accumulating the contributions of several connected elements. If $\mathbf{u_k}$ denotes the generalized displacement associated with the $k-th$ degree of freedom, then

$$[\mathbf{f}^{\,i}(\mathbf{u})]_k = \sum_{el=1}^{el=N_k} [\mathbf{f}^{\,i}(\mathbf{u})]_k^{(el)} \qquad (6)$$

where $[\mathbf{f}^{\,i}(\mathbf{u})]_k^{(el)}$ denotes the contribution of element $\mathbf{el}$ connected to the $k-th$ degree of freedom and is computed directly at the element level from a potential functional. It is this particular part of the computations that motivates us to develop two different domain decompositions for local and shared memory multiprocessors.

4. Implementation on Local Memory Multiprocessors An arbitrary finite element domain is automatically decomposed by a nonnumerical algorithm into a number of subdomains equal to the number of available processors, N_p. Each subdomain contains equal number of elements, so that load balance among the processors is achieved. Moreover, the algorithm for decomposition optimizes the number of interface nodes and hence minimizes the interprocessor communication requirements. Each processor computes first the components of $\mathbf{f}^{\,i}$ it can access, then communicates with its neighbors only for those components of $\mathbf{f}^{\,i}$ which correspond to interface nodes. After that, it updates the components of the generalized displacement vector $\mathbf{r}$ that are lying in its proper subdomain, without any communication with any other processor. As an example, consider the domain of figure 1.a. By setting N_p to 4, it is automatically decomposed into four subdomains as shown in figure 1.b.

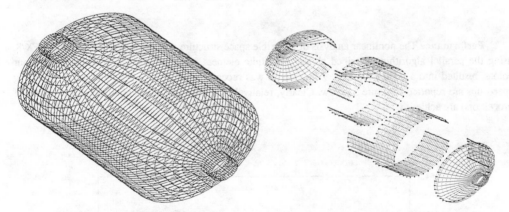

FIG. 1-a. Space Habitation Module   **FIG. 1-b. Decomposition into 4 Subdomains**

5. Implementation on Shared Memory Multiprocessors Within this class of hardware architecture, a processor can reference any location in the global memory. However, one needs to insure that no two memory locations are fetched then updated at the same time, because this will lead to wrong numerical results. Usually, to avoid memory conflict, the computing environment provides the user with software constructs such as *Critical Section* [4] which serialize the computations upon memory contentions in order to maintain consistency. However, if numerous, these serializations degrade the performance and ruin the sought-after speed-up. The domain decomposition we propose for this class of machines consists of a coloring technique, where no two adjacent elements get the same color. By mapping the processors successively on each color, the computation of $\mathbf{f}^{\,i}$ can be performed in a number of waves equal to the number of resulting colors, with synchronization occurring only once, at the beginning of a wave. The computations within a wave are carried out in parallel. Figure 2 illustrates the application of this coloring technique to the same previous finite element domain (for clarity, only one color is drawn).

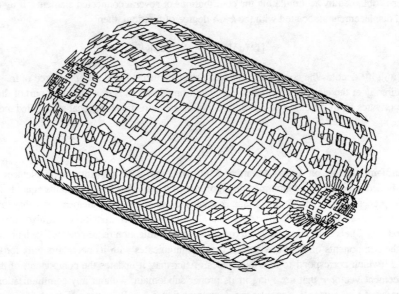

FIG. 2. Elements With a Same Color

Performance The nonlinear analysis of a flexible space structure was carried out on an Alliant FX/8, using the parallel algorithm described above. The finite element mesh, which was partitioned into four colors, resulted into 13,000 equations. The analysis was repeated with 1, 2, 3, ..,7 processors. Resulting speed-ups are reported in figure 3 below. Clearly, relatively high rates of efficiency (speed-up/number of processors) are achieved.

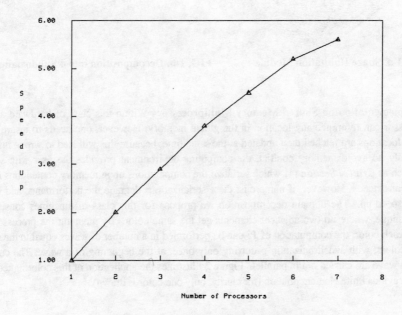

FIG. 3. Performance on an Alliant FX/8

REFERENCES

[1] C. Farhat and E. Wilson, A New Finite Element Concurrent Computer Program Architecture, *Int. J. Num. Meth. Engng,* Vol. 24 (1987), pp. 1771-1792

[2] C. Felippa, Dynamic Relaxation Under General Increment Control, *Innovative Methods for Nonlinear Problems,* (Eds. W. K. Liu, T. Belytschko and K. C. Park), Pineridge Press, Swansea, U.K. (1984), pp. 103-134

[3] P. G. Underwood, Dynamic Relaxation Techniques: A Review, *Computational Methods for Transient Analysis,* (Eds. T. Belytschko and T. R. J. Hughes), North-Holland, Amsterdam (1982), pp. 245-263

[4] H. Jordan, Structuring Parallel Algorithms in an MIMD, Shared Memory Environment, *Parallel Computing,* Vol. 3 (1986), pp. 93-110

CHAPTER 42

A Parallel Algorithm for Rapid Computation of Transient Fields*

B. J. Helland†
R. J. Krueger†

Abstract. The problem considered here is that of determining the internal electromagnetic field produced by an arbitrary incident pulse impinging on a one-dimensional medium with spatially varying permittivity and conductivity profiles. This is done by introducing a special set of Green's functions for the problem. These functions have the property that they can be computed for all time (at a fixed spatial location) via an algorithm which uses as input data the values of the functions for a fixed initial period of time. Although this technique is more computationally intensive than standard methods for computing Green's functions, it is ideally suited for a parallel environment because there is no sequential code involved. Numerical examples are presented comparing this approach with standard techniques on the Sequent Balance 21000 and the Intel iPSC Hypercube.

1. Introduction. The accurate computation of transient fields is a subject of increasing interest in areas ranging from geophysics to biology. In this paper a generic problem modeling electromagnetic wave propagation in an inhomogeneous, lossy medium is addressed. The field is induced by a rapidly changing pulse of short duration which impinges on the medium from free space. The interest in this type of problem is in quantitatively determining the transients in the medium and locating regions of high energy and current density. The goal in this paper is to provide an algorithm for accurately determining such fields.

A Green's function in the time domain is computed for the purpose of calculating the transient field in the medium. Thus, the burden of the numerical com-

* This work was supported by the Applied Mathematical Sciences subprogram of the Office of Energy Research, U.S. Department of Energy under contract No. W-7405-ENG-82. The numerical computations were performed at the Advanced Computing Research Facility, Mathematics and Computer Science Division of Argonne National Laboratory.

† Applied Mathematical Sciences, Ames Laboratory - USDOE, Iowa State University, Ames, IA 50011.

putation is in the construction of the Green's function. Although this can be done via any standard time-stepping algorithm, in this paper it is shown that an alternate algorithm, which is not competitive in a sequential computing environment, is in fact superior in a parallel environment because it minimizes computation time.

2. Green's Functions. A one-dimensional problem is considered in which a medium of length L is situated in free space. This slab has permittivity, ϵ, and conductivity, σ, varying with depth in the medium. A plane electromagnetic wave, E^{inc}, impinges normally on the medium from free space. Thus, the electric intensity, $E(z,t)$, inside the slab satisfies

$$\frac{\partial^2}{\partial z^2} E(z,t) - \epsilon(z)\mu_0 \frac{\partial^2}{\partial t^2} E(z,t) - \sigma(z)\mu_0 \frac{\partial}{\partial t} E(z,t) = 0 \tag{1}$$

where z denotes depth into the slab, t denotes time and μ_0 is the permeability of free space.

It is shown in [2] that Eq. (1) can be rewritten as follows via transformations of the independent and dependent variables:

$$\frac{\partial}{\partial x} u^+(x,s) + \frac{\partial}{\partial s} u^+(x,s) = - a(x)u^+(x,s) + b(x)u^-(x,s)$$
$$\frac{\partial}{\partial x} u^-(x,s) - \frac{\partial}{\partial s} u^-(x,s) = \quad a(x)u^+(x,s) - b(x)u^-(x,s) \tag{2}$$

for $0 < x < 1$, $s > 0$. The initial and boundary conditions are

$$u^+(x,0) = u^-(x,0) = 0, \tag{3}$$

$$u^+(0,s) \ \ given, \quad u^-(1,s) = 0. \tag{4}$$

Heuristically, the functions $u^\pm$ can be thought of as right and left moving waves, respectively, and their sum is equal to the electric intensity E. The variables x and s can be thought of as denoting space and time, respectively. The coefficients a and b are related to ϵ, σ, μ_0 and L. In Eqs. (4) $u^+(0,s)$ is the given incident field, E^{inc}, and $u^-(1,s) = 0$ denotes the fact that no incident field impinges on the medium from the right.

The Green's functions $G_1(x,s)$, $G_2(x,s)$ map any incident field over to the corresponding $u^\pm$ fields,

$$u^+(x,s) = c(x)\left[u^+(0,s-x) + \int_0^{s-x} u^+(0,s')G_1(x,s-s')\,ds' \right],$$
$$u^-(x,s) = d(x) \int_0^{s-x} u^+(0,s')G_2(x,s-s')\,ds' \tag{5}$$

where c and d are related to a and b. This is shown in [2], as is the system which the G's satisfy,

$$\frac{\partial}{\partial x}G_1 + \frac{\partial}{\partial s}G_1 = b(x)c(x)d^{-1}(x)\ G_2,$$
$$\frac{\partial}{\partial x}G_2 - \frac{\partial}{\partial s}G_2 = a(x)c^{-1}(x)d(x)\ G_1 \tag{6}$$

for $0 < x < 1$, $s > x$. The boundary and initial conditions for the G's are

$$G_1(0, s) = 0, \quad s > 0, \quad G_2(1, s) = 0, \quad s > 1$$

$$G_2(x, x) = -\tfrac{1}{2}a(x)c^{-1}(x)d(x) \quad 0 < x < 1.$$

Notice that since this latter condition is given on a characteristic, it specifies $G_1(x, x)$.

The goal of this paper is to accurately compute interior fields, $E(z, t)$. It may appear that conceptually there is no advantage in using Green's functions since the functions $u^{\pm}$ satisfy a hyperbolic system of equations, as do the Green's functions G_1, G_2. However, the Green's function approach enjoys two advantages over straightforward solution of the system (2). First, since the Green's function is independent of the particular choice of incident field, it follows that once the G's are computed, the interior fields due to a variety of incident fields can be examined using Eq. (5). In fact, it is possible to compute these fields at selected spatial and temporal points, without resorting to computing the entire field within the medium. Second, the grid size used to compute the G's depends only on the magnitude of the spatial gradients in the constitutive parameters of the medium, i.e., the grid size is independent of the temporal variation of the incident field. Thus, even if a rapidly changing incident field is impinging on the medium, it is not necessary to use a fine grid in the computational domain to accommodate this temporal variation. Rather, once the Green's functions are computed they can be interpolated as functions of s for fixed x to provide accurate field calculations using Eq. (5).

The system (6) provides a means of computing Green's functions. However, in [2] a new identity for Green's functions is derived. Begin by introducing the resolvent kernel, $W(x)$, of $G_1(1, s)$, i.e.,

$$W(s) + G_1(1, s) + \int_1^s W(s')G_1(1, s - s') \, ds' = 0, \quad s \geq 1.$$

It can be shown that $W(s) = 0$ for $s < 1$ and $s > 3$. Next, define the total Green's function, $G(x, s)$, by

$$G(x, s) = G_1(x, s) + c(x)d^{-1}(x) G_2(x, s).$$

Then the total field, $u(x, s)$, is given by

$$u(x, s) = u^+(x, s) + u^-(x, s) = c(x)\left[u^+(0, s - x) + \int_0^{s-x} u^+(0, s')G(x, s - s') \, ds'\right].$$

Finally, it can be shown that

$$G(x, s) + \int_{2+x}^s G(x, s')W(s + 1 - s') \, ds' = -\int_{s-2}^{2+x} G(x, s')W(s + 1 - s') \, ds' \quad (7)$$

for $s \geq 2 + x$. Notice that for fixed x this is a Volterra integral equation of the second kind for $G(x, s)$. This equation provides a basis for a new algorithm for computing Green's functions.

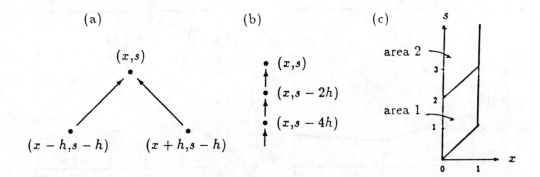

Fig. 1. (a) Computational grid for method of characteristics. (b) Computational grid for extension method. (c) Computational domain.

3. Algorithms for Computing Green's Functions. Two algorithms are considered here. The first consists of using the method of characteristics (see [1]) on the system (6). This is done by integrating along the characteristics using the trapezoid rule. This low order method easily accommodates the discontinuity in $G_2(x,s)$ along the line $s = 2 - x$. Mesh sizes of h and $h/2$ are used and the final results are obtained via passive polynomial extrapolation. The computational molecule is shown in Fig. 1a.

The second algorithm is called the extension method, since it is based on the extension of data result given in Eq. (7). The first step in this technique is to compute $G(x,s)$ in area 1, defined by $0 \leq x \leq 1$, $x \leq s \leq x + 2$. (See Fig. 1c.) This is done via the method of characteristics discussed above. Then in area 2 (defined by $0 \leq x \leq 1$, $x + 2 \leq s$) the Volterra equation for the G's is solved using Simpson's rule. By using the extrapolated values of the G's from area 1, this extension technique yields comparable accuracy in area 2 to that given by the method of characteristics. The computational grid for a typical point in area 2 is shown in Fig. 1b.

An operations count for each grid point in area 2 at which G is computed shows that the method of characteristics requires 65 multiplications and 33 additions per point. The extension technique requires $N + 3$ multiplications and $N + 1$ additions, where N is related to the step size h by $h = 1/N$. Thus, for $N = 64$ the two methods require roughly the same number of operations, for $N = 128$ the extension technique requires twice the number of operations, for $N = 256$ four times the number of operations, etc.

Based on this operations count, the extension technique is uncompetitive for large problems when sequential computation is employed. However, for parallel computation the method of characteristics requires that data from neighboring spatial points be communicated to a given grid point. This necessitates synchronization of processes, and in a non-shared memory environment, message passing (see [3]). The extension technique, on the other hand, requires neither synchronization nor message passing.

The characteristic and extension algorithms were implemented on an Intel iPSC Hypercube and a Sequent Balance 21000. The particular medium under study consisted of three lossy dielectric layers smoothly joined to each other. (This

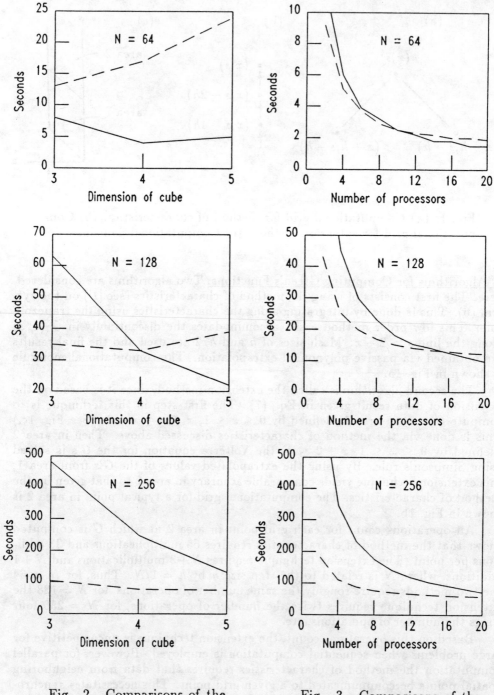

Fig. 2. Comparisons of the extension method (solid line) vs. the method of characteristics (broken line) in area 2 on the Intel iPSC hypercube as a function of problem size, N.

Fig. 3. Comparisons of the extension method (solid line) vs. the method of characteristics (broken line) in area 2 on the Sequent Balance 21000 as a function of problem size, N.

has no bearing on the execution times.) Green's functions were constructed for a time period of five round trips in the medium, meaning that area 2 consisted of four round trips. On the Hypercube, each node was dedicated to a specified spatial subinterval of $[0, 1]$. The method of characteristics was tuned somewhat to take advantage of the one-dimensional geometry of the problem by "time stepping" from one fixed value of $s - x$ to the next, i.e., stepping from one characteristic to the next, rather than stepping from one fixed value of s to the next. On the Balance which is a shared memory machine, the C$DOACROSS directive was invoked in the implementation of both algorithms. In the method of characteristics it was invoked at each fixed time step, with each processor taking the next available spatial step. Thus, there were no reduction, ordered or locked variables. In the extension method, each processor is responsible for an entire column of G data (i.e., fixed x), with each processor taking the next available column.

The results of these implementations in area 2 are shown in Figs. 2 and 3. Notice that on the Hypercube, message passing and synchronization overhead dominate the execution time for the method of characteristics (see the $N = 64$ case and, for 32 processors, the $N = 128$ and $N = 256$ cases). On the Balance, synchronization overhead ultimately causes the method of characteristics to become uncompetitive in the $N = 128$ case, even though the extension method performs twice the number of operations.

4. Conclusions. The results summarized in Figs. 2 and 3 show that the extension method is well suited to paralled computation in spite of its larger operations count. Both sets of figures seem to support the conjecture that as the size of the problem, N, increases and as the number of processors increases, the extension method will outperform the method of characteristics.

The extension method becomes even more competitive if interior fields are to be computed at a subset of the grid points. For example, computing the Green's functions at every second spatial grid point while maintaining temporal accuracy cuts the operations count in half for the extension method, while the method of characteristics requires the same number of operations as previously.

REFERENCES

[1] W. F. Ames, *Numerical Methods for Partial Differential Equations*, Academic Press, New York, 1977.

[2] R. J. Krueger and R. L. Ochs, Jr., *A Green's function approach to the determination of internal fields*, to appear.

[3] O. A. McBryan and E. F. Van de Velde,*Hypercube algorithms and implementations*, SIAM J. Sci. Stat. Comput., 8, (1987), pp. 227-287.

A Family of Concurrent Algorithms for the Solution of Transient Finite Element Equations
B. Nour-Omid and M. Ortiz

In this paper, we describe a solution method for problems in structural dynamics. This method makes effective use of the architecture of concurrent computers. The algorithm is based on partitioning the structure into substructures. Each substructure is then processed over a time step independently of the others. Thus, high levels of concurrency can be achieved at this phase of the analysis. The solution of the complete system is constructed by gluing the solution for the substructures. This involves averaging the solution at the interfaces between the subsystems. This averaging scheme is derived by enforcing the consistency condition on the underlying algorithm. The resulting two parameter algorithm is shown to be unconditionally stable.

CHAPTER 43

Mapping Large Scale Computational Problems on a Highly Parallel SIMD Computer

H. M. Liddell[*]
D. Parkinson[*]

Abstract. The mapping strategies employed to solve various size computational problems on an SIMD computer with p = 1000 or more processors are described. For 'small' problems, whose size n << p, many problems can be solved simultaneously, and when n is approximately equal to p, straightforward 'direct' mappings can be used. The most interesting situation arises for large problems, n >> p where various techniques have been applied to a wide variety of computational problems. These include sheet mapping, crinkled mapping, linear array, multi-serial or a combination of mappings. The problem of changing mappings is also considered. Finally, a summary of the application of these strategies to a number of problems will be given.

Introduction. An extensive subject of on-going research is the study of mapping strategies for solving various sizes of computational problems on highly parallel SIMD computers. In this paper we will attempt to summarise some of our experience to date. By 'highly parallel', we mean systems where the number of processors, p, is of the order of 1000 or more. At Queen Mary College, we have had experience of three generations of DAP systems [6], [11], [8]; the first is an array of 4096 processors attached to a conventional mainframe host computer, the second and third contain 1024 processors attached to scientific workstations. However, many of the results of our work apply to other highly parallel systems. 'SIMD' implies synchronous operation of the processors, which has the advantage of providing the user with no synchronisation problems! In the case of the DAP, each processor can be switched on or off, providing some degree of local autonomy; also the host and the DAP form an asynchronous system. We are primarily concerned with large scale 'fine grained' parallelism rather than 'task'parallelism, although some strategies used in the latter are also employed.

Features of the AMT DAP Architecture. The AMT DAP 500 series is the third generation of DAP products which has evolved as a result of more than ten years of experience by a community of several hundred users. Full details are included in [8], but it is relevant to include here a short description of some features which are particularly important from the user point of view.

---

*Queen Mary College, University of London. D Parkinson is also at Active Memory Technology Ltd, Reading, UK.

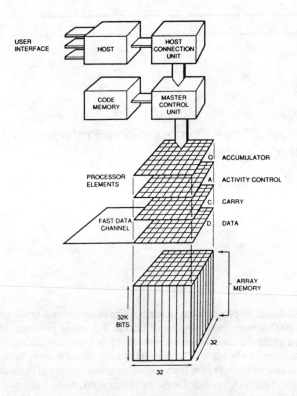

FIG.1. DAP 510 schematic

The DAP consists of 1024 Processing Elements (PEs), arranged as a 32 x 32 array , each with its own store of 32K or 64K bits; the architecture can support up to 1Mbit per PE, giving a total store potential of 128 MBytes. The D-plane provides a fast data channel for Input/Output, with a transfer rate of up to 50 MBytes/sec, a facility which is extremely useful for image processing, graphics and CAD applications. The mechanism for switching individual processors on or off is provided by the A (Activity) plane.

The DAP 510 can be attached to either a VAX or a SUN workstation with the corresponding operating system being either VMS or UNIX. Our experience with the SUN is that the high performance graphics workstation host with windowing facilities, combined with the computing power of the DAP make an excellent environment for interactive use. A fast frame store buffer and colour screen attached to the fast data channel provides a very powerful data visualisation tool. Another important feature from the user point of view is the good software support provided by the FORTRAN-PLUS compiler, APAL assembler, simulation facilities, run time support, the Subroutine Library (largely developed by Queen Mary College) and other applications software.

There are two systems of interprocessor communications: row and column highways which provide rapid data fetch and broadcasting to all elements in the array, and local connections to nearest neighbours. The functionality of these networks is reflected by functions in the FORTRAN -PLUS language. The overall result is an excellent balance between communication and processing speed with none of the communication bottlenecks experienced in some other multiprocessor systems.

Problem size. There are three cases to be considered: (1) the 'embarrassingly parallel' situation, where the size of an individual problem, n, is very much less than p, but where one needs to solve many such problems simultaneously (e.g. random number generation or rotation matrix multiplications, where one is manipulating sets of data of size 4x4); (2) problems whose size is approximately equal to the number of processors - either $n^2 = p$, which is the 'matrix processing' case, such as in the solution of a dense system of n linear equations, or n = p, where one may treat the system as a linear array (long vector) of processors, which is useful, for example, in sparse matrix applications. In both cases straightforward 'direct' mapping can be employed, and if the problem size is slightly less than that of the processor array, masking techniques can be used with very little loss of efficiency; (3) 'large' problems whose size n >> p (or n^2 >> p, n^3 >> p, depending on the dimensionality of the problem) where the upper limit for n is usually determined by the amount of store available. A hybrid approach of mixed parallel and serial algorithms is often used for this class of problem; this latter category is the most demanding of the programmer's ingenuity and could therefore be considered the most interesting; it includes many large scale problems - Finite Element applications, Computational Fluid Dynamics, Image Processing etc.

Mapping Strategies. The mapping strategies we are attempting to categorise in this paper have been developed for particular problems, but have been found applicable to many others. Although we concentrate on numerical problems here, the work has much in common with that described by Reddaway [10] for Image Processing. The first technique is 'sliced' or sheet mapping, illustrated below, in which the system is divided into sheets (slices) which match the size of the processor array.

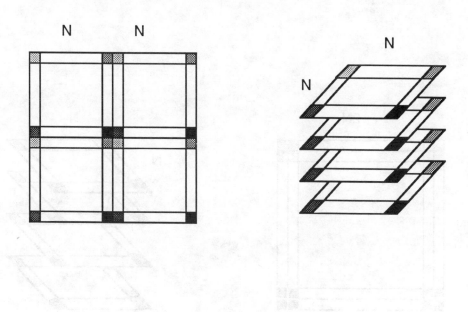

Fig. 2. Sliced mapping

In this case neighbouring matrix elements are stored in different PEs (see Fig.2). In image processing, the slices may be thought of as 'windows' displaying a section of the total image. Examples of applications where this type of mapping is used include multiplication of large matrices - the matrix can be partitioned into sub-matrices and the BIGSOLVE method described by Liddell et al for solving large dense systems of equations [4] applied; this is based on dividing the augmented matrix of the system into DAP size blocks. In both cases a mixture of serial and parallel techniques is used; in the former, the multiplication of any two submatrices employs a parallel matrix multiply strategy, but the overall block method is serial. For example, if one wishes to multiply two matrices A and B whose linear dimension is twice that of the processor array, one may partition them into submatrices, A_{ij}, B_{ij} $(i,j = 1,2)$ and the corresponding submatrices in the result C are given by

$$C_{ij} = \sum_{k=1}^{2} A_{ik} * B_{kj} \quad , \quad i,j = 1,2$$

(In practice, it is more efficient to use a modification of Strassen's algorithm [7]). In the BIGSOLVE method, a hybrid Gauss-Jordan, Gauss elimination technique is employed - the diagonal blocks of the system are converted to diagonal matrices using a parallel version of the Gauss-Jordan method, but the overall strategy is a block form of Gauss elimination.

For applications involving nearest neighbour operations, such as the solution of partial differential equations using finite different techniques, or for smoothing an image in an image processing application, it is often better to use 'crinkled' mapping, in which neighbouring elements of the matrix are stored in the same processor.

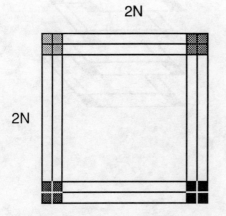
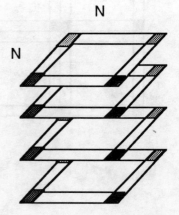

Fig. 3 Crinkled mapping

This 'crinkled' mapping is similar in concept to the 'domain decomposition' techniques popular for MIMD configurations with fewer processors. It has been employed by Wait [12] for FE Analysis and also by Pearmain et al for VLSI design [9]. Again a mixture of serial and parallel techniques is used, as serial preconditioning strategies are used within each processor prior to the application of a parallel Conjugate Gradient algorithm - this is similar to some of the domain decomposition methods used in MIMD systems. One disadvantage is that it is often necessary to have boundary values of the neighbourhood stored in more than one processor. Lai and Liddell [2] have also considered the application of sheet mapping for large FE problems and Mouhas [5] is comparing the application of both mappings to mesh generation.

In many cases, the mapping used is determined by the solution method and vice versa; for application of Conjugate Gradient techniques, a 'long vector' or linear array mapping is preferred - see Lai and Liddell [3]. This technique is used in the solution of tridiagonal systems and for storing the system matrix in Finite Element calculations. In both cases the system is mapped by diagonals, each being stored as a long vector.

All of these techniques can be extended fairly easily to 3D and higher dimensional problems; for example, the store of the processor array can be treated as a third dimension (see Fig. 1), which is a special form of 'sliced' mapping, or a 3D neighbourhood can be mapped into a single processor - again, it would be necessary to overlap the storage of data for the region boundaries.

Other calculations require that each item or element of data is treated independently and then one uses a 'multi-serial' approach, in which a serial algorithm is applied simultaneously to each processor to produce, say, 1024 (or a multiple of 1024) parallel results. An example of this is found in the calculation of the element stiffness matrices in an FE calculation, or the computation of a 1024 point FFT. An alternative method for calculating FFTs in parallel has been described by Flanders [1], who developed a novel technique called Parallel Data Transforms (PDTs) which can be used for many problems which involve reorganisation or sorting of the data held in the store of the processors. The mapping is specified by a one-dimensional binary mapping vector, and changes in the mapping are achieved by changing the mapping vector. This is a very important technique, which provides underlying support for many of the features in the FORTRAN-PLUS language and library routines.

Reddaway [10] addresses the problem of changing mappings during the course of the computation, and also describes the use of combined mappings where one dimension uses sheet and another crinkled mappings - this type of strategy could also be of use in mesh refinement. Many applications require various mapping strategies to be applied at different stages of the calculation.

Comparison of Sliced and Crinkled mapping techniques for nearest neighbour operations. In the previous section, it was suggested that for operations involving neighbouring elements, the 'crinkled' mapping is preferable to the 'sliced' mapping strategy. This may be quantified for a processor array such as the DAP, by considering skeleton codes for a simple 2D problem where the operations are of the form:

$$\phi_o = 0.25*(\phi_n + \phi_e + \phi_s + \phi_w)$$

and the value of some quantity ϕ at a central point o is given by the average value at neighbouring points, n, e, s, w. If t_c represents the time for communication from a neighbouring processor, and t_a the arithmetic time involved in the above operation (which includes three additions and one multiplication), we find that for a problem whose overall size is md x md (where the number of processors, $p = d^2$):

$$t_{sliced} = m^2 [t_a + 8 t_c]$$

$$t_{crinkled} = m^2 [t_a + \frac{4}{m} t_c]$$

Since the DAP is based on a bit serial architecture, the time for arithmetic operations varies with the precision used. In image processing, where one uses short fixed point arithmetic, the time for an arithmetic operation is approximately equal to that for communication between processors - i.e. $t_a = 4t_c$, so that if m is very large, crinkled mapping would give almost a factor of three performance improvement over sliced mapping. However, for numerical applications t_a is more than an order of magnitude greater than t_c, and the performance is approximately the same for both mappings, although the crinkled mapping is marginally preferable on performance grounds. A disadvantage of the crinkled mapping is that special processing over subdomains is more difficult to implement efficiently.

Summary. Only a few years ago, many researchers felt that there were few problems which would effectively use systems containing thousands of processors, but more recently it has become widely accepted that a large number of problems exhibit many thousands of degrees of parallelism, and the mapping of the problem onto the limited parallelism of even a highly parallel system is a major task. Table 1 shows the various mapping strategies which have been employed or are being investigated for a number of large scale application areas.

	Sliced	Crinkled	Long Vector (Linear Array)	Data Set	PDTs	Combination
Dense Linear Equations	√		√			
Banded Systems	√	√	√			
Finite Element	√	√	√		√	√
Mesh generation Mesh Refinement	√	√ √				√
ODEs	√		√	√		
Finite Difference		√				
Image Processing	√	√	√		√	√
FFTs				√	√	
Sorting					√	
Random Numbers			√			
VLSI Design		√				

Table 1 : Mapping Strategies used in Applications

It has been suggested that the mapping should be a task performed by the compiler; however, it is difficult to see how the appropriate choice can be made for complex problems, or how transformations between mappings can be done automatically as the computation proceeds. It is seen that many applications require a number of different strategies. For fuller details of the techniques of the techniques used in image processing, the reader should consult the survey paper by Reddaway [10].

The work is by no means complete; currently, further information is being gathered on the use of these and other strategies in a wide range of applications in science and engineering as part of an Alvey research project. We plan to produce more detailed results of this work in due course.

REFERENCES

[1] P. M. FLANDERS, A Unified approach to a class of data movements on an array processor, IEEE Trans. on Comput., C31-9, 809-819.

[2] C. H. LAI and H. M. LIDDELL, Preconditioned Conjugate Gradient methods on the DAP, paper presented at MAFELAP VI, April 1987, to be published in the conference proceedings.

[3] C. H. LAI and H. M. LIDDELL, Finite Elements using Long Vectors of the DAP, to be published in VAPPIII conference proceedings, North-Holland (1988).

[4] H. M. LIDDELL, D. J. HUNT and G. S. J. BOWGEN, The Solution of N linear equations on a P-Processor parallel computer, paper presented at the First SIAM Conference on Parallel Processing, Norfolk, Virginia, November 1983.

[5] C. MOUHAS, Automatic Mesh Generation on the DAP, presented at VAPPIII, Liverpool, August 1987.

[6] D. PARKINSON, The Distributed Array Processor (DAP), Comput.Phys.Comms, 28, (1983), pp 325-336.

[7] D. PARKINSON, The multiplication of matrices of size N x N on a P-Processor with P < N, DAPSU Technical Report 2.27 (1981).

[8] D. PARKINSON, D. J. HUNT and K. S. MACQUEEN, The AMT DAP 500, submitted to CompCom 88, (San Francisco).

[9] A. J. PEARMAIN, C. C. JONG and P. R. COWARD, Using the ICL Distributed Array Processor in a CAD system, presented at the EDA conference, Wembley, London, July 1987, to be published in the conference proceedings.

[10] S. REDDAWAY, Mapping images onto processor array hardware, paper presented at the BCS Parallel Architectures and Computer Vision Workshop, Oxford, March 1987, to be published in the conference proceedings.

[11] P. SIMPSON and J. B. G. ROBERTS, Speech Recognition on a DAP, Electronic Letters, 24, (1983)pp 1018-1020.

[12] R. WAIT, The solution of Finite Element equations on the DAP, paper presented at the International Conference on Vector and Parallel Processing, Loen, Norway, June 1986.

Analysis of a Parallelized Elliptic Solver for Reacting Flows

David E. Keyes and Mitchell D. Smooke

A parallelized finite difference code for systems of nonlinear elliptic boundary value problems in two dimensions, based on Newton's method, is analyzed in terms of computational complexity and parallel efficiency.

An approximate cost function depending on 15 dimensionless parameters (including discrete problem dimensions, convergence parameters, and machine characteristics) is derived for algorithms based on stripwise and boxwise decompositions of the domain and a one–to–one assignment of the strip or box subdomains to processors. The sensitivity of the cost function to the parameters is explored throughout a region of parameter space corresponding to a coupled system of nineteen equations with very expensive function evaluations (a detailed–kinetics diffusion flame). The algorithm has been implemented on parallel hardware and some experimental results for a flamesheet calculation with stripwise decompositions are presented and compared with the theory.

Performance of an Ocean Circulation Model on LCAP

Hsiao–Ming Hsu, Jih–Kwon Peir and Dale B. Haidvogel

A numerical ocean model based on the primitive equations has been parallelized and run on the LCAP system–an experimental parallel machine with an IBM host and ten FPS-x64 attached processors. Using coarse–grain parallelism approach, the computational domain is partitioned into a number of sub–domains according to one of the horizontal directions. Each attached processor computes one sub–domain. Inter–processor communication is accomplished through a set of shared bulk memories. Five synchronization (barrier) points are inserted into the program to guarantee a correct sequence of execution.

A set of ocean model benchmarks have been carried out. The best result shows that an impressive speed–up of 6.9 on a 10–processors scale can be achieved.

PART V

Languages

CHAPTER 44

Programming Parallel Architectures: The BLAZE Family of Languages

Piyush Mehrotra*

Abstract. Programming multiprocessor architectures is a critical research issue. This paper gives an overview of the various approaches to programming these architectures that are currently being explored. We argue that two of these approaches, interactive programming environments and functional parallel languages, are particularly attractive, since they remove much of the burden of exploiting parallel architectures from the user.

This paper also describes recent work by the authors in the design of parallel languages. Research on languages for both shared and nonshared memory multiprocessors is described, as well as the relation of our work to other current language research projects.

1. INTRODUCTION. One of the insights that has emerged in parallel computing research is that the design of portable and efficient software for parallel systems is a critical issue. Our success in inventing new high performance parallel architectures has not been matched by equal success in learning how to program them. We need mechanisms (languages, compilers, libraries, and tools) that allow parallel algorithms to be mapped to high performance computers, and that fully exploit the variety of kinds of parallelism supported by current architectures.

Today, each multiprocessor system typically comes equipped with its own dialect of C or FORTRAN, containing a variety of language extensions for exploiting parallelism. These extensions are generally specific to a given architecture, and may require programming techniques unique to that architecture. Thus, programmers wishing to use such a machine not only need to know the intricacies of their application, but must also become knowledgeable about the architecture and its programming environment.

While it is clearly useful to have parallel languages available for each new architecture, in the end we must get away from the idiosyncratic languages provided by manufacturers. For one thing, manufacturers face a host of hardware and software problems, and rarely have the resources or inclination to create carefully designed and well thought through programming environments. Of equal importance, parallel architectures continue to proliferate, and todays high-end machine will soon be replaced by new generations of machines, having new and different architectures. There is a clear need for programming environments which are portable across machines of different makes, and across the generations of architectures from each manufacture.

1.1. Parallel Programming Environments. We are still some distance away from having portable and user-friendly parallel programming environments, but a great deal of progress has been made towards achieving them. Four basic approaches to the construction of such programming environments have emerged.

---
*Dept. of Computer Science, Purdue University, West Lafayette, In. 47906. pm@cs.purdue.edu.
Research supported by NASA Contract No. NAS1-18107 while the author was in residence at ICASE, NASA Langley Research Center.

1. Explicit-tasking languages.
2. Direct compilation of existing sequential languages for multiprocessor execution.
3. Interactive program restructuring systems for existing sequential languages.
4. New high level parallel languages.

In practice, the distinctions between these four approaches are often quite blurred. For example, language constructs intended for high level parallel languages can easily migrate into interactive program restructuring systems. However, to keep the discussion here simple, we will treat these four approaches as distinct.

In the following paragraphs, we give a quick overview of current research in each of these areas. After that, we will focus more deeply on our own efforts in parallel language design.

Explicit-Tasking Languages. This approach is based on the view that it is the programmers responsibility to control and manage the resources of the underlying parallel architecture. Explicit-tasking languages generally follow the concept of "communicating sequential processes," advocated by C.A.R. Hoare, and embodied in the theoretical programming language, CSP [9]. In such explicit-tasking languages, the programmer defines and controls a system of interacting "tasks" or "processes." Depending on the underlying architecture, the interaction between tasks is either via synchronized sharing of data structures or via messages. In either case, the language provides mechanisms for the programmer to explicitly manage this interaction.

There is a clear advantage to this approach, since it allows complete control of the machine resources, and allows the programmer to fully exploit the target architecture. Efforts in this domain include the set of message-passing and synchronization primitives developed at Argonne National Laboratory [6], and a variety of new languages such as Occam [19], PISCES [20], FORCE [11], and LINDA [1].

The fact that explicit-tasking languages allow programs to exactly match the target architecture, carries with it an attendant disadvantage: loss of portability. Portability is lost even with "portable languages" like PISCES and FORCE; though such languages can run on a variety of machines, programs generally need to be structured differently for different architectures. Moreover, the visibility of the underlying architecture makes such programs quite difficult to design and debug. The programmer is faced with a variety of load-balancing, resource allocation, and communication and synchronization issues which are not present on sequential machines.

Direct Compilation of Conventional Languages. The second approach to programming multiprocessors, direct compilation of conventional languages for parallel execution, provides a number of important advantages. First, it allows programmers to continue using familiar languages as they move to newer and more complex machines. Second, there is a large body of existing programs which can be transported to parallel architectures without change. Third, the details of the target architecture are invisible to the programmer, so the complex load-balancing and program design issues, which must be faced with the explicit-tasking languages, are not present.

This approach is, in a real sense, a direct outgrowth of successful research in construction of vectorizing compilers. It is being actively explored by major groups at at Illinois [18], Rice [4], and IBM [2], and by smaller groups elsewhere. Since the millions of lines of existing sequential programs cannot be easily replaced, nor are they readily modifiable, there is clear importance to this approach, and it will surely continue.

There are, however, a number of difficulties with this approach. The major one is that the semantics of conventional languages strongly reflects the sequential von Neumann architecture, making the task of automatic restructuring very difficult. Aliasing effects in virtually all current languages obscure data dependencies, and severely limit the compiler's ability to extract parallelism. Moreover, existing languages, especially FORTRAN, encourage programming styles which make it extremely difficult for compilers to extract much parallelism. When arrays are freely "equivalenced," and passing of "pointers" is used to simulate dynamic allocation, the potential for parallel execution is quickly lost. The end result seems to be that direct compilation of sequential languages can extract only modest amounts of loop-level parallelism.

Interactive Restructuring Systems. The difficulties in direct compilation of sequential programs for multiprocessor execution has led to the exploration of a third approach to parallel programming, interactive restructuring systems. The problem with conventional sequential languages is that they do not provide the compiler with the "right" information for mapping programs to multiprocessors, and the information which they do provide is thoroughly hidden.

One way to alleviate this problem is to design a compiler which asks for "help" during the program transformation process. Programmers know far more about their programs than is directly visible in a program. For example, the typical number of invocations of a loop, the frequency with which a procedure is called, whether a procedures has "side effects," and so on, are all pieces of information a programmer might have. Interactive restructuring systems are systems which allow programmers to express this "deeper" knowledge to the compiler, and thereby guide the compilation process.

There are several mechanisms through which the programmer can provide this information to the compiler. First, the programmer can modify the source code, inserting pragmas and assertions, to help the compiler extract parallelism [8]. Second, the programmer can communicate interactively with the compiler, through a window-based graphics system, which allows the programmer to view and modify the program at various stages of the transformation process [3] [18]. Third, a variety of interactive performance analysis and debugging tools are possible, which can provide rapid feedback to the programmer.

There are two principal disadvantages to interactive program restructuring systems. First, the user of such a system has to be quite knowledgeable about the target architecture to be able to provide appropriate guidance. A naive user would not understand the nuances of the architecture and the program transformation process well enough to be of help. Second, the concept of a "program" as a file containing an algorithm expressed in a high level language is lost. Instead, the "program" is now the original source code, plus the sequence of hints and mouse clicks provided by the user during the interactive transformation process.

High Level Languages. The fourth approach to programming multiprocessors is to construct new high-level languages designed expressly for compilation to parallel architectures. There are a number of research projects focusing on the design of parallel languages which will hide most details of the parallel runtime environment. Examples of such projects include the Crystal [7], ParAlfl [10], VAL [16], SISAL [15], and BLAZE projects.

These language design projects, which are generally focusing on functional languages, are attempting to allow the programmer to concentrate on the *specification* of the algorithm rather than on its *implementation*. The goal has been to provide languages with simple and clean semantics, which make them easy for programmers to use, while also enabling the compiler to produce efficient executable code for parallel systems.

There are difficulties with this approach as well, though we tend to favor it over the other three approaches. One issue is that it is not yet clear whether these languages will enable programmers to extract the full potential of highly parallel architectures. All of the projects mentioned are in their infancy, and it is too early to declare any of them successful. Also, there is a problem with user acceptance. Programmers are, in general, reluctant to move to new a language, no matter how elegant, unless the benefits to doing so are overwhelming. Finally, none of these new languages mates well with existing languages, so it is difficult to combine the millions of lines of existing code with procedures or modules written in these newer languages.

1.2. Comparison of Approaches. In describing these four approaches to parallel programming, we have listed some of the advantages and disadvantages of each. Each of these approaches is being actively explored, because each offers advantages lacking in the others. However, in the longer term, we feel that the latter two approaches will come to dominate. The following are our reasons for this conjecture.

First, regarding explicit-tasking languages, such languages have been extremely useful in allowing users to experiment with parallel algorithm design. However, multiprocessor architectures are becoming increasingly complex, and now provide a variety of types of parallelism within the same system. As this happens, it is becoming increasingly unreasonable to expect programmers to manage the variety of parallel resources available in a complex multiprocessor systems. Further, multiprocessors are changing from laboratory curiousities to every-day work-horses. As this happens, higher level programming environments will become essential.

Second, regarding direct compilation of current sequential languages to parallel architectures, this approach has a continuing role to play in allowing exploitation of "dusty deck" programs. However, as multiprocessors become standard, and as the number of processors in a high performance systems grows, the limitations of this approach will become apparent. As Kennedy and other experts have pointed out, compilers cannot do everything, the programmer must help.

The implication of all this is that in order to obtain maximum performance from existing code, those programs will have to be extensively "massaged" with interactive program restructuring tools and compilers. Heavily used "kernels" may be reprogrammed in new languages, but the bulk of these programs will remain essentially unchanged. Users will have to add various pragmas and compiler directives either directly to the source, or indirectly through the program transformation system. However, the programs themselves will change relatively little.

The situation for new programs is different. In this case, both of the last two approaches to parallel programming seem viable. One will either use one of the current sequential languages, together with a sophisticated interactive compiler, or one will write the program in a new high level parallel language. Both of these approaches are able to fully exploit the performance potential of highly parallel multiprocessors, and both are also relatively user-friendly, hiding most of the complex details of the parallel runtime environment from the programmer.

The choice between these two approaches is complex. Our research has focused on the design of new languages, though we fully appreciate the merits of interactive compilers and program restructuring systems. In the end we expect these two research directions to merge, resulting in the eventual creation of elegant and friendly parallel programming environments combining the best aspects of both approaches.

2. THE BLAZE PROJECT. The focus of our research in the last few years has been the BLAZE Programming Environment. BLAZE [17] is a new functional programming language for scientific applications. The intention with BLAZE is to achieve highly parallel execution on a variety of shared memory multiprocessor architectures, while shielding the user from the details of parallel execution. In particular, neither the program structure nor the execution results will in any way reflect the multiple threads of control flow which may be present during execution. Our point of view is that such issues should be the responsibility of the compiler and runtime environment. In this section, we provide an overview of the BLAZE language and its compiler. We also briefly sketch our efforts in targeting compilers to nonshared memory machines.

2.1. The BLAZE Language. BLAZE is a parallel language for scientific programming with its roots in modern programming languages such as PASCAL, ADA, EUCLID, and MODULA 2. It contains extensive data structuring facilities and structured flow control constructs.

BLAZE is similar to data-flow languages in that it uses functional procedure invocation. That is, procedures in BLAZE act like functions which may return several values and operate without side-effects. In other words, they use value-result semantics for arguments, and have no access to nonlocal variables. This simple semantics makes restructuring of BLAZE programs for parallel execution much simpler than analogous restructuring of conventional languages. In particular, whenever there are no dependencies between the input and output values of two procedure calls, they can be executed in parallel. For example, the calls to the two procedures F and G in the following program fragment can be executed in parallel.

```
a, b  :=  F( c, d);
x  :=  G( c, y);
```

With conventional languages, determining when two procedures can safely be executed in parallel requires a complex and expensive global analysis of the program.

At the statement level, BLAZE differs from data-flow languages in that it uses traditional imperative semantics, rather than the "single assignment rule" used by data-flow languages. In BLAZE variables hold values that can be altered by assignment, just as in PASCAL, FORTRAN and other conventional languages. By contrast, in data-flow languages "variables" represent values rather than storage locations. Since names are bound to values rather than storage cells in these languages, the value of a "variable" cannot be altered once it has been set. Hence the idea of the single assignment rule.

Using traditional semantics at the statement level makes BLAZE programming natural to programmers accustomed to conventional languages. More surprisingly, the single assignment rule of data-flow languages does little to help compilation for multiprocessor architectures, and in some cases can severely hamper the compilation process. In particular, handling arrays and other large data structures in data-flow languages has proven quite awkward.

In addition to the functional procedure calling semantics, which allows procedures to be executed in parallel, there are two levels at which parallelism can be explicitly expressed in BLAZE. First, BLAZE provides extensive array manipulating facilities similar to those in ADA and FORTRAN 8x. Given the right hardware, these array operations can be executed in parallel.

Second, BLAZE contains an explicit parallel loop construct, called a **forall** loop. This provides a mechanism for the programmer to specify low-level parallelism not associated with vectors and arrays. Consider, for example, the loop:

```
forall   j   in  1 .. N      do
        . . .
     end;
```

In this **forall** loop, each of the N invocations of the body can run in parallel. The actual number of parallel threads of control at runtime depends on the number of processors available on the target machine.

Each invocation of the *forall* loop body is independent and cannot modify any variable which is being accessed by another invocation. The only interaction allowed between invocations is through reduction operators as shown below:

```
x   :=   0.0;

forall  i  in  1 .. 100  do
     x  +=  y[i];
     . . .
   end;
```

In the above loop the values in the array y are summed across the loop invocations. Other reduction operators provided by the language include: ***=**, **max=**, and **cat=** where the last is used for concatenation of one-dimensional dynamic lists. Properly implemented, these operators can be executed in "log-time" on a sufficiently parallel machine.

Below we give a BLAZE procedure which performs the forward elimination phase of Gaussian elimination without pivoting:

```
procedure gauss (A,b)  returns : (A,b)

param A: array [ , ] of real;
      b: array [ ]   of real;

const N = upper(A,1);                   -- upper bound of the first dimension
                                        -- of the array A

  begin

     for  k  in  1 .. N  do                    -- loop over pivot rows

          forall  i  in  k+1 .. N  do
               real   scale;                   -- local variable for each
                                               -- loop invocation

               scale       :=  -A[i, k] / A[k,k];   -- compute scale factor

               A[i, k..N]  +=  scale * A[k, k..N];  -- update row
               b[i]        +=  scale * b[k];        -- modify data vector
          end;
     end;
  end;
```

The outer loop is a sequential loop over all the rows of the array making each row the pivot row in turn. The inner loop is a parallel loop which updates all rows below the the pivot row. In this example, the reduction operator "+=" is used simply as a notational convenience. Since the variables on boths sides of these "reductions" are local to the current loop invocation, no parallel "tree-sum" is implied.

Structure of the BLAZE Compiler. The structure of BLAZE compilers is dictated by our desire to target this language to a number of sequential and shared memory systems, and by the necessity of performing extensive transformation and optimization during compilation. There is a machine independent front-end which performs, lexical analysis, parsing, and first few phases of optimization and machine-independent transformations. After this, further optimization and code generation is performed in machine specific compiler back-ends.

BLAZE source programs are first translated into an intermediate form representing the control-dependence between the statements of the program [12]. Extensive data-flow analysis is then performed to augment the control-dependence graph with the data dependencies between the variables. These include flow-dependencies, anti-dependencies and output-dependencies, as described by Kuck et al [14].

The functional procedure invocation semantics of BLAZE makes data-flow analysis much simpler and more "accurate" than it is with conventional languages, since no inter-procedural analysis is required. Even when complete inter-procedural data-flow analysis is performed for conventional languages, the resulting information is imprecise, because language features such as pointers and common blocks frequently obscure data-flow information.

An extensive set of program transformation techniques have been developed over the years for automatic vectorization of sequential code, most notably by the research groups at University of Illinois [14] and at Rice University [5]. The BLAZE compiler builds up upon this body of knowledge in an attempt to generate code for multiprocessor architectures.

The underlying goal of this analysis and transformation phase is to expose the parallelism available in the program. However, in most cases, the inherent parallelism of the algorithm does not exactly match that of the target architecture. Thus, the next step in the transformation process is to map the algorithm parallelism onto the architecture at hand. The independent threads of control in the program are 'bundled' into a set of concurrently executing processes, which can efficiently exploit the parallel architecture.

There currently exist implementations of the BLAZE compiler for Sequent, Alliant, and Butterfly multiprocessors systems. These have been implemented by students at Purdue University, Indiana University, and the University of Utah. These are experimental versions, and take a relatively naive approach to implementing parallel constructs such as the `forall` loop. We are currently exploring alternate implementations, and are beginning to study the effect that alternative implementations have on runtime performance [13].

2.2. Targeting Nonshared Memory Architectures. The BLAZE language is targeted primarily towards shared memory multiprocessors. While nonshared memory architectures having very high performance can be built, programming them is substantially harder than programming shared memory multiprocessors. This is primarily because issues such as data distribution and load balancing play a far more critical role on nonshared memory architectures than they due on shared memory machines. These issues make it very difficult for a compiler to automatically generate good code for nonshared memory machines. Instead, the user needs to explicitly specify data distributions, and must carefully plan load balancing strategies, in order to effectively utilize these machines.

The current approach to programming nonshared memory architectures is based on the use of explicit-tasking languages. Such languages seem to be ideally suited for some classes of algorithms, such as game tree searching and discrete event simulation, where the problem decomposes naturally into a system of cooperating processes. However, for algorithms relying on synchronous manipulation of distributed data structures, such languages have proven quite awkward.

KALI [2] is a research language designed to simplify the problem of programming nonshared

---

[2] The name "KALI" is taken from one of the Hindu goddesses with multiple arms, suggesting the idea of parallel execution.

memory architectures. It provides the semantic power of message-passing languages, such as CSP and Occam, while also providing a set of novel features for specifying and manipulating distributed data structures. The goal is to allow the user to retain control over data distribution and other issues critical to efficient parallel execution, while leaving the complex details of data transmission to the compiler and runtime environment. In the next section we give an overview of the parallel constructs of KALI.

2.2.1. Overview of KALI.
KALI is a high-level object-oriented language for distributed memory architectures. A KALI program is a sequence of *cluster* specifications, followed by an optional list of procedures. Clusters are a form of "object" or "process." That is, each cluster encapsulates a data structure, and has its own independent thread of control flow.

At program initiation, the unique cluster main begins execution. It may in turn dynamically create *instances* or *activations* of other clusters during program execution by sending create messages. Arbitrarily many instances of any cluster may be created, except for main, which has only one instance. Cluster instances do not share variables and can interact only via asynchronous message passing.

CLUSTERS. There are two kinds of clusters, *sequential* clusters and *distributed* clusters. A *sequential* cluster is a process or object having a single thread of control flow. Multiple instances of a sequential cluster may execute concurrently, but each executes as a sequential process. A *distributed* cluster, by contrast, supports SPMD-style (Single Program Multiple Data) parallel execution within each instance of the cluster. Since sequential clusters are quite conventional, we will focus only on distributed clusters, and on the data-parallel execution within them.

KALI assumes a nonshared memory architecture in which the programmer explicitly manages all critical resources. It further assumes that the architecture can support the idea of *processor arrays*, multidimensional arrays of physical processors, dynamically allocated by the user. This assumption is natural for hypercubes or mesh connected machines, and can easily be accommodated on a variety of other architectures. At the time of its creation, each instance of a distributed cluster is allocated a processor array, on which it will execute.

Syntactically, a cluster specification has a single level of static nesting. There is a sequence of declarations at the beginning of the cluster specification declaring variables and constants visible to procedures within the cluster. For a distributed cluster, this sequence of declarations also contains a declaration of a processor array as shown below:

```
procs    P[np, np];
integer  np ~ 10;
```

These statements allocate a square array P of np^2 processors, where np is an integer constant between 1 and 10 dynamically chosen by the runtime system.

The programmer controls the distribution of the data structures across the cluster's processor array. KALI currently supports only distributed arrays, though other distributed data structures will be allowed in future versions of the language. Arrays distributions are specified by a "distribution clause" in their declaration. This clause specifies a sequence of distribution patterns, one for each dimension of the array. Scalar variables and arrays without a distribution clause are simply replicated, with one copy on each of the processors in the processor array.

Data Distribution Primitives. Each dimension of a data array can be distributed across processors in one of two patterns, or can be left undistributed. The distribution patterns are block and cyclic. With a block distribution, each processor contains a contiguous block of elements of the array. Conversely, with a cyclic distribution, the array elements are distributed in a round-robin fashion across the processors. As an example, consider the following declarations:

```
procs    P[np];
integer  np ~ 10;
real     A[100]  dist [block];
real     B[100]  dist [cyclic];
```

Here, P is an array of up to ten processors and A and B are vectors having 100 elements. Assuming for simplicity that P contains exactly 10 processors, the subvector $A[1..10]$ would be assigned to processor $P[1]$, $A[11..20]$ would be assigned to processor $P[2]$, and so on. By contrast, with the cyclic distribution, processor $P[1]$ will have elements 1, 11, 21, ..., 91 of the vector B, processor $P[2]$ will have elements 2, 12, 22, ..., 92 of B, and so on.

The number of dimensions of an array that are distributed must match the number of dimensions of the underlying processor array. Hyphens are used to indicate dimensions of data arrays which are not to be distributed. Consider, for example, the following declarations.

```
procs    P[np];
integer  np ~ 10;
integer  C[100, 100]  dist [block, -],
                D[100, 100]  dist [-, block];
```

In this case, the row dimension of C is broken into blocks. Thus, each processor in the processor array P contains a group of rows of C, and each column is distributed across all processors in P. Conversely, in the case of the array D, each processor contains a group of columns, and the rows of D are split across processors.

Forall Loops. Data-parallel computation on distributed data structures is specified via forall loops. The forall loop header consists of a range specification and an on clause. The range specification specifies the number of invocations of the loop body, while the on clause specifies the processor on which each loop is to be executed. The most elementary way of doing this is to simply specify the processor explicitly:

```
procs   P[10];
real    A[100]  dist [block];
   . . .


forall  i  in 1 : 10 on P[i] do
   . . .

end;
```

Here the ith processor executes the ith loop invocation.

More generally, the execution of a forall loop can be tied to a distributed data structure through the use of a proc primitive in the on clause. Given an element of a distributed data structure, proc returns the processor on which it resides. This allows one to specify that a loop invocation be executed on the processor containing certain data, avoiding the necessity of messy index calculations. In the program fragment below, 100 loop invocations are performed, with the ith invocation executed on the processor owning the ith element the vector A.

```
procs P[10];
real A[100] dist [block];
   . . .

forall  i  in  1 : 100  on  proc( A[i] )  do
   . . .
end;
```

The compiler will strip-mine the above loop and convert it into a system of cooperating processes, one per processor. Each process will contain a sequential loop running over the elements of the distributed vector A local to that processor.

Data Movement. Each invocation of a forall loop can directly access only those data elements local to the processor executing that loop invocation; nonlocal parts of the data structure cannot be implicitly accessed. In KALI, access to nonlocal data must be explicitly specified via a set of high level primitives provided by the language. There are five communication primitives for data movement within a cluster: expand, <- (send), nbr, fetch and reply. Though KALI requires the user to manage communication within a cluster, these are relatively high-level primitives. The compiler translates these primitives into the system of sends, receives, and synchronization barriers that will actually be executed.

The first two of these primitives are used for sending data to other processors. **Expand** is used to broadcast data. It takes as argument data local to a processor, and broadcasts it to all processors in the processor array. Thus, in the following example, the processor owning the element $A[j]$ broadcasts it to all others.

```
forall  i  in  1 : N  on  proc( A[i] )  do
    real  x;
    x :=  expand ( A[j] );
        . . .

end;
```

Since x is declared within the loop, each loop invocation has its own copy of x.

The other form of "send" is denoted by an arrow: `<-`. It is used in place of the normal assignment operator `:=` to send data to a remote place. For example, consider the following program fragment:

```
forall  i  in  1 : 100  on  proc( A[i] )  do
    B[ f(i) ]  <-  A[i] ;
        . . .

end;
```

Here the values in the array A are "permuted" to form the array B.

The next two primitives, **nbr** and **fetch**, provide software simulation of shared memory semantics. **Nbr** is used for fetching data from adjacent processors in the processor array. This primitive is useful in a variety of numerical applications, such as relaxation schemes and "smoothing" algorithms, where the value at a point is computed from a set of neighboring values. As an example, consider the program segment:

```
forall  i  in  1 : 100  on  proc( A[i] )  do
    A[i] :=  A[i]  +  nbr( A[i-1] )  +  nbr( A[i+1] );
        . . .

end;
```

If A is distributed **block**, most of the accesses to $A[i-1]$ and $A[i+1]$ will be "local." However, communication is required at block boundaries, so the **nbr** primitive is needed.

The **fetch** primitive is more general, and can be used to access data from anywhere in the processor array.

```
forall  i  in  1 : 100  on  proc(A[i])  do
    real x;
    x  :=  fetch ( A[f(i)] );
        . . .

end;
```

Since this primitive amounts to direct software simulation of shared memory, it should be used with great caution; the overhead involved is likely to be quite high. On a non-shared memory architecture, the compiler must translate each fetch request into a system of sends and receives. In this example, if the function f is not a permutation, each processor would have to field an arbitrary number of fetch requests. In general processors must busy-wait for fetch requests until all outstanding fetches have been answered, before continuing on to the next computation.

In order to make this approach tractable, the runtime environment must be designed so that communication between processors occurs in "phases." Between communication phases, processors execute sequentially, without interference from other processors. A communication phase occurs whenever processors synchronize. Synchronization occurs:

    a. At the end of **forall** loops

    b. After any of the above communication operators, and

    c. When induced by a **reply** statements.

Whenever a processor encounters any of these synchronization points it blocks and handles pending fetch and send requests. The semantics of these synchronization constructs is subtle, but fortunately has no effect program correctness. If the synchronization is handled badly, only performance bugs results, not incorrect results or dead-lock.

Comparison of KALI and BLAZE. We have given above, a brief overview of KALI, concentrating on its most novel features. KALI was not designed to be an "elegant" language, in the sense of Modula, BLAZE or Sisal. Rather, KALI was driven by the needs of programmers trying to map numerical algorithms to nonshared memory architectures. After one has seen programmers struggling with the index arithmetic needed to implement `cyclic` and `block` distributions by hand enough times, it becomes apparent that language features such as those given here are clearly needed for nonshared memory architectures.

KALI has been designed only recently, and is not yet implemented. Thus, it is impossible to estimate accurately the cost of the communication primitives described. However, in many cases we can guarantee that the KALI implementation will execute as fast as the more laborious message-passing code. Consider for example the forward elimination phase of Gaussian elimination as shown below.

```
procedure  gauss(A, B, P)  returns(A, B);
procs    P[np];
real     A[n, n]   dist[cyclic, -];
         B[n]      dist[cyclic];
begin

    for  k  in  1:n  do                            -- loop over pivot rows

        forall  i  in  k+1:n  on  proc(A[i, -])  do
            real  PivRow[n],  scale;              -- these are local to each
                                                  -- loop invocation

            PivRow[k:n] :=  expand( A[k, k:n] );   -- broadcast pivot row

            scale       :=  -A[i, k] / PivRow[k];  -- compute scale factor

            A[i, k:n]  +=  scale*PivRow[k:n];      -- zero A[i, k]
            B[i]       +=  scale*expand(B[k]);     -- modify data vector
        end;
    end;
end;
```

This *gauss* procedure, as expressed in KALI, will perform as well as the analogous CSP or Occam procedure, since the output of the restructuring phase of the compiler is virtually identical to the analogous Occam procedure. Despite this, the above KALI procedure is shorter and much simpler than the analogous Occam code. In fact, it closely resembles the analogous BLAZE procedure given before. The precise differences are:

1. The processor array is explicitly present here.
2. Data distributions to accomplish load-balancing has been specified here.
3. Interprocessor communication was specified by the **expand** primitive.

However, despite this syntactic resemblance between KALI and BLAZE one should not overlook the dramatic semantic differences. We do not want to suggest that KALI procedures like this are as easy to write as comparable BLAZE code; there are subtleties and land-mines here not present in analogous shared memory code. We are merely arguing that this is a moderately high level way to specify algorithms for nonshared memory architectures, while retaining the full performance potential of these architectures.

3. CONCLUSIONS. Software technology has not kept pace with hardware technology in the domain of parallel processing. One of the major problems facing users of multiprocessor systems is the lack of adequate software tools. Until this problem is resolved, we will fail to effectively utilize parallel architectures on most problems.

In this paper we have briefly sketched the several directions that are being explored to provide portable programming environments for parallel machines. As architectures become more complex, tools such as the languages and compilers described will have to assume a greater role in making these architectures programmable and useful. Efficient utilization of parallel architectures requires a combination of good language design and advanced compiler technology.

Acknowledgement. The BLAZE project has been a joint research project with John Van Rosendale of University of Utah, currently serving as visiting faculty at Argonne National Laboratory. The author gratefully acknowledges his collaboration in this project, and his helpful comments regarding this paper.

REFERENCES

[1] S. AHUJA, N. CARRIERO, AND D. GELERNTER, *Linda and friends*, IEEE Computer, 19 (1986), pp. 26–34.

[2] F. ALLEN, M. BURKE, P. CHARLES, R. CYTRON, AND J. FERRANTE, *An Overview of the PTRAN Analysis System for Multiprocessing*, Research Report RC 13115 (#56866), IBM T. J. Watson Research Center, Yorktown Heights, NY, Sep. 1987.

[3] J. R. ALLEN, D. CALLAHAN, AND K. KENNEDY, *A parallel programming environment*, IEEE Software, 2 (1985), pp. 21–29.

[4] ——, *Program Transformations for Parallel Machines*, Technical Report TR85-20, Department of Computer Science, Rice University, Houston, TX, March 1985.

[5] R. ALLEN AND K. KENNEDY, *Automatic translation of Fortran programs to vector form*, ACM Trans. Prog. Lang. Syst., 9 (1987), pp. 491–542.

[6] J. BOYLE, T. D. R. BUTLER, B. GLICKFIELD, E. LUSK, R. OVERBEEK, J. PATTERSON, AND R. STEVENS, *Portable Programs for Parallel Processors*, Holt, RineHart and Winston, Inc., 1987.

[7] M. C. CHEN, *Very-high-Level Programming in Crystal*, Research Report YALEU/DCS/RR-506, Department of Computer Science, Yale University, New Have, Ct., Dec. 1986.

[8] H. DIETZ AND D. KLAPPHOLZ, *Refined FORTRAN: another sequential language for parallel programming*, in Proceedings of the 1986 International Conference on Parallel Processing, K. Hwang, S. M. Jacobs, and E. E. Swartzlander, eds., IEEE Computer Society Press, Aug. 1986, pp. 184–191.

[9] C. A. R. HOARE, *Communicating sequential processes*, Comm. ACM, 21 (1978), pp. 666–677.

[10] P. HUDAK, *Parafunctional programming*, IEEE Computer, 19 (1986), pp. 60–71.

[11] H. F. JORDAN, *The Force*, Report, Dept. of Electrical and Computer Engineering, University of Colorado, Boulder, Co., Jan. 1987.

[12] C. KOELBEL AND P. MEHROTRA, *The BIF Data Structures User's Manual*, in preparation, Purdue University, West Lafayette, IN, 1988.

[13] C. KOELBEL, P. MEHROTRA, AND J. V. ROSENDALE, *Semi-automatic domain decomposition in BLAZE*, in Proceedings of the 1987 International Conference on Parallel Processing, S. K. Sahni, ed., Pennsylvania State University Press, Aug. 1987, pp. 521–524.

[14] D. J. KUCK, R. H. KUHN, D. A. PADUA, B. LEASURE, AND M. WOLFE, *Dependence graphs and compiler optimizations*, in Conference Record of the Eighth Annual ACM Symposium on Principles of Programming Languages, Jan. 1981, pp. 207–218.

[15] J. McGRAW, S. SKEDZIELEWSKI, S. ALLAN, D. GALE, R. OLDEHOEFT, J. GLAUERT, I. DOBES, AND P. HOHENSEE, *SISAL: Streams and Iterations in a Single-Assignment Language: Reference Manual Ver. 1.2*, Manual M-146, Rev. 1, Lawrence Livermore National Labs, Livermore, Ca., 1985.

[16] J. R. McGRAW, *The VAL language: description and analysis*, ACM Trans. Prog. Lang. Syst., 4 (1982), pp. 44–82.

[17] P. MEHROTRA AND J. VAN ROSENDALE, *The BLAZE language: a parallel language for scientific programming*, Parallel Computing, 5 (1987), pp. 339–361.

[18] D. A. PADUA, D. J. KUCK, AND D. H. LAWRIE, *High-speed multiprocessors and compilation techniques*, IEEE Trans. Comput., C-29 (1980), pp. 763–776.

[19] D. POUNTAIN, *A Tutorial Introduction to Occam Programming*, Inmos, Colorado Springs, Colo., 1986.

[20] T. W. PRATT, *Pisces: an environment for parallel scientific computation*, IEEE Software, 2 (1985), pp. 7–20.

Automatic Decomposition of Fortran Programs for Execution on Multiprocessors
Ken Kennedy

It is clear that future generations of scientific supercomputers will employ multiple independent processors. What form of programming support software should be provided with such machines? Existing Fortran programs, written for sequential machines, are not well suited to parallel execution. If these programs are to run efficiently on a multiprocessor system, they must be decomposed into subproblems that can be executed in parallel.

The usual approach to decomposition is to provide language primitives or system calls that permit concurrent programming in Fortran. The programmer is thus responsible for handling all the synchronization. Unfortunately, concurrent programming is unnatural for many scientific programmers. It would be better to permit the programmer to design parallelism into his programs at an abstract level and have the programming environment automatically construct synchronization primitives.

In his talk, the author explores automatic techniques for uncovering parallelism in Fortran programs and for translating them to run efficiently on multiprocessor supercomputers. These techniques rely upon a sophisticated theory of dependence in programs developed by Kuck at Illinois and extended for our previous work on automatic vectorization. In addition, the methods rely extensively on intraprocedural and interprocedural data flow analysis.

In order to achieve enough parallelism to make the translation profitable, a number of very ambitious transformations are required, including loop distribution, loop interchange, loop alignment, and loop fusion. The problem of generating optimal code using these transformations has been shown to be intractable, but heuristic techniques have proved effective in practice. The talk will conclude with a discussion of our experience with a prototype translation system, called PFC Plus, developed at Rice.

CHAPTER 45

A Software Tool for Building Supercomputer Applications

Dennis Gannon[*]
Daya Attapattu[*]
Mann Ho Lee[*]
Bruce Shei[*]

Abstract. In this paper we describe a programming tool designed to help users of parallel supercomputers retarget and optimize application codes. In a sense, the system can be viewed as a tool to help users "fine-tune" the output of an automatic system or, if he or she has been so inspired, optimize the design of a new parallel algorithm.

1. Introduction. The software system described in this paper is an interactive program editing and transformation system that helps the user with this task of restructuring sequential programs for parallel execution. Each program that enters the system is completely parsed and all data dependences are recorded. The user then works with the system to restructure his code to a form suitable for a given target architecture. If the target is known to the system, it monitors the users transformations to the code. If the user attempts to transform the program in violation of the original semantics of the code he is warned that a change in the meaning of his program has taken place. At any time the user can ask the system to tell him what legal parallelizing transformation can be applied to a segment of selected code. More important, he can ask the system to make the program modifications the user desires. In this mode, the user is assured that of the correctness of the changes in his code.

In its current form, the system, known as the Sigma Editor or "Sigma", can support either FORTRAN (with Cedar and Alliant 8x extensions) or Blaze (a Pascal based functional language designed by Mehrotra and Van Rosendale [5]). In the future we plan to support C, C++ and Cedar Parallel C [4]. The target machines supported currently include the BBN Butterfly, the Alliant FX/8 and the Cedar System [2].

This tool is one part of a much larger programming environment known as the Faust Project. This effort, based at the Center for Supercomputer Research and Development in Urbana Illinois, has designed a common software platform for a number of programming tools including a performance analysis package, a program debugger, Sigma, and a graphics based program maintenance system. All the software has been written to use the X system from the MIT

[*]Department of Computer Science, Indiana University, Bloomington, Indiana, 47401. This research was supported by the Air Force Office of Scientific Research.

project Athena and, therefore, it will run on any Unix based workstation.

2. A Sample User Scenario. To illustrate how a user would interact with this system we shall step through a very simple example. The Blaze subroutine below is a simple matrix times vector routine.

```
Procedure MatVec(n,A,x) returns: y;
param A: array[1..n, 1..n] of real;
    x,y: array[1..n] of real;
    n: integer;
begin
    for j in 1..n loop
        for i in 1..n loop
            y[i] := y[i]+A[i,j]*x[j];
        end;
    end;
end;
```

The user loads this program into the system as if he were entering a text editor. The result is a window displaying the program and a list of menu headers (as illustrated in Figure 1).

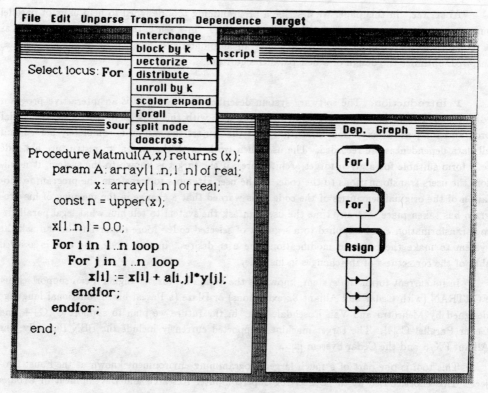

Figure 1. Sigma Screen After Program Load.
The Window on the Right is the Data Dependence Graph.

The first thing he may wish to do is to tell the system which machine is the intended target of this optimization. Currently this menu only lists three active choices: the BBN Butterfly, the

Alliant FX/8 and the Illinois Cedar. (We plan to extend this list to include the IBM RP3, the CRAY 2, the Connection Machine, and the ETA-10 during the next two years.)

The significance of having the user tell the system about the choice of target is threefold. First, and most obvious, we would like to have the system generate code for the given target. We will say more about this later. Second, it is important that program transformations that are inappropriate for the target machine be disabled. Third, and of most immediate concern to the user, selecting a target will enable the appropriate performance estimators which are described below.

To begin working with the system the user selects a segment of code, which we call a 'focus', on which to apply the tools BLED provides. For example, suppose the target is the BBN Butterfly and the user picks the inner most loop (by a mouse selection) as his focus. Among the menu headings he has at the top of the screen, is one called "Transforms". This menu contains a list of program restructuring transformations that can be used to expose concurrency or make parallelism explicit. For example, the user may have decided that he wishes the innermost "for" loop to be run in parallel. By selecting the transformation "forall" the system will first verify that the transformation can be legally applied. This requires a search of the data dependence graph to make sure that the appropriate conditions are satisfied so that the transformation can be legally applied. In this case the transformation is legal and the code now takes the form

```
for j in 1..n loop
    forall i in 1..n do
        y[i] := y[i]+A[i,j]*x[j];
    end;
end;
```

Following this operation, the user could invoke the code generator, but a better use of the system is to first invoke the analysis tool. This involves the selection of a menu item "analysis: Parallel Loop". Because the target machine, the Butterfly, executes such loops at the cost of a function call and an atomic increment to the index, the reply will be in the form

Loop Overhead to Body ratio > 50%.
Suggestion: increase granularity by blocking,
     merging or loop interchange.

Of the three suggestions, loop interchange is the easiest. After this operation the loop takes the form

```
forall i in 1..n do
    for j in 1..n loop
        y[i] := y[i]+A[i,j]*x[j];
    end;
end;
```

The "forall" loop body will now be large with respect to the loop overhead, but another problem that may inhibit performance is memory contention for shared data. In this case a second analysis tool called "analysis: Cache Management" can be invoked (see [3], [6] for theoretical background). This tool will report the following information when the program focus is the "forall" loop.

Cache/Local memory analysis for iterate i:
Suggest local copies of: A[i,1..n], x[1..n]
for n=100 hit ratios will be 0.99, 0.9999

This (too cryptic) message suggests that local copies of these variables be made in each processor. A major shortcoming of this form of the analysis is that the use is not informed of the penalty for failing to take this advise. The resulting program will take the form

```
forall i in 1..n do
var: x_local, A_local: array[1..n] of real;
    A_local[1..n] := a[i,1..n];
    x_local[1..n] := x[1..n];
    for j in 1..n loop
        y[i] := y[i]+A_local[j]*x_local[j];
    end;
end;
```

Clearly *x_local* need only be initialized once per processor but in Blaze we have no way to express this, so this task is left to the code generation step.

3. Generating Target System Code. Once a program has been optimized for a given architecture a programmer will want to have it run. This means that code will need to be generated for the target. Our approach to this problem follows the tradition of other restructuring systems. Rather than generate object code, we output source code that utilizes the language extensions and special function calls that are specific to that machine. The advantage is that we can take advantage of the native code generators and optimizers that exist for that machine.

In the case of FORTRAN, we generate Alliant vector-concurrent FORTRAN 8X for that machine. For Blaze programs we generate C code for the Alliant. In the case of the Butterfly we generate a C program that utilizes the BBN "uniform system" runtime environment for both FORTRAN and Blaze.

The first problem that one confronts in mapping the computational model defined by a parallel programming language like Cedar Fortran or Blaze to a target like the Alliant or the Butterfly is the following:

How does one provide efficient parallel execution of a language that is historically rooted in sequential stack based semantics?

The first instance of where this problem arises is in the execution of concurrent loops where the body of the loop contains references to names defined outside the loop. We would like to have a runtime stack which, at the point of a parallel loop invocation, can branch so that each processor has a private stack branch, but they share the part of the stack before the parallel execution call.

With the Alliant FX/8 the solution to this problem is built into the hardware. There are special instructions to set up this type of "cactus stack" and have each processor start execution at the appropriate place and time.

Unfortunately, the Butterfly presents a different computational model. Using the "uniform system" on that machine each processor gets a copy of the C program static data and has its own stack in private memory space. All shared data must be explicitly allocated in global memory. Consequently, if one processor updates a static variable or pushes an item on the stack it is invisible to all other processors. The solution to the problem for the Butterfly is to make the code generator allocate a copy of the activation record in global memory that is reachable by all processors. Any reference to data outside the scope of the loop is made through this record.

It would seem that the Alliant solution is by far superior, but there are other problems when one tries to extend the semantics in directions suggested by Schedule [1] and other portable concurrency packages. Dongarra and Sorrenson have argued that parallel programmers should have the ability to write code that permits tasks to be generated dynamically and scheduled for execution when the appropriate data is available. It is important that these tasks be "light weight", i.e. unlike a Unix process, the creation and scheduling of a task should involve no more overhead that a typical function call.

The bulk of our current research involves the automatic generation of light weight task and optimization of code that respects the memory hierarchy of the multiprocessor.

REFERENCES

(1)   J. Dongarra, D. Sorensen, "SCHEDULE: Tools for Developing and Analyzing Parallel Fortran Programs," in **Characteristics of Parallel Algorithms,** Jamieson, Gannon, Douglas, eds. MIT Press, 1987, pp.363-394.

(2)   D. Gajski, D. Kuck, D. Lawrie, A. Sameh, "Cedar - A large Scale Multiprocessor", Proc. of the 1983 International Confreence on Parallel Processing. Aug. 1983.

(3)   D. Gannon, W. Jalby, "Strategies for Cache and Local Memory Management by Global Program Transformation," Proc. of 1987 International Conference on Supercomputing, Athens, Greece, June 1987. Springer-Verlag Lecture Notes in Computer Science.

(4)   V. Guarna, "VPC - A Proposal for a Vector Parallel C programming Language," June 1987, Center for Supercomputer Research and Development, Technical Report no. 666.

(5)   P. Mehrotra, J. R. Van Rosendale, "The BLAZE Language: A Parallel Language for Scientific Programming," Report No. 85-29, ICASE, NASA Langley Research Center, Hampton, Va. (May 1985). (to appear in Journal of Parallel Computing).

(6)   Wang, K.-Y., Gannon, D., "Applying AI Techniques to Program Optimization for Parallel Computers," To appear in "AI Machines and Supercomputer Systems", Hwang, DeGroot, eds. McGraw Hill, 1987.

Parallel Fortran: Why You Can't. How You Can.
Clifford Arnold

With the advent of a variety of parallel processing systems, computational investigators are looking for convenient programming models and tools for executing their favorite codes. The first choice of investigators in the physical sciences and engineering appears to be access to parallel processing within Fortran. Unfortunately, the basic rules and requirements of an operational Fortran product strongly conflict with the most obvious primitives for parallel processing. I will discuss reasons why Fortran cannot be formally extended for parallel processing without transforming the language beyond recognition (and beyond its operational requirements). I will then discuss an alternative for altering Fortran for parallel processing without any formal extensions. This is done with a set of a dozen compiler directives. The goal is to invent a set of directives that will be portable and allow efficient code to be generated for a large variety of parallel processing architectures.

CHAPTER 46

Scientific Parallel Processing with LGDF2*

Robert G. Babb II[†]
David C. DiNucci[†]

Abstract. We describe a programming model, Large-Grain Data Flow (LGDF), and runtime support environment, that together provide parallel programmers with a high-level tool for parallel computation. The method uses a graphical language for specifying parallelism, combining dataflow-like programming constructs with blocks of ordinary sequential code. Macro expansion automatically generates source code containing efficient scheduling mechanisms for a particular (parallel) computer system. An earlier version of the LGDF model proved successful for programming several parallel scientific computers, including the CRAY X-MP, Denelcor HEP, and the Sequent Balance Series, as well as for debugging parallel programs on sequential computers via simulated parallel execution. An improved version of the programming model, referred to as LGDF2, allows automatic generation of more efficient scheduling mechanisms. Results of parallel scheduling performance experiments for the Sequent Balance 21000 are given and software tools are described to assist in the reliable construction, tuning, debugging, and porting of large scientific parallel programs.

1. Introduction. Programming a parallel computer is quite different from programming a sequential computer. Added complexity results from the need to coordinate and control the interactions of a number of concurrently active processes.

* This work was supported in part by Los Alamos National Laboratory under contract 9-Z34-P3915-1, Lawrence Livermore National Laboratory subcontract 1506703 and by the NSF Office of Advanced Scientific Computing under grant number ASC-8518527.

† Department of Computer Science and Engineering, Oregon Graduate Center, 19600 NW Von Neumann Drive, Beaverton, Oregon, 97006.

Programmers must learn a variety of parallel processing concepts that had little or
no relevance for either scalar or vector processing. For example, recognizing the end
of a computation is usually a trivial concept in ordinary sequential processing. However, program termination takes on new, subtle nuances in parallel processing, where
some processes may continue to run long after the "answer" has been computed, and
need to be explicitly killed.

The most obvious way to code parallel programs has been termed the "bare-
knuckles" method. In this approach, the basic low-level parallel programming
mechanisms provided by a particular parallel architecture are invoked explicitly by
the programmer. Programmers tend to underestimate the problems involved in this
approach, since the amount of explicit synchronization code required is typically
small and tends to be repetitive. However, keeping track of synchronizing variables
(locks) can be tedious and surprisingly error-prone, even in small codes at low
degrees of parallelism. The number of possible interaction patterns, even for very
simple parallel programs, is astronomical. Process synchronization and shared
memory access, especially for initial, termination, and boundary cases, are easily
mishandled.

Most parallel programming errors seem to result from the unconscious (and unin-
tended) application of sequential reasoning for parallel processing. McGraw and
Axelrod [5] give a summary of the issues involved in programming parallel supercomputers, illustrated by case studies of parallel bugs encountered in actual experience
with developing scientific parallel programs. They also point out that if a parallel
program is producing correct results, there is no guarantee that the code is in fact
correct. In fact, even if the program and the supporting parallel computer system
remain unchanged, there is in general no assurance that it will *continue* to produce
correct results!

A more reliable approach to parallel programming is to use a higher-level, more restricted model of parallelism than that provided by the basic machine facilities. Programmers use the higher-level model (language) to design applications, and a preprocessor or compiler is provided for implementation on a particular target parallel
machine. The effect is quite similar to the difference between programming in assembly language versus using a higher-level language. Although a lower-level language
may be more powerful, that power is most useful when harnessed via easy-to-reason-
with higher level concepts.

This paper briefly describes one such higher-level model approach, an improved version of the original Large-Grain Data Flow model [1,2].

2. Large-Grain Data Flow 2.

In LGDF2, a program consists of sequential
processes, which can be programmed in the language of the user's choice without
regard to the usual pitfalls of parallel programming. This provides a natural and
familiar environment for programmers to exercise their sequential programming
skills. Each process when executing can access one or more data areas called *datapaths*. A separate graphical language is used to specify how the processes will share
access to the values on the datapaths. This *network*, provides a simple basis for reasoning about parallel algorithms, and also serves as a source of information that the
scheduler can use to efficiently implement the LGDF2 program. Since all allowed
process/datapath interactions are expressed declaratively in the network, they can
be implemented in a shared- or distributed-memory parallel environment, or in a
uniprocessor environment, without requiring modification of the source code of the
processes.

An example of an LGDF2 network is shown in Figure 1. Each circle represents a process (i.e. a sequential user program) and each rectangle represents a datapath (e.g., corresponding to a Fortran COMMON block or a Pascal record structure). The edges connecting datapaths and processes specify the type of access each process has to the values on the datapath. An arrowhead represents a permitted direction of data flow. An arrowhead toward a process means that the process can read the data from the datapath while an arrowhead the other direction means that the process can write new values onto the datapath.

Problems in parallel processing often arise when one process attempts to change the data in a memory area while another process is also reading or writing it. LGDF2 execution semantics serve to protect against such occurrences. Each datapath has a datapath state which can activate processes on either the left or right side of the datapath. A process can begin execution only when it is connected (via edges) to the active side of all of its datapaths. When the process has finished accessing the data on a datapath, it informs the scheduler via a high level command (macro or subroutine call). When relinquishing the datapath, the process can also change the active side of the datapath. When a process suspends, it automatically relinquishes its access rights to all of its datapaths that have not already been explicitly relinquished.

Toggling the datapath state serves as a signal to the process(es) on the other side that they may wake up (fire) and access the data on the datapath (assuming all of their other datapaths are in the proper state to allow execution). All conflict resolution is handled implicitly by the scheduler. For example, the scheduler will not allow two processes to access a datapath simultaneously if one of them might write

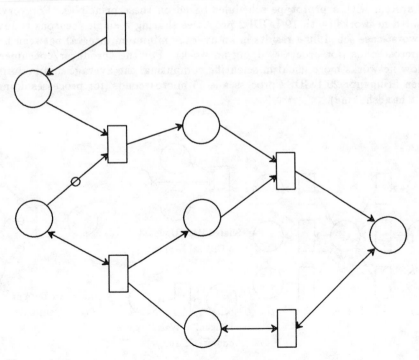

Figure 1. Example of an LGDF2 network.

(modify) its contents. The semantics of the latter case have been defined in order to facilitate efficient scheduler implementations for both shared- and distributed-memory multiprocessors.

These facilities should prove quite natural for expressing parallelism. We hope that LGDF2 will be useful as a way to retain the parallelism inherent in most algorithms, rather than forcing the programmer to linearize code which is not logically ordered simply because that is the only way to easily express it. Primitives in other parallel programming languages can be illustrated by relatively simple diagrams in LGDF2, as shown in Figure 2.

The LGDF software toolset does not currently support automatic transformation of "dusty deck" programs into LGDF form. However, we are currently engaged in a cooperative research project to enable use of a Fortran version of PREFINE [3] in order to provide automatic assistance for this process.

3. Scheduler Design and Performance. The heart of the strategy we employed for the scheduler code for the Sequent (which should generalize well to other shared memory parallel processors) is to have each process keep a count of how many "reasons" there are that it cannot fire. A "reason" might be that the process is currently executing, or that a number of its datapaths are not in the appropriate state (*left* or *right*). Whenever a process performs some action that affects the "reasons count" of another process, the scheduler updates it (safely). If the "reasons count" has become zero, the scheduler causes the other process to fire and places it on a ready queue of executable processes.

We have run several small experiments using 8 CPUs of a 16 CPU Sequent Balance 21000 system with a prototype scheduler based on these principles. For very highly contended networks (with 19 LGDF2 processes sharing a single "reasons count" variable) worst-case scheduling results in an average minimum interval between firings of 125 microseconds (for processes doing no work). For the "producer/consumer" style dataflow networks more usual in scientific computing, the average minimum interval between firings for 20 LGDF2 processes is 55 microseconds (for processes doing nothing but handshaking).

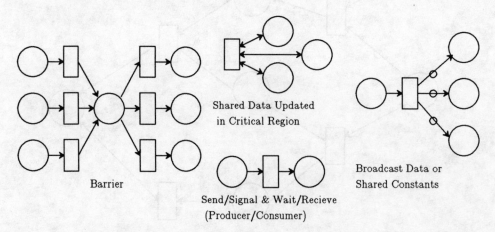

Barrier

Shared Data Updated
in Critical Region

Send/Signal & Wait/Recieve
(Producer/Consumer)

Broadcast Data or
Shared Constants

Figure 2. Expressing traditional parallel constructs in LGDF2.

More details on the LGDF2 model and a scheduler implementation strategy for a particular parallel processor, the Sequent Balance 21000, are given in [4].

4. Conclusions and Future Work. We have been using LGDF techniques until now mainly to investigate ways to achieve portability of scientific application codes between shared memory parallel processors. Our current research is concentrating on refinements to the computation model that should aid in implementation of LGDF2 programs on distributed memory multiprocessors.

Another active research direction is in design and construction of tools to aid in debugging parallel applications designed using the model. Our hope is to minimize the parallel debugging problem by breaking it into several parts. The discipline required in the analysis of programs to cast them into LGDF2 form should eliminate almost all of the usual species of parallel bugs. In addition, the model provides a way to graphically depict the parallel strategy chosen, which should aid in visualizing and avoiding possible deadlock and performance bottleneck problems. We are developing a prototype multilevel LGDF2 monitor/debugger that will integrate standard sequential debugging tools with the "structured" high-level debugging capabilities made possible by use of the model.

REFERENCES

[1] R. G. Babb II, "Programming the HEP with Large-Grain Data Flow Techniques", in *MIMD Computation: HEP Supercomputer and Its Applications,* (ed. by J. S. Kowalik). Cambridge, MA: The MIT Press, 1985.

[2] R. G. Babb II, "Parallel Processing with Large-Grain Data Flow Techniques," *Computer*, vol. 17, no. 7, July 1984, pp. 55-61.

[3] H. Dietz and D. Klappholz, "Refined FORTRAN: Another sequential language for parallel programming", in *Proc. 1986 Int. Conf. on Parallel Processing*, Aug. 1986, pp. 184-191.

[4] D. C. DiNucci and R. G. Babb II, "Practical Support for Parallel Processors", in *Proc. 21st Hawaii Int. Conf. on System Sciences 1988*, vol. 2: *Software Track,* Jan. 1988, pp. 109-118.

[5] J. R. McGraw and T. S. Axelrod, "Exploiting multiprocessors: Issues and options", in *Programming Parallel Processors,* (ed. by R. G. Babb II). Reading, MA: Addison-Wesley, 1988, pp. 7-25.

CHAPTER 47

An Overview of Dino—A New Language for Numerical
Computation on Distributed Memory Multiprocessors

Matthew Rosing[*]
Robert B. Schnabel[*]

Abstract.   We briefly discuss the design of a new language, called Dino, for programming parallel numerical algorithms on distributed memory multiprocessors. A significant difficulty with most current approaches to programming such computers is that interprocess communication and process control must be specified explicitly through messages, thereby making the parallel program difficult to write, debug, and understand. Our approach is to add several high level constructs to standard C that allow the programmer to describe the parallel algorithm to the computer in a natural way, similar to the way in which the algorithm designer might informally describe the algorithm. These constructs include the specification of a data structure of virtual processors that is appropriate for the problem, and the ability to map data and procedures to this virtual parallel machine. Parallelism is achieved through a concurrent procedure call that utilizes these data and procedure mappings. All the necessary interprocess communication and process control results implicitly through these constructs.

1. Introduction.   Dino ("**DI**stributed **N**umerically **O**riented language") is a new language for writing numerical programs for distributed memory multiprocessors. By distributed memory multiprocessor we mean any computer consisting of multiple processors with their own memories and no shared memory, which communicate by passing messages. Examples include hypercubes, and networks of computers used as multiprocessors.

The main goal of our work is to make parallel numerical programs for such computers easy to write and understand. The approach we take is to try to make the programs similar to the natural descriptions of parallel algorithms that algorithm designers often use in explaining their methods. Inherent in this approach is that interprocess communication and process control should be implicit in the language constructs.

The constructs of Dino use the fact that many numerical algorithms are highly structured. The main data structures used in these algorithms are arrays and possibly trees. The processes that execute in parallel usually are also highly structured; sometimes the algorithm consists of "single program, multiple data" segments where the same code executes on different processors and different parts of the data structure simultaneously. In conjunction, the distribution of data among processors, and the communication between them, typically follows regular patterns.

The Dino language allows such parallel numerical algorithms to be described in a natural,

* Department of Computer Science, University of Colorado, Boulder CO 80309. Research supported by AFOSR grant AFOSR-85-0251, and NSF cooperative agreement DCR-8420944.

top down manner. It provides both a mechanism for efficiently distributing data over the processors of a distributed memory machine, and the ability to easily operate on that distributed data concurrently. Interprocess communications is implicit in the data mapping constructs, and is therefore less subject to programmer errors than sending and receiving messages. The key to these capabilities is the ability of the user to define a data structure of virtual processes (called environments) that fits the algorithm and data structures.

Dino consists of extensions to standard C. We have chosen C for several reasons. It is available on all the target parallel machines we have considered; it is a structured language, which complements the new, highly structured characteristics of DINO; and there are a wealth of compiler and other tools associated with C which considerably ease the task of implementing the new language. In addition, by choosing C we have been able to use C++ [6] for our initial, prototype implementation.

To our knowledge, relatively little high level language design has been done for distributed memory multiprocessors. Languages such as Linda [1,2] and Pisces [5] support distributed numerical programming, but have a more low level orientation to issues such as communication. We were partially motivated by languages such as the Force [3,4] which have greatly facilitated parallel numerical programming on shared memory multiprocessors; we would like to bring a similar level of ease to programming distributed memory multiprocessors.

The rest of this paper briefly and informally introduces the main ideas in Dino, gives one simple example of a Dino program, and gives a very brief synopsis of the status and directions of this work. Subsequent papers will describe Dino in more detail and discuss our experience with using it more thoroughly.

2. Dino Overview.

The goal of Dino is to allow the programmer to communicate a distributed parallel algorithm to the computer in a way that is similar to the natural way that we often observe algorithm designers informally describing their methods. To facilitate this, Dino provides two important new capabilities, distributed data structures and composite procedures. Both are in turn based upon an underlying data structure of virtual concurrent processors (environments) that is provided by the user. We now briefly describe the characteristics of each of these three fundamental aspects of Dino.

2.1. Environments.

The key construct that allows Dino to provide a natural, high level description of a parallel algorithm is a user defined structure of environments. An environment consists of data and procedures. It may contain multiple procedures, but only one procedure in an environment may be active at a time. Thus each environment can correspond to a process, and this is how environments are implemented in our Dino prototype. A more general possibility is mentioned in Section 3.

To create a parallel algorithm, the user declares a structure of environments which best fits the number of processes, and the communication pattern between processes, in the parallel algorithm. This structure can be viewed as a virtual parallel machine constructed for the particular algorithm. In our experience, the most common structures of environments are one, two, and higher dimensional arrays. This is because the parallelism in numerical algorithms often derives either from the partitioning of physical space into neighboring regions, or from the partitioning of arrays, both of which result in parallel algorithms whose data mappings and procedural parallelism are naturally described in terms of arrays of processors. It is possible, however, to use any data structure in defining a structure of environments.

2.2. Distributed data.

Dino allows the user to specify mappings, either one to one or one to many, of data structures to the underlying virtual machine structure given by the user-defined structure of environments. These mappings, which are specified as part of the declaration of the data structure, are selected according to how the processors will access and share the data. They are the key to making interprocess communication natural and implicit.

An example of distributed variables is illustrated in Fig. 1 and in the Dino program in the Appendix. Suppose we wish to solve Poisson's equation with zero right hand side,

$$\frac{\partial^2 U}{\partial x^2} + \frac{\partial^2 U}{\partial y^2} = 0 \; ,$$

on a square domain with some given boundary condition by a simple finite difference method. In this method we discretize the variable space into $U_{i,j}$, $i=0,\cdots,N$, $j=0,\cdots,N$, and then iteratively apply the formula

$$U_{i,j} = (U_{i+1,j} + U_{i-1,j} + U_{i,j+1} + U_{i,j-1}) / 4 \qquad (2.1)$$

to calculate the new value at each grid point except the border points, until the values converge. The natural topology for this problem is a two dimensional grid. If we assume for simplicity that we create a unique environment $e_{i,j}$ for each grid point except the border points, then the natural structure of environments is also a two dimensional array, $e_{i,j}$, $i=0,\cdots,N-1$, $j=0,\cdots,N-1$. Now from equation (2.1), each variable $U_{i,j}$ (except border variables) is used in the environment where it resides, and the environments directly to the north, south, east, and west. Thus the distributed data mapping function is to map each $U_{i,j}$ to $e_{i,j}$, $e_{i+1,j}$, $e_{i-1,j}$, $e_{i,j+1}$, and $e_{i,j-1}$.

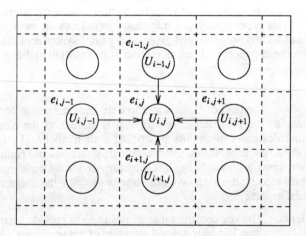

Fig. 1. Distributing a partial differential equation calculation

The variable $U_{i,j}$ is an example of a distributed variable that is mapped to multiple environments. In this case a local copy of that variable will exist in each of the environments to which it is mapped. A procedure in any of these environments can then access the variable either locally or remotely. A local access, which uses standard syntax, affects just the local copy and is the same as any standard reference to a variable. A remote access, which uses the syntax *variable name#*, is used to generate interprocess communication. A remote assignment to a distributed variable generates a message that is sent to every other processor to which that variable is mapped, while a remote read of a distributed variable will receive such a message and update the variable's value, in one of two ways that are described below. Thus one to many mapping functions can be thought of as defining locally shared variables, where certain data elements are shared between certain subsets of the processors. This mapping provides the information required to automatically generate the necessary sends and receives when these variables are accessed remotely.

Dino distributed variables may be either *synchronous* and *asynchronous*. The default is synchronous, but a distributed variable may be made asynchronous by placing "asynchronous" before "distributed" in its declaration. In either case, a remote write causes a message, with a new value of the distributed variable, to be sent to a buffer in each other environment to which that variable is mapped. A remote read of a synchronous variable causes it to overwrite its local copy with the *first* value that has been received since the last remote read; if no new value is present, it blocks until one is received. A remote read of an ansynchronous variable causes it to overwrite its local copy with the *last* value that has been received since the last remote read; if no value has been received, it retains its current local value, and does not block. A nice consequence of this construction is that by changing a distributed variable's declaration from synchronous to asynchronous, a program can be changed from a data synchronous parallel program to an asynchronous ("chaotic") program.

Dino provides an extensive collection of standard mapping functions, such as the *NSEWoverlap* mapping used in the example. The user can also create arbitrary mapping functions.

2.3. Composite procedures.   A composite procedure is a set of identical procedures, one residing within each environment of a structure of environments, which is called concurrently. Its parameters typically include distributed variables. A composite procedure call causes each procedure to execute, utilizing the portion of the distributed parameters, and possibly other distributed data structures, that are mapped to its environment. This results in a "single program, multiple data" form of parallelism. There is no need to explicitly send code to each processor, to initiate execution on individual processors, or to explicitly distribute or collect data among processors.

Distributed variables that are parameters to composite procedures can be declared as input, output, or input/output parameters. Upon invocation of the composite procedure, a input parameter is distributed based upon its mapping function. That is, the value of each of its elements is sent from the calling procedure to each environment to which that element is mapped. Upon terminatation of the composite procedure, the value of each output parameter is sent from the environment to which it is mapped back to the calling procedure. If the mapping is one to many, the mapping function specifies from which environment the value should be returned.

In the Poisson solver example, the composite procedure *Poisson* performs *itns* iterations of equations (2.1) at grid point $U_{i,j}$. The distributed variable U is a one to many, input and output parameter. The mapping function specifies that upon process termination, the value of each $U_{i,j}$ should be retrieved from the environment $e_{i,j}$.

In summary, the main advantage of composite procedures, together with distributed data and environments, is that they permit a natural, high-level description of many parallel algorithms, while making the details of interprocess communication implicit in the language.

3. Current Status and Future Directions.   We have implemented a prototype of Dino using C++ [6]. It runs on the Intel hypercube, the hypercube simulator, and on our network of Sun workstations. We have written parallel programs for a variety of numerical algorithms in Dino/C++, as well as paper programs in standard Dino. Our opinion is that these programs are usually considerably easier to write and understand than the same parallel programs in existing languages for distributed memory multiprocessors; subsequent papers will give a larger number of examples that help support this claim. Another benefit of the prototype has been to enable us to understand better the low level issues in implementing Dino.

The Dino research is leading in a number of interesting directions, which we are beginning to pursue. First, we continue to consider new features for Dino. Mainly these are enrichments which would allow Dino to express a broader class of parallel algorithms easily. Examples include multiple distributed mapping functions for a single data structure, multiple environments in a single program, and facilities for supporting dynamic process structures and dynamic distributed data structures. A second area is tools for optimizing parallel programs. The information about interaction between environments that is readily available in a Dino program provides the opportunity to make good mappings of environments to processes, and of processes to the processor topology of the target computer. The latter is already done in a simple way in the C++ prototype. The former might allow the programmer to specify environments at a finer grain than there are processors, when this is the natural way to describe the algorithm, and then have a tool make a good decision about how best to bundle the environments into processes. Finally, the graphical nature of the basic Dino constructs, such as the structure of environments and the data mapping functions, makes in natural to consider a graphical interface to Dino.

Appendix -- Dino Program for Parallel Solution of Poisson's Equation

```
#define N 128
environment node[N:xid][N:yid]{
        composite Poisson(U, in itns)
                distributed float U[N+2][N+2] : NSEWoverlap;
                int itns; /*number of iterations*/
                {
                int i,x,y;
                x = xid + 1;
                y = yid + 1;
                        /*calculate using local values to start cycle*/
                U[x][y]# = 4*U[x][y] - U[x+1][y] - U[x-1][y] - U[x][y+1] -
                        U[x][y-1];
                        /*calculate remaining iterations using remote values*/
                for (i=1; i<itns; i++)
                        U[x][y]# = U[x][y]*4 - U[x+1][y]# - U[x-1][y]# -
                                U[x][y+1]# - U[x][y-1] ;
                } /*poisson*/
        } /*node*/
environment host{
        float G[N+2][N+2];
        main(){
                initPoisson(G);
                Poisson(G[][], 250)#;
                display(G);
                } /*main*/
}/*host*/
```

REFERENCES

(1)   N. CARRIERO and D. GELERNTER, *The S/Net's Linda kernel,* ACM Transactions on Computer Systems 4, 1986, pp. 110-129.

(2)   D. GELERNTER, N.CARRIERO, S. CHANDRAN, and S.CHANG, *Parallel programming in Linda,* in Proceedings of the 1985 International Conference on Parallel Processing, IEEE Press, 1985, pp. 255-263.

(3)   H. F. JORDAN, *The Force,* in The Characteristics of Parallel Algorithms, L. H. Jamieson, D. B. Gannon and R. J. Douglass, Eds., MIT Press, 1987, pp. 395-436.

(4)   H. F. JORDAN, *Structuring parallel algorithms in an MIMD, shared memory environment,* Parallel Computing 3, 1986, pp. 93-110.

(5)   T. PRATT, *The Pisces 2 parallel programming environment,* in Proceedings of the 1987 International Conference on Parallel Processing, IEEE Press, 1987, pp. 439-445.

(6)   B. STROUSTRUP, *The C++ Programming Language,* Addison-Wesley, Reading, Massachusetts, 1986.

CHAPTER 48

The Design and Implementation of Parallel Algorithms in ADA

Richard Sincovec[*]

Abstract. Ada contains a number of new language features that support parallel computing and a new way of designing mathematical software components. These features include tasks, packages and generics. This paper explores several issues relating to the use of Ada in the design and implementation of reusable mathematical software components for parallel computers.

1. Introduction. Ada contains numerous features to support scientific computing including parallel computing. Ada also supports a new approach to the development of mathematical software for scientific computations. This new approach is based on data abstraction and information hiding implemented using the generic, package and tasking features of Ada. Software systems that use this approach are expected to be more reliable and more maintainable than systems designed using other approaches.

This paper examines several issues related to the use of Ada in mathematical software for scientific computing with an emphasis on the design and implementation of parallel algorithms. A more comprehensive discussion is presented by Sincovec (3).

2. Abstract Data Type Components. Ada represents a significant leap in computer language support for problem abstraction. The software developer can create new abstract data types (ADTs) along with operations on objects of the types thereby extending the language to the problem space. To preserve the integrity of the abstraction, the representation of the type along with the implementation of the operations are hidden from the client program, i.e., the user of the abstract data type. This assures consistency in the use of the type and eliminates maintenance costs for the client program if the representation or algorithms are changed at a later time.

The implementation of an abstract data type consists of two parts: a visible/public part and a hidden/private part. The visible part defines the interface to the operations available on the type. The hidden part implements the

_____
*Department of Computer Science, University of Colorado at Colorado Springs, Colorado Springs, Colorado 80903-7150

operations promised in the visible part. The visible part and the implementation part are separately defined and compiled. This permits computational resources to be made available without revealing the implementation details and permits multiple implementation parts corresponding to a single interface part.

In Ada, the package structure is used to support abstract data types. It has the following form:

```
package Abstract_Data_Type is
     -- Types and objects are identified.
     -- Subprogram interfaces are specified.
private
     -- Internal data structure of the abstract data type.
     -- This information is for the compiler.
end Abstract_Data_Type;
```

```
package body Abstract_Data_Type is
     -- Declaration of local variables and types.
     -- Implementation of subprograms defined in package specifications.
end Abstract_Data_Type;
```

Libraries of abstract data types are significantly different from usual subroutine libraries. When using a subroutine library the client program is often required to allocate storage for data structures used in the implementation of the routine. The client program may also be required to initialize these data structures prior to calling the routine. This results in the client program becoming tightly coupled to the representation and implementation details of the routine. If the representation of the data structures are later changed to reflect an improved algorithm or a parallel computer architecture, then the client program must be modified.

Some obvious candidates for abstract data types in scientific computing are complex numbers, polynomials, Taylor series, rational number arithmetic, vector and matrix arithmetic, and sparse matrices, to name a few. Several examples are presented by Sincovec (3).

3.   Parallel Computing in Ada.   The tasking features of Ada support the development of algorithms for parallel computers. Ada tasks are entities whose execution proceed in parallel. Different tasks proceed independently except at points where they synchronize, called a rendezvous. Tasks can be considered to execute on their own logical processor which is assigned by the Ada run-time environment. Ada tasking may have various actual implementations for multicomputers, multiprocessors, and interleaved execution on a single processor.

Each task depends on at least one master, called its parent. A task is activated after the declarative part of its parent is elaborated and before the first statement of its parent is executed. A task terminates when it is completed and all its children tasks are terminated. The syntactical structure of a task is:

```
task Task_Name is
     -- Entry declarations and representation clauses.
end Task_Name;
```

```
task body Task_Name is
     -- Declarations
begin
     -- Statements including accept statements.
     -- Accept statements can define a set of statements to be
     -- executed during rendezvous with a calling task.
end Task_Name;
```

To illustrate these features, we present a simple matrix operations package that contains only one operation, a matrix vector multiply. We present a package specification and then show three package bodies (implementations) for scalar computers, distributed memory parallel computers and shared memory parallel computers. The goal is to design a single interface (package specification) but with multiple implementations (package bodies) that take advantage of the underlying computer architecture. This permits a client program to be moved to a new computer environment without any changes.

The package specification is given by:

```
package matrix_operations is
        type vector is array( integer range <>) of float;
        type matrix is array( integer range <>) of vector;
        function "*"( a :  matrix; x : vector ) return vector;
    end matrix_operations;
```

A client program that uses this package would need to "with" and "use" the package, declare appropriate matrices and vectors, and call the matrix vector multiply function. For example, if we make the declarations: x: matrix(1..40); u, p: vector(1..40); then p := x * u; is the actual call to the matrix vector multiply.

The client program can use the package without knowledge of the implementation details. For a von Neumann computer, the usual implementation of matrix vector may be appropriate but for a parallel computer, a different implementation that takes advantage of the computer architecture may be provided.

A parallel implementation of the package body uses the tasking features of Ada. The basic idea is to create separate tasks to multiply each row of the matrix by the vector. The Ada run-time environment allocates these tasks to the available processors. We present the following implementation, see Booch (2), that is suitable for a distributed memory parallel computer and then briefly describe its operation.

```
    package body matrix_operations is

        task type partial_product is
            entry receive_value( p       : out float );
            entry send_values( v_1, v_2 : in vector );
        end partial_product;
        task body partial_product is separate;

        function "*"( a : matrix; x : vector ) return vector is
            b                  : vector (x'range);
            parallel_product : array (x'range) of partial_product;
        begin
            for i in x'range loop
                parallel_product(i).send_values( a(i), x );
            end loop;
            for i in x'range loop
                parallel_product(i).receive_value( b(i) );
            end loop;
            return b;
        end "*";
    end matrix_operations;
```

The package body defines a task type called partial product with two entries. The entry send_values receives two vectors from some parent task. The entry receive_value sends the result of the scalar product of the two vectors back to the parent task. Since this is a type definition, we may declare objects of this type. Each object is then a task that can potentially execute in parallel with other tasks.

The implementation of the multiply function declares an array, parallel_product, of partial product tasks. The size of this array is the same as the size of the matrix and vector to be multiplied. The body of the function sends the vector and the ith row of the matrix to the ith partial product task. When the ith task completes its operations, the body of the function is ready to receive the ith component of the resultant vector.

The actual task body is presented below:

```
separate ( matrix_operations )
task body partial_product is
    type pointer is access vector;
    product   : float;
    vector_1 : pointer;
    vector_2 : pointer;
begin
    accept send_values( v_1, v_2 : in vector ) do
        vector_1 := new vector' (v_1);
        vector_2 := new vector' (v_2);
    end send_values;
    product:= 0.0;
    for i in vector_1.all' range loop
        product := product + ( vector_1(i) * vector_2(i) );
    end loop;
    accept receive_value( p : out float ) do
        p := product;
    end receive_value;
end partial_product;
```

The client program (parent task) interacts with a task by making entry calls. When an entry call to send_values is made by the client program, the task makes copies of the two vectors in its local processor's memory using the dynamic allocation features of Ada. Then the scalar product of the two vectors is formed and made ready to be sent to the client program when a receive_value entry call is made.

To implement this algorithm on a shared memory parallel computer, we make the vector and matrix global variables to the tasks. This permits each task to directly access these variables without making local copies as in the previous implementation. We also redefine the type matrix to be a two-dimensional array as opposed to an array of vectors which was required in the previous implementation due to the strong typing of Ada.

```
type vector is array( integer range <>) of float;
type matrix is array( integer range <>, integer range <> ) of float;

function "*"( a : matrix; x : vector ) return vector is

task type partial_product is
    entry form_component( i : in integer );
    entry receive_component( r : out float );
end partial_product;

b : vector (x'range);
parallel_product : array(x'range) of partial_product;
```

```
task body partial_product is
        row     : integer;
        product : float;
begin
        accept form_component( i : in integer ) do
            row := i;
        end form_component;
        product := 0.0;
        for j in x'range loop
            product := product + a(row,j) * x(j);
        end loop;
        accept receive_component( r : out float ) do
            r := product;
        end receive_component;
end partial_product;

begin
    for i in x'range loop
        parallel_product(i).form_component(i);
    end loop;
    for i in x'range loop
        parallel_product(i).receive_component( b(i) );
    end loop;
    return b;
end "*";
```

This implementation is similar to the preceding implementation except that the entry form_component receives the subscript of the matrix vector component it should form.

The tasking features of Ada are quite extensive. The preceding example gives only a small indication of the total capabilities included in the Ada language. We have also not attempted to cover the semantics associated with Ada tasking. The reader is referred to the Ada Language Reference Manual (1) for additional details.

There are a number of issues that need to be considered in using the tasking features of Ada for the development of software for scientific computations on parallel computers. These include communication/synchronization requirements of the parallel algorithm, efficiency, speedup, tasking overhead with respect to storage, Ada tasking constructs and their use on various parallel computer architectures, and the granularity of the parallel algorithm, to name a few.

4. Conclusions. Ada supports a new way of designing mathematical software. The package construct permits a single interface definition with multiple implementations depending on the computer architecture. The impact of data abstraction on efficiency and a number of issues relating to the use of the tasking features of Ada for mathematical software are currently under investigation.

REFERENCES

(1) ADA PROGRAMMING LANGUAGE, Military Standard. ANSI/MIL-STD-1815A, January 1983.

(2) G. BOOCH, Software Engineering with Ada, Second Edition, Benjamin/Cummings, 1987.

(3) R. F. SINCOVEC, "The Design and Implementation of Mathematical Software in Ada", UCCS Technical Report, January 1988.

CHAPTER 49

The BF (Boundary-Fitted) Coordinate Transformation Technique of DEQSOL (Differential EQuation SOlver Language)

Chisato Konno*
Michiru Yamabe*
Miyuki Saji*
Yukio Umetani*

Abstract. DEQSOL is a high-level programming language specially designed to describe PDE problems in a way quite natural for numerical analyses. The DEQSOL translator automatically generates highly vectorizable FORTRAN simulation codes from the mathematical-level DEQSOL descriptions. The BF facility has been developed in addition to the existing FDM and FEM facilities. The BF facility is implemented as a preprocessor for the existing FDM facility. The PDE problem, described by theextended DEQSOL language, is transformed into a problem on rectangular regions in transformed space by this facility, and the existing FDM facility is applied to generate FORTRAN simulation codes. This facility has made DEQSOL applicable to the development of simulators on curved and moving boundary regions, such as the impurity diffusion simulator in semiconductor devices.

Despite this extension, programming productivity is still an order of magnitude higher than FORTRAN programming, and generated codes achieve extremely high vectorization ratios (over 90%) on the Hitachi S-810 vector processor.

1. Introduction. As the widespread use of supercomputers indicates, the demand for numerical simulations of physical phenomina is rapidly increasing. However, despite the innovations in hardware technology and computer architecture, the programming environment itself has remained at a relatively elementary level. DEQSOL (Differential EQuation SOlver Language)[5,6,7] is a very high level language specifically designed to describe PDE (Partial Differential Equation) problems in a quite natural way for numerical analyses. This language has two main objectives:

(1) To increase programming productivity by establishing a new, architecture-independent language interface between numerical analysts and vector/parallel processors.

(2) To generate highly vectorizable FORTRAN codes from DEQSOL descriptions by using their intrinsic parallelism, thus achieving efficient execution.

The structure of the previously developed DEQSOL system and its processing flow are shown in Fig.1. The DEQSOL description is automatically translated into a FORTRAN simulation program by the DEQSOL translator. The DEQSOL translator has two discretization facilities, namely FDM (Finite Difference Method) and FEM (Finite Element Method).

Several numerical simulation language systems , such as SALEM[1], PDEL[2] and ELLPACK[3,4], have been developed. Compared with these systems, DEQSOL has the

---

* Central Res. Lab., Hitachi Ltd., Kokubunji, Tokyo 185, Japan

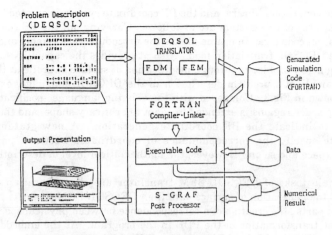

Problem Description
(DEQSOL)

DEQSOL
TRANSLATOR

FDM   FEM

Generated
Simulation
Code
(FORTRAN)

FORTRAN
Compiler·Linker

Output Presentation

Executable Code

Data

S - G R A F
Post Processor

Numerical
Result

DEQSOL : Differential EQuation SOlver Language
SGRAF : Scientific GRAphing Facilities
FDM : Finite Difference Method
FEM : Finite Element Method

FIG.1 DEQSOL System and its Processing Flow

following significant advantages:

* It can flexibly describe various numerical algorithms within the language.
* It can describe spatial regions and shapes with their automatic meshing facilities.
*It generates FORTRAN codes which allow supercomputer to achieve high acceleration.

Thanks to these features, DEQSOL is being applied to a wide range of practical and complex problems[6,7].

FDM is the most straightforward and widely-used method for analyzing PDEs on the rectangular regions. The matrix composed by FDM has the regular form. Using this regularity, fast matrix solution algorithms, efficient on supercomputers, can be applied with rather small memory. However, this method requires some technical interpolation between grid points for shapes with complex boundaries. And this may cause significant numerical errors. By contrast, FEM has geometrical advantages for matching shapes with complex boundaries. However, it requires much more memory to store and calculate the matrix because of its irregular meshes.

Another approach to deal with geometrical complexity is the BF (Boundary Fitted) coordinate transformation technique. This technique is based on automated curvilinear coordinate system having a coordinate line coincident with each boundary of the arbitrary shape domain, which was originated by Thompson[8].

In this paper, the new BF-facility(BFM) of DEQSOL is presented.

2. DEQSOL-BFM System.

An outline of BF technique is as follows: A PDE problem defined on an arbitrary-shaped region in physical space is mapped by a transformation to a problem defined on a rectangular region in transformed space. The ordinary FDM is applied to the problem to solve it numerically in transformed space. The calculated numerical solution is then transformed to physical space by reverse mapping. FDM can solve problems defined on various-shaped regions in physical space using this procedure.

As DEQSOL already includes a FDM solver, this procedure can be realized by providing a preprocessor for the existing DEQSOL translator. The extended DEQSOL system with BF facility is shown in Fig.2 . A DEQSOL-BFM program is written in the extended DEQSOL language, allowing arbitrary-shaped regions to be described in physical space. The DEQSOL-BFM translator automatically generates DEQSOL-FDM program from DEQSOL-BFM program. The existing DEQSOL translator then generates FORTRAN codes. Finally,the generated codes are compiled, linked and executed in the usual way. At execution time, the surface data file, which is generated by DEQSOL-BFM translator to describe the boundary and the structure information of the arbitrary shape, is inputted by the embedded

BF coordinate generation library, and the BF coordinate system is created.

The features of this system are as follows: Carrying out the BF coordinate generation at the simulation code execution time makes the procedure vectorizable and makes it possible to reduce the DEQSOL translation load. Also, it provides the capability to apply the method to moving-boundary problems, through modification of surface data file at the execution time.

An example of a problem description in the DEQSOL language, extended for the BF facility, is shown in Fig.3. Compared with the existing language specification, the capability for describing space regions is extended to adapt to arbitrary shape, and the MESH statement is extended to indicate the BF coordinate generation. A new statement (NCORD) is introduced to specify the correspondence of coordinates between physical space and transformed space and so on. However, the mathematical level of description is high-level as before.

The main tasks of DEQSOL-BFM translator are the generation of DEQSOL-FDM program and the generation of surface data file. DEQSOL-FDM program generation consists mainly of two parts : the creation of a correspondence between physical space and transformed space, and the transformation of the PDE in the program. As the embedded library, the BF coordinate function is added, which creates a BF coordinate system from the information in the surface data file, by Thompson's method[8].

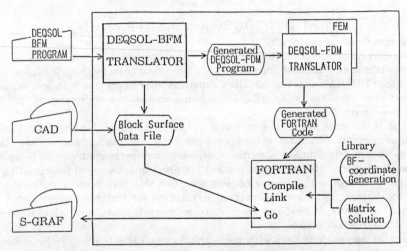

FIG.2 Extended DEQSOL System with BF Facility

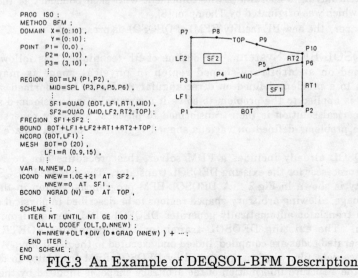

```
PROG ISO :
METHOD BFM ;
DOMAIN  X = (0:10) ,
        Y = (0:10) :
POINT  P1 = (0,0) ,
       P2 = (0,10) ,
       P3 = (3,10) ,
          :
REGION  BOT = LN (P1,P2) ,
        MID = SPL (P3,P4,P5,P6) ,
           :
        SF1 = QUAD (BOT,LF1,RT1,MID) ,
        SF2 = QUAD (MID,LF2,RT2,TOP) :
FREGION SF1 + SF2 :
BOUND  BOT + LF1 + LF2 + RT1 + RT2 + TOP :
NCORD (BOT,LF1) :
MESH  BOT = D (20) ,
      LF1 = R (0.9,15) ,
         :
VAR  N,NNEW,D :
ICOND NNEW = 1.0E+21 AT SF2 ,
      NNEW = 0 AT SF1 ,
BCOND NGRAD (N) = 0  AT  TOP ,
         :
SCHEME :
   ITER NT UNTIL NT GE 100 :
      CALL DCOEF (DLT,D,NNEW) :
      N = NNEW + DLT * DIV (D * GRAD (NNEW) ) + ··· :
   END ITER :
END SCHEME :
END :
```

FIG.3 An Example of DEQSOL-BFM Description

3. PDE transformation.

PDE transformation is one of the key requirements of the DEQSOL-BFM translator. Since DEQSOL can describe arbitrary PDEs of second order, the transformation should be adaptable to any PDEs. And the transformation should such that the transformed PDEs are suitable for numerical calculation.

There are two methods for PDE transformation. One, which is often used, is the non conservative method, which replaces the first and second derivative with a fixed transformed formula. The other is the conservative method, which transforms each first derivative from the inner part recursively, using the fixed transformation rules. The latter method works as follows. A second derivative is assumed as the composite derivative, the inner derivative is transformed, and then the outer derivative is transformed, as below:

$$f_{xx} \rightarrow (f_x)_x \rightarrow [\frac{1}{J}\{(Y_\eta f)_\xi - (Y_\xi f)_\eta\}]_x$$

$$\rightarrow \frac{1}{J}[[(Y_\eta \frac{1}{J}\{(Y_\eta f)_\xi - (Y_\xi f)_\eta\}]_\xi - [(Y_\xi \frac{1}{J}\{(Y_\eta f)_\xi - (Y_\xi f)_\eta\}]_\eta]$$

These two methods are equivalent mathematically, but sometimes cause crucial differences in numerical result because of the discretization procedure. It has been verified that the conservative method is better than non-conservative one in numerical accuracy. In the DEQSOL-BFM translator, the conservative method has been realized.

The processing flow of PDE transformation is shown in Fig. 4. The processing mainly consists of symbolic manipulation, refering to a transformation rule dictionary. First, the translator extracts a term that includes differential operators from the original program, and then expands the differential operator recursively according to the operator rule, and finally transforms derivatives in the expanded equation from the inner part recursively. Using this procedure, any PDEs can be transformed conservatively and naturally.

4. Evaluation.

The BF facility makes it possible to simulate such complex phenomena as the impurity distribution analysis in semiconductor devices. The section of the simulation model is shown in Fig.5. This numerical analysis has the following difficulties:
* Both the silicon and the oxidized silicon layers have curved shapes.
* During diffusion, the borders of the layers move according to the oxidation process. This is the moving boundary problem.
* This is non-linear and simultaneous problem coupled by plural impurities.

A part of the evaluation are summarized from a viewpoint of DEQSOL's effect in Table.1. The descriptive efficiency of DEQSOL (lines-of-code) and the runtime efficiency of the generated FORTRAN code (vectorization ratio, acceleration ratio by vector processor) are illustrated. Here, acceleration ratio refers to the elevation of the machine speed for the programs execution on the vector processor versus the scalar processor. From these data,

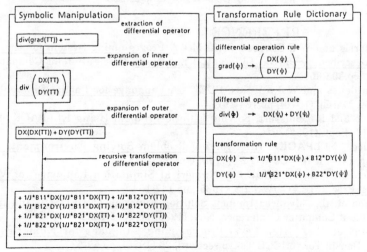

FIG.4 Recurcive Transformation Technique of PDE

despite this extension, the following conclusions can be deduced:
* Programming productivity is still an order of magnitude higher than FORTRAN.
* The generated codes achieve over 90% vectorization ratios on the Hitachi S-810 vector processor, thus exploiting most of its processing power.

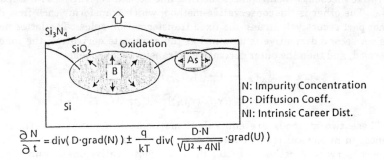

N: Impurity Concentration
D: Diffusion Coeff.
NI: Intrinsic Career Dist.

$$\frac{\partial N}{\partial t} = \mathrm{div}(\,D \cdot \mathrm{grad}(N)\,) \pm \frac{q}{kT}\,\mathrm{div}(\,\frac{D \cdot N}{\sqrt{U^2 + 4NI}} \cdot \mathrm{grad}(U)\,)$$

FIG.5 Simulation Model of Impurity Distribution Analysis

TAB.1 Evaluation of DEQSOL-BFM

Program Name		SHT	PCAD	THD
Discretization Method (dim.)		2	2	3
Scale-of -Problem	Mesh	20*20	50*50	10*10*10
	Node	441	2601	1331
Lines-of Codes	DEQSOL	33	68	69
	Generated FORTRAN	814	1,003	2,602
	Ratio (FORT./DEQ)	24.6	14.7	37.7
Vectorization Ratio (S-810) [%]		90.5	95.3	96.6
Acceleration Ratio (S/V)		3.9	5.7	7.5

SHT : Electric Field Analysis of Sheet Resistor
PCAD: Impurity Distribution Analysis in LSI Process CAD
THD : Thermal Diffusion Analysis in 3D Block

REFERENCE

(1) S.M.Morris and W.E.Sciesser, SALEM - A Programming System for the Simulation of Systems Described by Partial Differential Equations, Proc.Fall Joint Computer Conference, 33 (1968), pp.353-357.
(2) A.F.Cardenas and W.J.Karplus, PDEL - A Language for Partial Differential Equations, Com. ACM, March, (1970), pp.184-191.
(3) J.R.Rice and R.F.Boisvert, Solving Elliptic Problem Using ELLPACK, Purdue Univ., CSD-TR414, Sept.,(1982)
(4) J.R.Rice, ELLPACK - An Evolving Problem Solving Environments for Scientific Computing, Proc. IFIP TC2/WG 2.5, (1985), pp.233-245
(5) Y.Umetani et al., DEQSOL - A Numerical Simulation Language for Vector/Parallel Processors, Proc. IFIP TC2/WG 2.5, (1985), pp.147-164
(6) C.Konno et al., Advanced Implicit Solution Function of DEQSOL and its Evaluation, Proc.Fall Joint Computer Conference, Nov, (1986) pp.1026-1033
(7) C.Konno et al, A High Level Programming Language for Numerical Simulation: DEQSOL, Denshi Tokyo(IEEE Tokyo section), 25 (1986), pp.50-53
(8) J.F.Thompson, Numerical Solution of Flow Problems Using Body-Fitted Coordinate Systems, Lecture Series in Computational Fluid Dynamics, Hemisphere (1980)

Performance Based Distributed Programming
Phillip Q. Hwang

Writing parallel and distributed programs will be very difficult. A number of approaches have been investigated to look at the viability of writing normal, sequential programs without consideration for the hardware architectures or the environmental characteristics. A separate decomposition configuration or specification is then written that will partition the software at runtime. By analyzing the software and hardware as separate steps with an automated specification generator, the program can be optimally applied to a larger number of architectures. The automation techniques will be very important especially for the development of large systems for hard real–time systems.

CHAPTER 50

A Programming Aid for Message-passing Systems

Min-You Wu[*]
Daniel D. Gajski[*]

Abstract. This paper describes an interactive tool for program development in message-passing multiprocessors. The tool provides performance estimates and program quality measures to help programmers improve their parallel programs. Quality measures are derived from a computation flow graph which shows the dependencies, parallelism, load distribution, and critical paths of a program. The tool is expected to increase programming quality and productivity, and reduce the number of errors by an order of magnitude.

1. Introduction. Partitioning, scheduling, and synchronization are the most important issues for parallelization [1]. There are two extreme approaches in solving these problems. One school of thought believes that these problems are complex and that they should be left to programmers [2]. The other school of thought believes in restructuring compilers that will extract parallelism and restructure sequential programs into parallel programs automatically [3]. We advocate an approach that falls between the above two extremes. We believe that parallelization is a very complex problem which can only be fully solved by a human expert. However, some of parallelization chores can be automated, especially scheduling and synchronization. A programming aid, that will automatically schedule processes on different processors and insert synchronization primitives where needed will increase programming productivity. Moreover, it will generate performance estimates and quality measure for programmers to improve their programs and algorithms.

Several research efforts have demonstrated the usefulness of programming aids for multiprocessing. DAPP [4] accepts a code in which synchronization has already been inserted and produces a report of parallel access anomalies, that is, pairs of statements that can access the same location simultaneously. PTOOL [5] performs sophisticated dependency analyses, extracts

---

* Department of Information and Computer Science, University of California at Irvine, CA 92717

This work was supported in part by National Science Foundation grant No.8700738

global variables, and provides a simple explanation facility. However, PTOOL only tests loops for independence. Neither DAPP nor PTOOL provide partitioning and synchronization mechanisms for non-parallel loops. CAMP [6] partitions both parallel and non-parallel loops, inserts synchronization primitives, and estimates performance for each partitioning strategy. However, it does not deal with serial segments of programs. Moreover, it performs partitioning and synchronization for shared-memory systems only.

This paper describes an interactive tool which is capable of analyzing programs consisting of serial and parallel code segments. This tool provides performance estimates and generates a *computation flow graph*, which displays process dependency, parallelism, load distribution, and critical path, so that the programmer can improve the program.

2. Issues in using message-passing systems. To write a program for message-passing systems a programmer must partition the problem into processes, group these processes into tasks, assign each task to a processor in a given topology, and insert synchronization primitives for proper execution.

A program usually consists of serial and parallel segments of code. It can be partitioned according to these natural boundaries. The parallel parts can be further partitioned along iteration boundaries. If there is no dependency between iterations they can be executed on different processing elements (PEs). In case they are dependent, a partitioning method with synchronization insertion must be used.

Those program partitions that are not partitioned further are called processes. They can be thought of as units of computation. When two processes executing on different PEs must exchange data, the data communication time is an overhead that slows down the computation. If this overhead is sizable, the maximum parallelism found in the program may not generate the optimum speedup. For this reason, several processes are merged into tasks which can be thought of as units of allocation. The amount of computation in each task defines the granularity of partitioning. Since message-passing systems usually have higher overhead than shared-memory systems, the tasks tend to have cruder granularity.

After merging processes into tasks, each task is assigned to a PE in a given topology. For the best speedup, two tasks that exchange data should be mapped to the neighboring PEs. Since this is not always possible, the assignment heuristic should attempt to minimize communication distance along the critical path in the program. Thus, an optimal speedup for a given number of PEs can be obtained by selecting proper granularity and proper mapping of tasks for the given topology.

In a message-passing system, there are basically two synchronization primitives: *send* and *receive*. These primitives must be inserted into a program for proper execution. The insertion may be automatic.

3. Development tool. The system diagram of our development tool is shown in Fig.1. First, a designer develops a proper algorithm, performs partitioning, and writes a program as a set of procedures. Second, the tool performs scheduling, and inserts synchronization primitives with users help. The tool then generates performance estimates, including execution time, communication time, and suspension time for each PE, and network delay for each communication channel. The explanation facility displays data dependencies between PEs, as well as parallelism and load distribution in any time interval. It also determines the critical path through the program. If the designer is not satisfied with the result, he will attempt to rewrite the program using the information provided by the performance estimator and the explanation facility. If the improved program still does not satisfy required performance, the designer must redesign the algorithm.

4. Computation flow graph. The main part of the explanation facility is the computation flow graph, as shown in Fig.2. It consists of computation, communication, suspension, and data transmission segments. Computation segments indicate the time spent on executing the program while

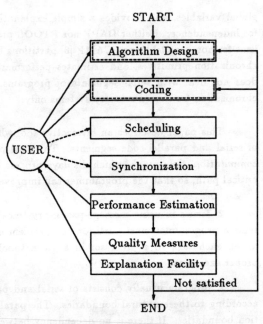

Fig.1 The development tool.

"send" and "receive" constitute communication segments. Suspension segments indicate PEs idle time while waiting for data. Data transmission segments show the dependences among PEs and indicate the time spent for data transfers. Thus, the graph can be used to determine several program quality measures such as parallelism explored during the computation and load distribution among PEs. Furthermore, the computation flow graph also identifies critical paths through the computation, indicated by arrows in Fig.2.

The computation flow graph also can be used for debugging. Since the graph shows all communications among PEs, synchronization errors can be found easily.

The main strategy for improving programs is speeding up the critical path by using the following optimization rules:

(a) move the computation segments forward which do not depend on a "receive", and postpone the segments which do not prepare data for a "send";

(b) move computation segments from overloaded PEs to suspended PEs;

(c) eliminate unnecessary data and information transmission;

(d) combine several communication segments into one;

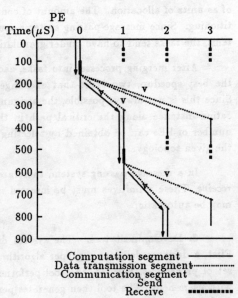

Fig.2 The computation flow graph.

(e) use asynchronous primitives whenever possible; and

(f) reduce the communication distance.

5. An example and results   We use the following example to illustrate dramatic improvement in program performance after the above optimization has been applied. Table 1 and Fig.3 represent the performance estimation and the computation flow graph of a matrix multiplication program. The size of the matrices is 8, and the number of available PEs is 4. From the table, we see that the communication time is short, but the suspension time is long. Since several computation segments on the critical path do not prepare data for "sending" they can be postponed. Results of the improved program are shown in Table 2 and Fig.4. The suspension time has been reduced by 89%.

Table 3 is a larger example with matrices of size 16 and 16 available PEs. This example demonstrates the interactive use of our tool. First, we used algorithm #1, which divided matrices by rows. For program #1(a), the speedup was small. After improving the critical path and rewriting the program as #1(b), we found that the communication time was still too long. We then used a different algorithm which required less communication. Algorithm #2 divided matrices by both rows and columns. The initial code, #2(a), showed improved performance over the previous program #1(b). After the second optimization step, code #2(b) substantially reduced the suspension time. This code exhibited the best performance among all tried in this experiment. This demonstrates how our tool helps programmers to improve both programs and algorithms.

Table 1. Performance estimation for the example.   (mS)

PE	0	1	2	3
Comp. time	8.7	8.7	8.7	8.7
Comm. time	1.0	1.0	1.0	1.0
Susp. time	4.6	4.6	3.9	4.8
CHANNEL	0	1	2	3
Trans. delay	0.9	0.9	0.9	0.9

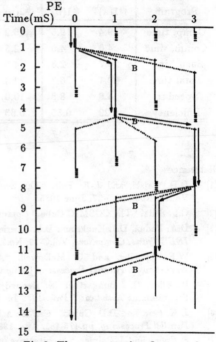

Fig.3 The computation flow graph for the example.

6. Conclusion   An interactive development tool for message-passing systems has been described. The tool used a computation flow graph for debugging and improving programs and algorithms. This tool drastically reduced development time on selected set of examples.

Table 2. Performance estimation after improving the critical path.   (mS)

PE	0	1	2	3
Comp. time	8.7	8.7	8.7	8.7
Comm. time	1.0	1.0	1.0	1.0
Susp. time	0.0	0.4	0.7	0.9
CHANNEL	0	1	2	3
Trans. delay	0.9	0.9	0.9	0.9

Table 3. Comparison   (mS)

Algorithm	#1		#2	
Program	#1(a)	#1(b)	#2(a)	#2(b)
Comp. time	19.4	19.5	16.2	16.3
Comm. time	4.0	4.0	2.5	2.5
Susp. time	26.7	2.8	4.6	0.9
Total time	50.1	26.3	23.3	19.7
Speedup	4.6	8.8	10.0	11.8
Efficiency	0.29	0.55	0.63	0.74

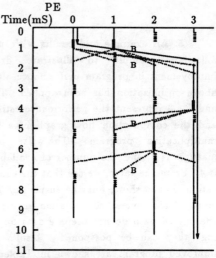

Fig.4 The computation flow graph after improving the critical path.

References

[1]   D.D. Gajski, and J.-K. Peir, "The Essential Issues in Multiprocessor Systems," *IEEE Computer*, Vol.18, No.6, pp.9-27, June 1985.

[2]   C.L. Seitz, "The COSMIC Cube," *Communications of the ACM*, Vol.28, pp.22-33, Jan. 1985.

[3]   D.A. Padua, D.J. Kuck, and D.L. Lawrie, "High Speed Multiprocessor and Compilation Techniques," *IEEE Trans. Computers*, Vol.C-29, No.9, pp.763-776, Sep. 1980.

[4]   W.F. Appelbe, and C. McDowell, "Anomaly Detection in Parallel Fortran Programs," *Proc. Workshop on Parallel Processing Using the HEP*, May 1985.

[5]   R. Allan, D. Baumgartner, K. Kennedy, and A. Porterfield, "PTOOL: A Semi-Automatic Parallel Programming Assistant," *Proc. 1986 Int'l Conf. on Parallel Processing*, pp.164-170, Aug. 1986.

[6]   J.-K. Peir, and D.D. Gajski, "CAMP: A Programming Aide for Multiprocessors," *Proc. Int'l Conf. on Parallel Processing*, pp.475-482, Aug. 1986.

[7]   S. Colley and J. Palmer, "Architecture of a Hypercube Supercomputer," *Proc. Int'l Conf. on Parallel Processing*, pp.653-660, August 1986.

[8]   L. Welty and P. Patton, "Hypercube Architecture," *AFIPS Conference Proceeding*, Vol 54, pp.495-501, 1985.

PART VI

Software Systems

CHAPTER 51

The PARTY Parallel Runtime System

J. H. Saltz[*]
Ravi Mirchandaney[*]
R. M. Smith[*]
D. M. Nicol[†]
Kay Crowley[*]

Abstract. There exists substantial data level parallelism in many applications. We are developing an automated system that is intended to organize the data and computation required for solving data parallel problems in ways that optimize multiprocessor performance. By capturing and manipulating representations of a computation at runtime, we are able to explore and implement rather general heuristics for partitioning program data and control. These heuristics are directed towards dynamic identification and allocation of concurrent work in computations with irregular computational patterns. In problems which involve repetitive patterns of computation, such as iterative methods seen in scientific computations; we calculate an optimized static workload partitioning.

The system is structured as follows: An appropriate level of granularity is first selected for the computations. A directed acyclic graph representation of the program is generated. Parallelization is identified and various workload clustering or aggregation techniques are employed in order to generate efficient schedules. These schedules are then mapped onto the target machine. When computations are irregular, this graph and schedule generation can proceed throughout the course of the computation.

1 Introduction

There exists substantial data level parallelism in scientific problems. The Parallel Automated Runtime Tookit at Yale (PARTY) is an attempt to obtain efficient parallel implementations for scientific computations, particularly those where the data dependencies are

[*]Department of Computer Science, Yale University, New Haven, CT 06520.

[†] Department of Computer Science, College of William and Mary, Williamsburg, VA. 23185,
Work supported in part by the Office of Naval Research under Contract No. N00014-86-K-0564, as well as NASA Grant NAS1-18107 for portion of work performed at ICASE, NASA Langley Research Center, Hampton VA 23185

manifest only at runtime. The two main areas where PARTY can provide performance benefits are the following:

- Automatic detection of parallelism that cannnot be determined by a compiler due to data dependencies that become manifest only at runtime.

- Partitioning and mapping of the computation in a manner that is able to take advantage of the multiprocessor architecture.

PARTY allows for the high level specification and control of computational granularity along with a robust load balancing mechanism for automated scheduling and mapping. The system is to be coupled with compiled languages. The coupling appears to be reasonably strightforward with functional or equational languages such as Crystal [2], [3], we are also actively pursueing the coupling of PARTY to standard imperative languages such as C or Fortran. We are endowing PARTY with a C interface so that the PARTY system can be interfaced directly with programs.

It should be emphasized that when utilizing our system, the user can specify the strategy to be used for parallelization. PARTY detects parallelism between computational grains specified by the user, and then clumps or aggregates those grains so as to optimize performance on different multiprocessor architectures. We thus provide a system that allows for a combination between high level user specification of parallelization strategy along with the automated aggregation and mapping of the computation.

2   Objectives

The goals of the runtime system effort are:

- to exploit the parallelism in programs which only becomes apparent at runtime and consequently is not detected by compile time techniques,

- to manipulate the resulting data dependency directed acyclic graphs (DAGs) into more efficient schedulable units,

- to automate the entire process and be able to generate performance data from a large variety of problems much more quickly than if the mapping and scheduling were specified by the programmer,

- use the above mentioned empirical data to generate realistic performance models of parallel systems.

3   Methodology

3.1   Motivation

Various research efforts have been underway to write compilers for both imperative as well as applicative languages in order to extract parallelism from user programs written in a sequential form [5], [13], [8], [6], [16], [12], [9], [5], [9], [12], [16]. While there has been considerable success in this endeavor, it has been recognized that

- compile time data is not always adequate to exploit hidden parallelism that may manifest itself only at runtime, and

- sophisticated aggregation and mapping techniques may be needed once this runtime parallelism is detected.

Experience has shown that a very substantial programming effort is required to obtain correct solutions and good performance on many currently available architectures. The complexity involved in partitioning the data and computation causes programmers to take advantage of problem specific information in mapping and scheduling work. Furthermore, when problem partitioning must be explicitly specified, the number of lines of code required to solve a problem using current methods often increases dramatically on a parallel machine.

There have been several efforts in which problem specific information is utilized at run-time to detect parallelism and perform appropriate mappings of these computations onto parallel machines. Fox [4] has utilized a scatter decomposition strategy for the Caltech Hypercube. This method partitions the problem domain into a fine lattice of tasks (which are far more numerous than the processors) and then scatters each processors work assignment throughout the domain. Baden [1] has used a method that takes advantage of the locality of interactions in certain types of problems in fluid dynamics and implemented a dynamic load balancing scheme for the Intel iPSC. Nicol and Saltz [11] have implemented the triangular solve and a battlefield simulation program on the Intel iPSC and the Encore Multimax. Lusk and Overbeek [10] implement a self scheduled mechanism to dynamically allocate work to processors. While this method has the advantage of simplicity, there are many potential problems, especially with distributed memory machines.

In many of the examples we have cited, each user ends up designing their own specific runtime system or adapting a runtime system package for either a single problem or for a relatively narrow class of problems. We believe that this approach hinders the speedy development and understanding of parallel systems and applications. Thus, our goal is to provide the runtime detection of parallelism and mapping using an automated system which is able to abstract away the unnecessary details of different computations.

3.2   Mechanism

We utilize a C based interface for the control of the runtime system, this provides us a very clean interface to functional languages such as Crystal and leaves the possibility open of linking the runtime system to other programming environments. In its present form, the syntax requires that the user specify the appropriate granularity of computation as well as portions of code that need to be manipulated by PARTY. Subsequently, we will enhance PARTY to include automated procedures for these functions, in case the user does not specify any problem specific information. With little additional work, it will be possible to generate performance data from applications already developed in C as well as from Crystal programs. Using these data, we hope to design performance models of parallel programs/machines. It is important to remember though that our goals are achieved with a well-defined, but functionally limited interface. The rules we impose on the interface specification are ones deemed expedient for our larger goal of building a working runtime system. The problems being studied in this context include:

- solution of sparse linear systems by Preconditioned Conjugate Gradient Methods. These methods require forward and back substitution of very sparse triangular systems, incomplete matrix factorizations, matrix vector multiplies and inner products,

- discrete event and time-stepped simulations

- realistic simulation of neural networks,

- adaptive mesh solutions of fluids problems,

- string matching problems.

The two main areas where a runtime system can provide performance benefits are the following:

- Automatic detection of parallelism that cannnot be determined by a compiler due to data dependencies that become manifest only at runtime.

- Partitioning and mapping of the computation in a manner that is able to take advantage of the multiprocessor architecture.

We illustrate the above issues with example C programs.

One crucial issue is the ability to extract sufficient parallelism from a computation despite ambiguities in data dependencies that may be present during compilation. Take for example the example of the sparse triangular solve as written in C and presented in Figure 1. Imagine that we parallelize the computation by allocating all work associated with a given row to a single processor. If the matrix is dense, the solution must proceed in a strictly sequential manner, with the solution of row 0 followed by row 1 and so on as the structure of the matrix necessitates this schedule. But, in many sparse matrix computations, each row has only a few non-zero elements. For a given row R, solution values from the rows corresponding to the columns of the non-zero elements of R are needed before R may

```
               Data  Structures
               ----------------
          struct st_matrix
            {
              int *col;
              float *value;
            }
          struct st_matrix *matrix;
          double soln;

               Code Segment
               ------------
          for (i=0; i < nrows; i++)
             for (j=0; j < MAXCOLS; j++)
               {
                 soln[i] = b[i] - matrix[i].value[j]*soln[matrix[i].col[j]];
               }
```

Figure 1. Triangular Solve

produce its solution value. This implies that the solution for row R need only wait for solutions from a few other rows. Once these solutions are available, a solution for row R can be obtained.

The structure of matrices used in many scientific problems are defined at runtime. In the example of the triangular solve, the compiler must sequentialize the solution of matrix rows. A parallelizing compiler could detect parallelism obtainable from the inner products involved in solving individual rows, however, in many cases there are very few non-zero elements in a row so that the amount of parallelism available from intra-row parallelization is not substantial.

Results presented in [7] describe results obtained in the solution of two triangular matrices on 16 processors of an Encore Multimax. Parallel efficiencies of 0.77 and 0.55 were obtained when runtime data dependencies are used to allow several processors to work on different rows at once. When only the inner product parallelism is exploited, efficiencies of 0.12 and 0.063 were obtained. Parallel efficiency on P processors is defined here as the execution time for a separate, optimized sequential code on a single processor, divided by the product of P times the execution time on P processors.

Another aspect of PARTY deals with partitioning and mapping of the computation in a manner that is able to take advantage of the multiprocessor architecture. In many instances,

```
              Data Structures
              ---------------

    struct st_matrix
       {
       int *col;
       float *value;
       }
    struct st_matrix *matrix;

              Code Segment
              ------------

    for(i=0;i<num_iters;i++)
    {
      for(i=0;i<numrows;i++)
      {
        for(j=0;j<matrix[i].ncol;j++)
        {
          newsoln[i] -=matrix[i].value[j]*
                  oldsoln[matrix[i].col[j]];
        }
      }
      for(i=0;i<numrows;i++)
      {
        oldsoln[i] = newsoln[i]
      }
    }
```

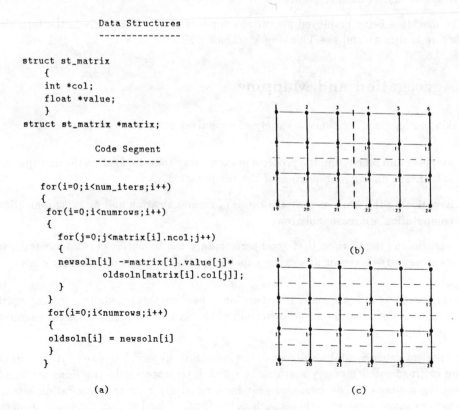

(a)

(b)

(c)

Figure 2. Sparse Jacobi Iterations

the canonical representation of a computation is specified at a very fine computation grain. As a result of this, various performance problems can occur, as described in [15]. Using the data dependencies discovered at runtime, the system uses this information to aggregate work into units which take into account the characteristics of machine architecture.

An example of the relevance of aggregation may be seen in the code for the sparse Jacobi iteration, shown in Figure 2. One of the natural methods for partitioning the work related to this function is based upon the distribution of one or more rows onto each processor. A compiler can detect the parallelism here and distribute this work among the processors. However, data dependencies affect the frequency of communication between the processors. A compiler based mapping is unable to take advantage of this important fact because the structure of the sparse matrix (and consequently the data dependencies) is not available to the compiler. To illustrate this aspect of the problem, we depict a 6x4 mesh with two possible partitions onto a 4 processor system. A partitioning that takes into account geometry is shown in Figure 2b, with the mesh having been divided into four quadrants and each of these being assigned to a processor. The property of this partition is that the amount of data transfer between iterations grows as $N^{1/2}$, where N is the size of the mesh. Depending upon the way the mesh points are numbered, a compiler, because it does not possess any knowledge of the problem structure, might assign rows 1-6 to processor 1, 7-12 to processor 2 and so on, as shown in Figure 2c. Asymptotically, this method incurs a communication cost of order N, which is significantly worse. When the problem is less regular than the one shown, the effect of not knowing problem geometry tends to become even more significant. In the next section, we present experimental results on the Intel iPSC hypercube that emphasize the importance of effective aggregation in problems that involve sparse matrix data structures.

The machines being employed for various portions of this work include the Intel iPSC, the Encore Multimax and the Thinking Machines CM-2.

4   Aggregation and Mapping

In making assignment decisions, a variety of objectives must be taken into account:

1. mapping and scheduling the problem in such a way that the load on the multiprocessor is balanced during each portion of the computation,

2. partitioning the work so that the ratios of communication and or synchronization to computation are reasonably low,

3. partitioning the work so that good performance can be obtained from any fast cache, local memory or vector processing capabilities that a target machine may have.

A variety of methods are under investigation to perform this scheduling and aggregation. All of these methods require a representation of the data dependency relations manifested by the problem.

We present experimental results using the Intel iPSC hypercube. The iPSC is a message passing or fragmented memory machine in which interprocessor interactions are handled via sending messages. The messages sent have relatively high communication latencies. We regard message passing machines such as the iPSC as having very pronounced memory hierarchies; there is a very large performance penalty to be paid when programs must move data between processors frequently and/or in large quantities. As has been discussed above, the efficient implementation of PCGPAK hinges on obtaining satisfactory performance for the sparse triangular solve.

High performance multiprocessor architectures differ both in the number of processors, and in the delay costs for synchronization and communication. In order to obtain good performance on a given architecture for a given problem, an appropriate choice of granularity is essential. The partitioning and mapping of the computation should be performed in a way that is able to take advantage of the multiprocessor architecture.

Experimental results measuring the performance of triangular solves on the Intel iPSC hypercube are presented in this section. As has been previously discussed, the triangular solve proves to be a crucial factor in determining the multiprocessor performance achievable by Krylov space methods when preconditioning using incompletely factored matrices is employed. The efficiencies obtained from the triangular solve represent a pessimistic lower bound on the performance that can be obtained by the iterative portion of the algorithm. We present here only the results pertaining to the triangular solve, the work on implementing and benchmarking other portions of the algorithm on the iPSC is still in progress.

Clustering or aggregation methods can be used to reduce the number of communication startups and the volume of information that must be communicated between processors. The effective use of these methods is essential for obtaining satisfactory results on machines such as the iPSC. The method used to aggregate is discussed in detail in [15] and involves two steps. In the first step, a coordinate system is obtained for the DAG. This coordinate system is obtained through a process of peeling off layers of the DAG. It is then straightforward to map the problem to a multiprocessor in a manner that restricts the fan-in and fan-out of data between processors. To the extent allowed by the data dependencies in the algorithm, it also becomes possible to map work so that only nearby processors have to communicate. Finally, the coordinate system is used to allow the specification of work clusters in a parametric manner.

In the applications discussed here, the clustering of work is controlled by two parameters, the *block size* which describes the number of consecutive DAG layers assigned to a processor and the *window size*, or number of wavefronts per block. The reduction in communication overhead is however, achieved at the risk of load imbalance, making this the critical tradeoff. The reader is referred to [15] for further details regarding this issue.

In Tables 1 and 2 we present results from two moderate sized problems solved on a 32 node Intel iPSC hypercube. Matrices were generated from 200×200 grids and 300×300 grids using a five point template. Reduced systems were formed and a zero fill incomplete factorization was carried out. The triangular systems have 20000 and 45000 rows in the smaller and larger versions of the problem, note that formation of the reduced system also alters the the inter-row data dependencies.

We depict the total parallel efficiency, the total execution time, the number of computational phases and the estimated communication time. Note that the best efficiencies we are able to obtain increase with the size of the problem. In the smaller of the two problems the best efficiency is 31% while in the larger problem the best efficency increases to 53%. These efficiencies in the absence of aggregation were 12% and 16% respectively.

Table 1: Matrix from 200x200 mesh, 5pt. template reduced system

window block size	efficiency	total time	phases	communication time
1- no grey code	0.06	4.58	398	-
1	0.12	2.34	398	1.92
2	0.21	1.35	200	1.06
4	0.31	0.90	100	0.45
6	0.24	1.21	67	0.65
8	0.18	1.54	50	0.57

Table 2: Matrix from 300x300 mesh, 5pt. template reduced system

window block size	efficiency	total time	phases	communication time
1- no grey code	0.08	7.61	598	-
1	0.16	3.99	598	3.45
2	0.31	2.07	299	1.57
4	0.53	1.21	149	0.88
6	0.44	1.44	99	0.65
8	0.33	1.95	75	0.45

The ability to aggregate work plays a central role in extracting increased efficiency from larger problems. When matrix rows are assigned to processors in an unaggregated manner, each row corresponding to an interior mesh point must communicate its value to another processor. Consequently, the communication volume does not tend to decrease as problems become larger. The above data did reveal a small improvement in efficiency when the larger problem was solved even in the absence of aggregation; the amount of computational work per phase does decrease as problem size grows, even in the absence of aggregation.

The aggregation here is performed using graph techniques on sparse matrices. In the absence of a method that is able to capture the geometrical relationship between matrix rows, it is not possible to optimize the *mapping* of the unaggregated problem. In Tables 1 and 2 we also depict the performance figures when the problems were executed without the use of grey coding. This was a conservative attempt to estimate the effect of not taking the problem geometry into account. In this case we still map in a manner that ensures that the assignment of work to processors was such that each processor needed to communicate with only one other processor in each phase. In the larger of the two problems our mapping and **aggregation** methods made the difference between an 8% efficiency and a 53% efficiency.

Aggregation allows a tradeoff between communication costs and time wasted due to load imbalance. For a given size machine, as problem size grows, one can obtain the same load balance by aggregating work into increasingly large chunks. This leads to increasingly favorable ratios of computation to communication costs.

It is natural to inquire as to the degree to which one might expect to improve those efficiencies though improved programming, mapping and scheduling. There is clearly a tradeoff made between costs of communication and the costs of load imbalance as granularity is increased. While one can reduce communication costs by aggregating work in small triangular solves, load imbalances increase rapidly with aggregation. These effects are documented in detail in [15] and will be examined below in the context of a model problem.

A great deal of care needs to be taken in ensuring that the actual code on message passing machines functions efficiently. Without the appropriate techniques, high overheads can result from managing and coordinating the execution of a single irregular problem on a number of processors with separate address spaces. One way of determining the role of these inefficiencies is to compute an *estimated optimal runtime*, which approximates the runtime that would be observed with the actual work distribution but in the absence of any other multiprocessing overheads. We perform an operation count based analyses at the time the workload is aggregated, this yields an estimated speedup. The execution time of the separate sequential program on one processor divided by this estimated speedup gives us this estimated optimal runtime. One can also estimate the time required for communication, by maintaining the pattern of communication and specifying the correct message sizes but deleting all computation and all actual data movement required to pack messages. In an efficient code, the communication costs and the estimated optimal runtime should approximately add up to the actual runtime.

We consider below the results of these analysis performed on a small model problem. This problem will will be presented using data from a triangular system generated from a zero fill incomplete factorization of a sparse matrix generated by a 120x120 five point template. The tradeoffs between load imbalance and communication costs in this model problem have been formally analyzed in some detail [14], [15]. In Table 3 we depict parallel efficiency, estimated optimal time, the total time required to solve the problem on a 32 node Intel Hypercube and the estimated communication time. The communication time estimate is obtained by running problems in which computation is deleted but communication patterns are maintained. We note that when we employ a very fine grained parallelism (window and block size equal to one), we pay a very heavy communication penalty relative to the computation time. The completion of this non-computationally intensive problem requires 240 phases, each one of which requires processors to both send and receive data. We can **reduce the number of computational phases, and hence reduce the communication time but we do this at the cost of increasing** the computation time since the available parallelism is degraded due to load imbalances. While appropriate choice of computational granularity is essential for maximizing computational efficiency, the nature of the triangular solve limits the performance that can be obtained in small to intermediate sized problems. The example here illustrates this well, we obtain a three fold improvement in efficiency through a moderate increase in granularity, i.e. speedup increases to 5.8 from 1.9. The absolute efficiency obtained, even given appropriate choice of granularity is still quite limited.

Table 3: Matrix from 120x120 mesh, 5pt. template

window block size	efficiency	total time	estimated optimal time	communication time
1	0.06	1.25	0.09	1.09
2	0.12	0.60	0.11	0.49
4	0.18	0.40	0.15	0.25
8	0.15	0.48	0.31	0.10
10	0.13	0.56	0.36	0.08

The communication times added to the estimated optimal times yield quantities that are reasonably close to the total time indicating that the code is not likely to contain gross inefficiencies that exaggerate the difficulties involved in solving this small problem. We note that only rough correspondence is expected as the assumptions made in the operation count analysis are clearly overly simplistic.

5   Status of the Project

We have provided some initial results that indicate the crucial performance role that can be played by a runtime system. We have seen that for sparse matrix computations on the Encore Multimax, the efficiencies generated by runtime parallelization were several times higher than those obtained without runtime parallelization. We have tested aggregation or clustering strategies on the Intel iPSC multiprocessor [15] and demonstrated that a graph based clustering of work can also lead to several fold improvements in efficiency. We can thus claim that an efficient, automated runtime system is desirable and that it can be built using the principles outlined in this paper.

The status of the various parts of the runtime system is as follows: We have preliminary designs of the intermediate C syntax as well as the design of the DAG encoding mechanism for the prescheduled system. We are in the process of developing the system that encodes a DAG from the intermediate C program. The designs for the self scheduled systems are still being developed. As regards running software is concerned, we have a schedule execution program, a version of which runs on the Intel iPSC and another the Encore Multimax. Preliminary DAG manipulation routines as well as the self-scheduled system are currently being developed and tested in the context of sparse matrix computations and discrete event simulation problems.

References

[1] S. B. Baden. *Run-Time Partitioning of Scientific Continuum Calculations Running on Mutiprocessors*. PhD thesis, Mathematics Dept., University of California, Berkeley, June 1987.

[2] Chen. *Can Data Parallel Machines be Made Easy to Program*. Technical Report YALEU/DCS/RR-556, Department of Computer Science, Yale University, August 1987.

[3] M. C. Chen. Very-high-level parallel programming in crystal. In *The Proceedings of the Hypercube Microprocessors Conf., Knoxville, TN*, September 1986.

[4] B. Fox and P. Glynn. *Computing Poisson Probabilities*. Report CALT-68-1343, University of Montreal, 1986.

[5] P. Hudak. Denotational semantics of a para-functional programming language. *Proceedings of Symposium on Principles of Distributed Systems*, 1981.

[6] P. Hudak. Para-functional programming. *Computer*, Aug 1986.

[7] Saltz J.H., Mirchandaney Ravi, Smith Roger, Nicol D.M., and Crowley Kay. *The Automated Crystal Runtime System: A Framework*. Technical Report YALEU/DCS/TR-588, Yale, January 1988.

[8] K. Kennedy. Compilation for n-processor architectures. In *Proceedings of the IEEE International Conference on Computer Design: VLSI in Computers*, page 15, October 1985.

[9] D. J. Kuck, R. Kuhn, D. Padua, B. Leasure, and M. Wolfe. Dependence graphs and compiler optimizations. *Proceedings of the 8th ACM Symposium on Principles of Programming Languages*, 1981.

[10] E. Lusk, R. Overbeek, and et. al. *Portable Programs for Parallel Processors*. Holt, Rinehart and Winston Inc., 1987.

[11] D. M. Nicol and J. H. Saltz. *Principles for Problem Aggregation and Assignment in Medium Scale Multiprocessors*. Report 87-39, ICASE, September 1987.

[12] D. Padua, D. Kuck, and D. Lawrie. High speed multiprocessors and compilation techniques. *IEEE Trans. Computers*, 1980.

[13] D. A. Padua and M. J. Wolfe. Advanced compiler optimizations for supercomputers. *CACM*, Dec 1986.

[14] Y. Saad and M. H. Schultz. *Topological Properties of Hypercubes*. Department of Computer Science YALEU/DCS/RR-389, Yale, June 1986.

[15] J. Saltz. *Automated Problem Scheduling and Reduction of Communication Delay Effects; submitted for publication*. Report 87-22, ICASE, May 1987.

[16] M. J. Wolfe. *Optimizing Supercompilers for Supercomputers*. PhD thesis, University of Illinois, Urbana-Champaign, 1982.

Parallel Systems: Performance Modeling and Numerical Algorithms
Daniel A. Reed

Historically, there have been two major techniques for modeling parallel systems: discrete event simulation and mathematical analysis. Although simulation models can mimic a real–world system as closely as understanding permits and needs require, highly detailed simulation models can be computationally taxing. In contrast, analytic techniques often can quickly provide mathematical insight into the behavior of systems over a broad range of parameter values. Their major limitation is the number of restrictive assumptions that often must be satisfied to insure tractability and accuracy. For many parallel systems, evaluation of realistic analytic models is as complex or even more complex than equivalent simulations.

Transient behavior, complex access patterns, and large state spaces all make analysis difficult. Eliding unnecessary detail while retaining important features is the essence of the modeler's art. The author presents a set of techniques for analyzing the performance of parallel architectures. These techniques permit the comparative evaluation of interconnection networks for message–based processors (e.g., hypercubes) as a function of communication pattern, including communication locality. In addition, the author shows how simple analytic techniques can capture the salient details of network traffic and memory conflicts in shared memory processors.

CHAPTER 52

Concurrent Management of Priority Queues for Adaptive Algorithms

J. Kapenga[*]
E. de Doncker[*]

Abstract. A study of the demands of several types of parallel numerical algorithms, which are synchronized by a task pool, is presented. This is used to evaluate different methods of concurrent priority queue management on various shared memory architectures. Both an analytic model and empirical results from the Denelcor HEP, Sequent Balance, Alliant FX-8 and Encore Multimax are considered.

This work is part of our effort to develop a set of macros, layered over the Argonne macro package, which provide a means of producing portable programs for adaptive partitioning problems. Portable efficient concurrent priority queue management is critical to this endeavor.

1. Introduction. Adaptive algorithms for ray tracing, numerical integration, function approximation, searching and optimization can often be viewed as an iterated four-step process: selecting a task from a task pool, subdividing the task into subtasks, testing for termination conditions and inserting subtasks back into the task pool. The selection of tasks is based on a priority, which usually corresponds to an error estimate or a value indicating how much effect the task might have on the final result. In [9] a meta-algorithm was presented which allowed semi-automatic parallelization of serial versions of algorithms in this class.

This conversion from serial to parallel algorithm is supported by a set of macros called ADAPT-MACS, which contain code for performing concurrent operations on a priority queue used to maintain the task pool. The operations required on the queue are: the insertion of one or more tasks into the queue, the removal of one or a group of tasks from the queue and the merging of two queues. The synchronization of these operations is hidden from the application program within macro invocations, such as *get-tasks*(), *get-space*(), *put-tasks*() and *merge-pools*() , producing parallel application programs which are provably correct, given that the original serial versions were correct.

Figure 1. presents a skeleton of the *work* module in the parallelized version of a program, which is executed by all processes in attacking the *task pool*. Not shown is the initialization by the master process, the starting of the slave processes, the end of stage processing (where separate queues of tasks may need to be merged), and the final processing, all of which are done in a similarly structured manner (see [9]). We are considering adjusting the system so that after initialization and process creation there is no special status for the master process. This can be done with little change to the current programs and macros and may produce important gains.

Using the m4 preprocessor, ADAPT-MACS was initially constructed as a macro layer, over the Argonne Macros package ([12,13,7]). This was running on a Denelcor HEP. One of the goals of the

* Western Michigan University, Computer Science Department, Kalamazoo MI 49008. kapenga@anl-mcs.arpa

FIG. 1. work *Module Executed By All Processes*

work: REPEAT
　　　　get-tasks(I)
　　　　get-space(NS,J)

　　　　⋮

　　　　(* Partitioning step code from the serial program to process the tasks
　　　　(* pointed to by I, placing the subtasks in the NS spaces pointed to by J.

　　　　⋮

　　　　update-globals(update-code)
　　　　IF *termination conditions* THEN *probend*
　　　　put-tasks(J)
　　　UNTIL doomsday

initial Argonne Macros package, and ADAPT-MACS, was to allow the creation of efficient portable parallel programs. As new MIMD shared memory machines reached the market it became clear this goal was partially met. With some small changes in the user interface to the macros the people at Argonne were able to port their macros to new machines. These included, among others, the CRAY XMP, Sequent Balance 21000, Encore Multimax and Alliant FX/8. The effort in porting the Argonne macros was said to be far less than their creation, in some cases requiring only a couple days. We were then able to follow, porting ADAPT-MACS to various systems with very little effort. Versions of the Argonne Macros were originally written in both FORTRAN and c. Currently, the c versions are receiving the most active development ([3]).

Two moderate size application programs were coded on the Denelcor HEP with the ADAPT-MACS: a two dimensional numerical integration program ([4]) and a parallel ray tracing program, with the serial version referred to in [14] and the parallel version mentioned in [10]. These ran basically unchanged on the new machines tried. For a discussion of multidimensional integration using this approach see de Doncker and Kapenga ([5]).

2. Concurrency in Standard Priority Queue Management Algorithms. There is an extensive literature on serial management of priority queues (e.g. [11]) and even some work on VLSI implementation of concurrently accessible priority queues. Much less is known about the concurrent management of priority queues on actual present day MIMD shared memory architectures by application programs. This is an important topic for study in connection with the improvement of performance of programs using the ADAPT-MACS package as well as for direct use. There are two areas to consider in evaluating queue management algorithms, the profile of an application's use of queues and the specific architecture of the target machine.

The first step is to consider the nature of the application program's use of the three queuing primitives *get-tasks*(), *add-tasks*() and *merge-pools*(). Following work by Rice ([15]) on parallel numerical integration, we propose a model for the behavior of adaptive processes. To describe the model select: a dimension d; a means of subdividing a d-dimensional hypercube; a set S of affine subspaces of dimension less then d (which will be called the set of singularities). We also require two distributions f_1 and f_2. The distribution f_1 is used to model the drop in the estimated error over a subproblem which intersects S, while f_2 is used to model the drop in the estimated error over regions which do not intersect S.

These selections are used in a natural manner. Initially the unit d-cube with an error of 1.0 is placed in the work pool. Tasks in general consist of a d-dimensional hypercube and an error estimate. Using the subdivision rule and the drop in error distributions, the algorithm in Figure 1. is applied until the total error in the task pool falls below a cut off value.

Rice's model corresponds to $d=1$, $S=$ a set of points, the subdivision is by bisection of each interval and the support of the densities can be assumed to be bounded from below by 0 and above by constants α and β.

Useful results can also be obtained for the cases when f_1 and f_2 are uniform or gamma distributions. Moving to at least dimension 2 and considering line singularities is important as well. These are reasonable and provide the basis for good simulation studies.

FIG. 2. *Serial Priority Queue Management Complexities*

	removal	insertion	merge
linked list	1	n	n
heap	$\log n$	$\log n$	$\log n$
leftish tree	$\log n$	$\log n$	$\log n$
B-tree	$\log n$	$\log n$	n
DAQ	1	1	1

The considerations of the various architectures include: the overhead of the synchronizations available, the number of synchronizations which can be specified, the existence of local memory and the blocking which may occur when accessing shared memory. Of course these must all be considered relative to the time the arithmetic, logical and nonshared memory units will require.

Once the nature of the architecture and the concurrent use of the priority queue primitives is specified, various queue management schemes can be compared. The standard data structures of linked lists, heaps, leftish trees and B-trees can all be parallelized reasonably for some applications by using a single lock to limit access to the structure to one process at a time. This method is acceptable when the synchronization primitives are fast, the maximum queue size is small and the computation required in the subdivision step is large; all relative to the number of processes.

Often a single lock on a queue turns into a bottleneck in the applications we have in mind. The queues can get very large so that the queue management time becomes significant, relative to the computation in the subdivision step. The use of more than one lock on a queue can help. On a Denelcor HEP each memory location had an extra bit that was used by machine instructions to allow asynchronous access to memory with virtually no overhead. This allows a fairly direct translation of the standard serial queue management algorithms into efficient parallel versions. Current commercial parallel processors do not provide this feature, making the management of large queues for a large number of processors difficult. The parallelization of B-trees is investigated in [1], while Heaps and Banyans are discussed in [2].

It should be noted that all the mentioned queue management methods return the task with the highest priority, though in the unlikely case, in these applications, of ties unusual things can happen. For the numerical applications we have in mind this restriction can be relaxed to requiring the return of a task from a class of tasks which all are close to the highest priority, and an insurance that all tasks in this class will be returned in a first in first out, FIFO, manner. The next section contains a description of a Distributive Array Queue, DAQ, which provides this behavior. Figure 2. is a table of the complexities for several serial priority queue management methods, including the DAQ.

Following the earlier work done on performance modeling ADAPT-MACS ([9]), as well as the task model above extended from Rice and Jordan's [8] ideas on parallel performance, the behavior of "parallelizable" adaptive processes can be specified. This results in an *effective parallelizability*, which is the maximum speedup which can be expected, for a given serial algorithm and class of problems. Effective parallelizability depends parametrically on the synchronization overheads associated with specific architectures.

Using the model so produced, DAQ and hybrid DAQ type methods are best on some architectures, in particular when the number of available processors is close to or greater than the effective parallelization and the task pool becomes large.

3. Distributive Array Queues. We now define a Distributive Array Queue, DAQ. A DAQ is:

- A set of NQ disjoint linked lists, $l_1, l_2, \ldots l_{NQ}$, which provide FIFO queues $q_1, q_2, \ldots q_{NQ}$.
- An array of dimension NQ of pointers, $q.head()$, which point at the heads of the corresponding linked lists.
- An array of dimension NQ of pointers, $q.tail()$, which point at the tails of the corresponding linked lists.
- An integer $q.first$ which is the index of the first non-empty queue.
- An optional array of dimension NQ of counts, $q.count()$, which gives the lengths of the linked lists.

FIG. 3. *Priority Queue Management System Effects on Execution Time on a Sequent 21000*

	Number of Processes									
	1	2	3	4	5	6	7	8	9	10
linked list	1.00	.71	.59	.73	.96	1.22	1.61	2.12	2.40	2.53
Heap	.55	.30	.24	.22	.17	.19	.23	.24	.37	.35
BAQ	.46	.25	.18	.14	.11	.09	.10	.11	.10	.12

- An optional array (or two) of dimension NQ of links, $q.link()$, which may be used to form a linked list (or doubly linked list) of some of the nonempty queues.
- A nonincreasing function $f : E \rightarrow I$, where E is the set of possible priorities and $I = \{1, 2, 3, \ldots, NQ\}$

The intention is simply to use $f(e)$ to select a queue, $q_{f(e)}$, to put all tasks in which have priority e. Each queue, q_i, is to be maintained as a FIFO structure. The function f could be a truncated linear function; however, letting $f(e) = -c * \log_2(e/s)$, where c and s are chosen positive scaling constants, has the advantage that the ratio of the largest to the smallest error mapped into a bucket can be controlled (if $c = 2$ then the ratio is $\sqrt{2}$). In the c language, in a nonportable manner, or in the proposed FORTRAN-8x standard ([16]), in a fully portable manner one can access the exponent and fraction in a floating point representation. This allows the construction of a log like function f, as far as the current usage requires, without a multiplication or division.

With this definition of a DAQ, implementations of the desired queue operation, (insertion, removal and merging) and the complexities claimed in the previous section are clear. The optional count array is intended for use in load balancing between disjoint DAQs or fast profiling of the total work in a task pool. A work profile may be sent out to determine which tasks to transfer form one DAQ to another without removing any tasks from the pool, this is important on loosely coupled architectures.

Figure 3. contains normalized execution times for an example run on a SEQUENT BALANCE 21000. The effective parallelizability of the problem was about 6. There were 4 point singularities on the unit interval. The task pool was about 1,000 at termination for the serial algorithm and up to 6,000 for the parallel algorithms with 10 processes. A dummy work load was set at 50 FLOPS. The distributions $f_1()$ and $f_2()$ were uniform with means 0.3 and 0.5, each had variance 0.15. There was a system limit of 10 processors per job and it appeared that a full 10 were applied to each job.

The conclusion is that the effect of too many processors, more then the parallizability of the problem allows, produces a loss in performance because of the extra size of the queue. This extra size is caused by unneeded work being done, producing singularity loss ([9]). This singularity loss does not appear in the DAQ because of the independence of the DAQs overhead with respect to queue size. Similar results were obtained on the Alliant and Encore, by setting different parameters. In all cases, with large queues, the DAQ performed as well as or better than any other method tried.

The fact that the function $f()$ can be evaluated in parallel and the only operations on insertion, removal and merging needed inside critical sections make it very appealing for concurrent use. This is true both in the case where there are many locks and when a single lock is used for the entire structure.

The optional array $q.link()$ is intended to provide a short linked list of the leading nonempty queues, say those before $q.thrash$. This can be used to prevent thrashing in the case where there is a very strong point singularity. For an automated system a hybrid system should be used, initially using a single ordered linked list, keeping some profile information, until the queue reached a critical size. Then the information would be used to adjust the parameters in the DAQ and a single process would start converting the queue over to the DAQ. Such a conversion operation, just like the merge, can be implemented concurrently with normal process access to the DAQ.

In general as the speed and number of processors increases in forthcoming shared memory machines, for individual user programs to take advantage of these gains, it will be necessary to use new types of higher level synchronization methods, such as DAQs.

4. Future Work. When the hybrid DAQ system is operating it can be adapted automatically either by using internal control or under application program control to help ensure effective parallelization. This control can be placed outside critical sections and can be very effective. The model mentioned here needs more work, both analytical and empirical, to determine the best control strategies.

The DAQ also holds great promise for use on loosely coupled architectures, such as a hypercube.

5. Acknowledgment. We would like to thank people and groups at Argonne National Laboratories, in particular R. Lusk and R. Overbeek in the Mathematics and Computer Science Division, who gave advice and produced the Argonne Macros and the Advanced Computing Research Facility, which provided excellent access, system support and processor time.

REFERENCES

[1] R. BAYER AND M. SCHKOLNICK, *Concurrency of operations on B-trees*, Acta Informatica, 9 (1977), pp. 1–21.

[2] J. BISWAS AND J. BROWNE, *Simultaneous update of priority structures*, in PROCEEDINGS OF THE 1987 INTERNATIONAL CONFERENCE ON PARALLEL PROCESSING, S. Sahni, ed., The Pennsylvania State University Press, August 17-21 1987, pp. 124–131.

[3] J. BOYLE, R. BUTLER, T. DISZ, B. GLICKFEILD, E. LUSK, R. OVERBEEK, J. PATTERSON, AND R. STEVENS, *PORTABLE PROGRAMS FOR PARALLEL PROCESSORS*, Holt, Rinehart and Winston, 1987.

[4] E. DE DONCKER AND J. KAPENGA, *Parallelization of adaptive integration methods*, in NUMERICAL INTEGRATION; RECENT DEVELOPMENTS, SOFTWARE AND APPLICATIONS, P. Keast and G. Fairweather, eds., Reidel, 1987, pp. 207–218.

[5] ———, *A portable parallel algorithm for multivariate numerical integration and its performance analysis*, 1988. this conference.

[6] L. DEVROYE, *LECTURE NOTES ON BUCKET ALGORITHMS*, Birkhäuser, 1985.

[7] C. HOARE, *MONITORS: an operating system structuring concept*, Comm. ACM, 17 (Oct. 1974), pp. 549–557.

[8] H. JORDAN, *INTERPRETING PARALLEL PROCESSOR PERFORMANCE MEASUREMENTS*, Tech. Rep., Department of Electrical Engineering and Computer Engineering, Univ. of Col., 1985. CSDG 85-1.

[9] J. KAPENGA AND E. DE DONCKER, *A parallelization of adaptive task partitioning algorithms*, Parallel Computing, (1988). to appear.

[10] J. KAPENGA AND R. PULLEYBLANK, *Raytracing bicubic patches on a hypercube*, Sept. 29 - Oct. 1, Knoxville, TN, 1986, Abstracts of the Second Conference on Hypercube Multiprocessors.

[11] D. KNUTH, *THE ART OF COMPUTER PROGRAMMING, VOLUME 3: SORTING AND SEARCHING*, Addison Wesley, 1973.

[12] E. LUSK AND R. OVERBEEK, *IMPLEMENTATION OF MONITORS WITH MACROS: A PROGRAMMING AID FOR THE HEP AND OTHER PARALLEL PROCESSORS*, Tech. Rep., Argonne National Laboratory, 1983. Report MCS ANL-83-97.

[13] ———, *USE OF MONITORS IN FORTRAN: A TUTORIAL ON THE BARRIER, SELF-SCHEDULING DO-LOOP AND ASKFOR MONITORS*, Tech. Rep., Argonne National Laboratory, 1984. Report MCS ANL-84-51.

[14] R. PULLEYBLANK AND J. KAPENGA, *A vlsi chip for raytracing bicubic patches*, IEEE Computer Graphics and Applications, 7 (March 1987), pp. 33–45.

[15] J. RICE, *Parallel algorithm for adaptive quadrature III - program correctness*, TOMS, 2 (1976), pp. 1–30.

[16] X3J3, *AMERICAN NATIONAL STANDARDS INSTITUTE FOR INFORMATION SYSTEMS PROGRAMMING LANGUAGE FORTRAN*, Tech. Rep., American National Standards Institute Inc., June 1987. S8 (X3.9-198x) version 104, revision of X3.9-1978.

CHAPTER 53

On the Placement of Parallel Processes

Michael R. Leuze[*]
Stephen R. Schach[*]

Abstract. An important problem of parallel computation is to determine the best placement in a multiprocessor system of each of a set of related processes. An earlier approach to this problem attempted to maximize the number of communicating processes which lie on adjacent processors. We demonstrate that this approach is not suitable for large problems through a comparison of mappings generated by this approach with random mappings and with mappings generated by heuristics which attempt to minimize the distance from each process to processes with which it communicates. The various approaches are experimentally compared with respect to a specific problem.

Introduction. The use of parallel processors is increasing as researchers in diverse application areas pose computational problems which are intractable on uniprocessor systems. When solving a problem using a parallel system which communicates by passing messages, a number of issues arise in addition to those that must be considered when using a system with a single processor. First, a researcher must determine how a problem is to be *partitioned*, that is, how a problem is to be divided into relatively independent modules, each of which will be executed by a single process requiring occasional communication with other processes. Next, it must be determined how these processes are to be assigned, or *mapped*, to processors. The mapping problem is of primary interest to us in this work.

---

[*] Department of Computer Science, Vanderbilt University, Nashville, TN 37235.

Bokhari [1] has considered a simple form of the mapping problem. Given a problem graph, in which a vertex corresponds to a process and an edge represents interprocess communication, and a machine graph, in which a vertex corresponds to a processor and an edge to a physical link between processors, Bokhari searched for mappings which maximize the number of pairs of communicating processes which lie on adjacent processors. He noted that this form of the problem is equivalent to the graph isomorphism problem, for which there are no known polynomial time algorithms for general graphs. To find good mappings, he used a heuristic which exchanges pairs of processes to increase the number of adjacent communicating processes. Occasional random exchanges were made in an attempt to avoid convergence to a local maximum. Bokhari tested his heuristic on structural problems with from 9 to 49 vertices and obtained "acceptable" results but speculated that the algorithm would not be suitable for much larger problems.

This paper reports on a process-placement heuristic which, for structural problems with 256 vertices, improves upon Bokhari's heuristic, both in the quality of mappings produced and in the speed with which they are found.

Placement Heuristics. Bokhari's only criterion used in finding or evaluating a mapping was the number of communicating processes which lie on adjacent processors. We felt this criterion to be insufficient either for finding good mappings or for evaluating mappings in the general case, when perfect mappings of the problem graph to the machine graph do not necessarily exist. Bokhari's criterion is insufficient for *finding* good mappings because it does not consider the placement of communicating processes which cannot be placed on adjacent processors. We felt it was important to place communicating processes as close to each other as possible. Consequently, we examined a modification of Bokhari's algorithm which attempts to minimize the total distance from each process to all processes on adjacent levels of the problem precedence graph with which it communicates. Bokhari's criterion is, furthermore, insufficient for *evaluating* mappings because it does not account for the communication-link contention which results when messages between communicating processes on non-adjacent processors must be routed through intermediate processors. As a result, to evaluate mappings we developed a model which simulates the execution of Gaussian elimination on a hypercube multiprocessor. The model assumes that at each of the hypercube nodes, a single Gaussian elimination process, responsible for maintaining and updating one row of the coefficient matrix, is executing. During each time step, the model passes as many messages as possible along appropriate links between processors. If two or more messages attempt to use the same link during the same time step, one message is chosen randomly.

A 256×256 matrix arising from a 16×16 finite element mesh was tested. The matrix was first ordered using a variety of techniques: random (RAN), reverse Cuthill-McKee (RCM) [2] [3], minimum degree (MIN) [6] [7], nested dissection (NED) [4], and two independent set orderings [5], one using a greedy algorithm to find maximal independent sets (IGR), the other using a vertex cover complement approach (IVC).

Three different mapping heuristics were then applied to each set of processes responsible for maintaining and updating the rows of a filled matrix. The placement heuristics examined were:

PH0: Random assignment of processes to processors.

PH1: Bokhari's heuristic [1], which attempts to maximize the number of communicating processes which lie on adjacent processors.

PH2: The modified heuristic, which attempts to minimize the total distance from each process to all processes with which it communicates on adjacent levels of the problem precedence graph.

Results. The model was applied to each of the six filled matrices with processes mapped to the processors using each of the three placement heuristics. Each of the 18 cases was tested 10 times, since choosing a message at random when contention arises causes some variation from test to test. Means and standard deviations of the total number of time steps required to complete each parallel computation appear in Table 1.

Table 1. Means and standard deviations of number of time steps required to complete parallel Gaussian elimination

	PH0	PH1	PH2	Percentage improvement of PH2 over PH1
RAN	1528.0 ± 34.8	1441.0 ± 53.5	951.6 ± 38.9	51.4%
RCM	1505.3 ± 19.8	1136.1 ± 28.6	811.8 ± 12.1	39.9%
MIN	263.2 ± 11.1	187.9 ± 7.8	153.7 ± 5.5	22.2%
NED	279.4 ± 9.9	212.9 ± 10.5	172.8 ± 7.6	23.2%
IGR	474.1 ± 17.9	427.3 ± 16.1	338.5 ± 16.7	26.2%
IVC	221.9 ± 9.6	183.4 ± 6.9	137.9 ± 8.2	33.0%

It is obvious that both the way in which a matrix is ordered and the way in which processes are mapped to processors are important considerations. Bokhari's heuristic (PH1) is, as he predicted, not satisfactory for large problems. PH1 performs only slightly better than random placement (PH0). PH2, which causes communicating processes to be placed close together even if they cannot be placed adjacently, consistently and significantly outperforms both PH0 and PH1.

Computation time is an important consideration in choosing a mapping. The times required on a Sun 3/50 workstation to determine mappings using PH1 and PH2 for the MIN, NED, and IVC orderings are listed in Table 2. The execution time of PH2 is much faster, better than two orders of magnitude, than the execution time of PH1. This increase in speed is due to the fact that in PH2's evaluation function, only communicating processes which lie on adjacent levels of the problem precedence graph must be considered, whereas in PH1, all communicating processes are considered.

Table 2.	Sun 3/50 execution times	
	PH1	PH2
MIN	2463.1	102.9
NED	5589.4	171.7
IVC	5147.6	110.0

A Non-Iterative Placement Heuristic. In an effort to reduce the computation times for mappings even further, a non-iterative placement heuristic (PH3) was developed. In PH3 once a process is assigned to a processor, no subsequent change is made. The steps of the placement algorithm are as follows:

1. Order processes according to the number of neighbors each has one level away in the precedence graph ("one-neighbors"), from largest to smallest. Break ties by choosing the process with the largest number of neighbors two levels away in the precedence graph ("two-neighbors"). Break remaining ties by choosing lowest numbered node in the ordering.

2. Map each process in order to that processor which will minimize the sum of the distances to previously placed one-neighbors. Break ties by minimizing the sum of the distances to previously placed two-neighbors.

3. If no one-neighbors or two-neighbors have been placed, map the process to that processor which will maximize the sum of distances to previously placed (non-neighbor) processes.

A comparison of PH2 and PH3 is shown in Table 3.

Table 3. A comparison of heuristics PH2 and PH3

	PH2		PH3	
	Computation (Sun CPU secs.)	Performance (time steps)	Computation (Sun CPU secs.)	Performance (time steps)
MIN	102.9	153.7	41.0	144.5
NED	171.7	172.8	38.8	169.0
IVC	110.0	137.9	43.0	133.8

Not only is PH3 two or three times faster than PH2, it also produces better mappings with respect to the simulation model. Consequently, it appears on the basis of this experimentation that non-iterative placement heuristics are to be preferred over iterative approaches.

It is difficult to determine how close to optimal the mappings of PH3 are. Further work is needed in the development and testing of process-placement heuristics, particularly for problems in which the number of processors is limited.

REFERENCES

[1] S. H. BOKHARI, On the mapping problem, *IEEE Trans. Comp.* C-30(3) (1981) pp. 207-214.

[2] E. CUTHILL and J. McKEE, Reducing the bandwidth of sparse symmetric matrices, *Proceedings of the 24th National Conference of the Association for Computing Machinery*, ACM Publications (1969) pp. 157-172.

[3] A. GEORGE, Computer implementation of the finite element method, Tech. Rep. *STAN-CS-208*, Stanford University (1971).

[4] A. GEORGE, Nested dissection of a regular finite element mesh, *SIAM Journal of Numerical Analysis* 10 (1973) pp. 345-363.

[5] M. R. LEUZE, Independent set orderings for parallel Gaussian elimination, Tech. Rep. *CS-86-12*, Department of Computer Science, Vanderbilt University (1986).

[6] D. J. ROSE, A graph-theoretic study of the numerical solution of sparse positive definite systems of linear equations, in **Graph Theory and Computing**, edited by R.C. Read, Academic Press, New York (1972).

[7] W. F. TINNEY, Comments on using sparsity techniques for power system problems, in *Sparse Matrix Proceedings*, IBM Research Rep. RAI 3-12-69 (1969).

CHAPTER 54

Iteration Space Tiling for Memory Hierarchies

Michael Wolfe*

Abstract. Researchers have found that block algorithms can improve the performance of computers via better utilization of virtual memory, cache memory, local memory or registers. Iteration space tiling is presented here as a compiler optimization that can be used to automatically generate block algorithms. Iteration space tiling is shown to be a combination of strip mining and loop interchanging. These, together with index set splitting and automatic concurrency detection, can be used to automatically optimize linear algebra algorithms for parallel machines with memory hierarchies or private memories.

1. Introduction. Block algorithms have been the subject of a great deal of research and study recently; as evidence, this conference alone included many papers and posters on different block algorithms for different machines. The advantage of block algorithms is that while computing within a block, there is a high degree of data reuse, allowing better register, cache or memory hierarchy performance. This research is now well developed enough to include some of the principles in compilers.

We have developed a retargetable restructuring processor, called KAP, which can discover vector and concurrent loops and represent these in a form suitable for execution on a wide variety of machines. KAP also includes a powerful vector optimization facility which is used to decide which of several methods to execute a loop will produce the best performance on the target architecture [7]. Naturally, vector optimization is very specific to a particular machine.

Development of optimizing compilers began with the study of the techniques used by intelligent programmers to improve the performance of their programs. From this study came theory and rules which are now applied in compilers automatically. Vectorization, for instance, is exactly one such optimization. Initially, programmers began to rewrite their programs and rethink their algorithms to fit onto new vector machines. This developed into the theory of data dependence and rules for automatic vectorization of loops. Now we have powerful automatic vectorizers which, while not perfect, are well accepted enough that every vendor supplies a vectorizing compiler. Other optimizations which were derived from manual algorithmic modifications are loop interchanging [2], [5] and loop skewing [6].

This approach to designing compiler optimizations can now be applied to optimizations for memory hierarchies. Previous work in automatically optimizing programs for virtual memory machines [1] can also be applied to other memory hierarchies. Rather than looking at the algorithm, these optimizations look at the structure of the program and modify the loop structure to produce the same benefits of manual algorithm enhancement. The methods involved break the "iteration space" defined by the loop

---

* Kuck and Associates, Inc., Champaign, IL 61874

structure into blocks or "tiles" of some regular shape, and traverse the tiles in an order designed to improve the memory hierarchy utilization. We shall see that some of the methods used in vectorizing compilers are used to implement loop tiling.

2. Block Algorithms. Much work has been done on block algorithms with the goal of reducing memory traffic. As an example, the block matrix multiply algorithm is shown here. A basic matrix multiply computes an inner product of a row and a column of matrices B and C for each element of the result matrix A:

```
DO 100 I = 1,N
DO 100 J = 1,N
DO 100 K = 1,N
100   A(I,J) = A(I,J) + B(I,K)*C(K,J)
```

If the arrays are stored columnwise (as in Fortran) in cache lines or pages, then the accesses to B in the loop above will likely refer to a different cache line or page of B in each iteration of the inner loop, causing much unwanted cache or page traffic (unless the cache or memory is large enough to hold the whole B array). The algorithm can be reformulated using subblocks of the arrays:

$$A^{1,1} = B^{1,1}*C^{1,1} + B^{1,2}*C^{2,1}$$
$$A^{1,2} = B^{1,1}*C^{1,2} + B^{1,2}*C^{2,2}$$
$$A^{2,1} = B^{2,1}*C^{1,1} + B^{2,2}*C^{2,1}$$
$$A^{2,2} = B^{2,1}*C^{1,2} + B^{2,2}*C^{2,2}$$

$A^{1,1}$	$A^{1,2}$
$A^{2,1}$	$A^{2,2}$

$B^{1,1}$	$B^{1,2}$
$B^{2,1}$	$B^{2,2}$

X

$C^{1,1}$	$C^{1,2}$
$C^{2,1}$	$C^{2,2}$

This formulation exhibits the advantage that the blocks can be made of a size to fit into the fastest level of the memory hierarchy, and that during each submatrix multiplication, the data in each submatrix is used many times. For this reason, block formulations have been developed for many algorithms. Our work tries to duplicate the benefits of block algorithms via program transformations.

3. Iteration Space. The basis of our work deals with tiling the iteration space. We define the iteration space as a discrete bounded cartesian space traversed by the DO loops in loop nest. For instance, the doubly-nested loop below traverses a 2-dimensional iteration space:

```
DO 100 I = 1,10
DO 100 J = 2,5
100   ...
```

A single loop traverses a one dimensional iteration space, triply nested loops define a three-dimensional iteration space, and so on. Triangular loops (where inner loop bounds depend on outer loop indices) define a triangular iteration space:

```
DO 100 I = 1,10
DO 100 J = 1,I
100   ...
```

4. Iteration Space Tiling. Once the iteration space is defined, we try to find a program transformation that will implement iteration space tiling. We define iteration space tiling as dividing the iteration space into tiles (or blocks) of some size and shape (typically squares or cubes), and traversing between the tiles to cover the whole iteration space. Optimal tiling for a memory hierarchy will find tiles such that all the data for each tile will fit into the highest level of the memory hierarchy and will exhibit high data reuse, reducing the total memory traffic. Also, the traversal between tiles will be done in an order that will reduce the amount of data that needs to be moved when going to the next tile.

Tiling is also a good paradigm for multiprocessor computers. If tiling can be done so that different tiles are independent of each other, then different tiles can be assigned to different processors. The data locality of tiling will also help multiprocessors by reducing memory contention.

Tiling can be implemented by a combination of strip mining and loop interchanging [2], [5]. Strip mining divides each loop into strips; each loop becomes two loops:

```
DO I = 1, N          DO IT = 1, N, IS
                     DO I = IT, MIN(N,IT+IS-1)
```

We call DO IT the "tile loop", and DO I the "element loop", since DO IT traverses between tiles and DO I traverses between elements of the iteration space. The tile size, IS, may be different for each loop in a loop nest, since tiles need not be square. Simple minded strip mining of each loop would produce the following code for a matrix multiply:

```
      DO 100 IT = 1, N, IS
      DO 100 I = IT, MIN(N, IT+IS-1)
        DO 100 JT = 1, N, JS
        DO 100 J = JT, MIN(N, JT+JS-1)
          DO 100 KT = 1, N, KS
          DO 100 K = KT, MIN(N, KT+KS-1)
100         A(I,J)=A(I,J)+B(I,K)*C(K,J)
```

Loop interchanging is then used to move the tile loops outwards and the element loops inwards:

```
      DO 100 IT = 1, N, IS
      DO 100 JT = 1, N, JS
      DO 100 KT = 1, N, KS
        DO 100 I = IT, MIN(N, IT+IS-1)
        DO 100 J = JT, MIN(N, JT+JS-1)
        DO 100 K = KT, MIN(N, KT+KS-1)
100       A(I,J)=A(I,J)+B(I,K)*C(K,J)
```

In the example of matrix multiply, the loops can easily be interchanged; in the general case, the loops cannot always be interchanged (due to data dependence constraints [2]). In those cases, automatic iteration space tiling may not produce an optimal program.

The final step is to optimize the order of the tile and element loops. The tile loops should be ordered so that the tiles are traversed in such a way as to reduce the amount of data moved between tiles. For instance, the tile loop ordering shown above refers to different blocks of B and C for each iteration of the KT loop:

```
      DO 100 IT = 1, N, IS
      DO 100 JT = 1, N, JS
        load block A^{IT,JT}
        DO 200 KT = 1, N, KS
          load blocks B^{IT,KT}, C^{KT,JT}
200       compute A^{IT,JT} = A^{IT,JT}+B^{IT,KT}*C^{KT,JT}
100     store block A^{IT,JT}
```

Each block of A is loaded and stored only once, while the B and C blocks are loaded multiple times (N/JS loads of each B block and N/IS loads of each C block). For computers with private memories on each processor, explicit data moves must be added in the code. For computers with hardware-managed caches, the data motion will be done automatically, though the same data motion patterns will apply. A different tile loop ordering has different characteristics:

```
             DO 100 IT = 1, N, IS
             DO 100 KT = 1, N, KS
                load block B^IT,KT
                DO 100 JT = 1, N, JS
                   load blocks A^IT,JT, C^KT,JT
                   compute A^IT,JT = A^IT,JT+B^IT,KT*C^KT,JT
        100      store block A^IT,JT
```

This tile loop ordering requires multiple loads and stores of the blocks of A, while reducing the number of loads required for each block of B. Generally, the first tile loop ordering would be preferable for most systems; if for some reason loading a block of B was expensive, the second tile loop ordering may be more effective. Tiling of the iteration space creates a tiling or blocking of the data arrays. The order in which the tiles are traversed induces the order in which the data blocks are accessed.

Another consideration is the concurrency in the algorithm, for multiprocessor computers. In general, it is more efficient to make outer loops concurrent (instead of inner loops) due to reduced overhead cost of forking and joining.

The element loop ordering must also be optimized. For vector computers, the element loops should ordered so that the inner loop is vectorizable, with appropriate memory strides (most vector computers work best with unit stride memory fetches). The element loop ordering may be important even for non-vector computers; when the loops can be ordered to increase the loop invariant computation or memory addresses in the inner loop, more code floating or better register assignment may be possible by the compiler.

5. Tiling of a Non-Square Iteration Space. Using simple minded strip mining to tile triangular loops may produce the wrong result. The desired tiling pattern for the two dimensional triangular loop is:

```
        DO 100 I = 1, N
        DO 100 J = I, N
100 ...
```

However, simple minded strip mining cannot generate this pattern. The proper way to strip mine the loops is:

```
             DO 100 IT = 1, N, IS
             DO 100 JT = IT, N, IS
             DO 100 I = IT, MIN(N, IT+IS-1)
             DO 100 J = MAX(JT,I),MIN(N, JT+IS-1)
        100    ...
```

Note that this produces some triangular tiles (along the diagonal) and some square tiles (throughout the rest of the iteration space). Note also that the tile size must be the same for both loops. For a triply-nested triangular loop, the situation gets even more complicated:

```
    DO 100 K = 1, N        DO 100 KT = 1, N, KS
    DO 100 I = K, N        DO 100 IT = KT, N, KS
    DO 100 J = K, N        DO 100 JT = KT, N, KS
                           DO 100 K = KT, MIN(N, KT+KS-1)
                           DO 100 I = MAX(K,IT),MIN(N, IT+KS-1)
                           DO 100 J = MAX(K,JT),MIN(N,JT+KS-1)
```

The original iteration space is a right pyramid (see fig. 1); the tiled iteration space produces some right pyramid tiles (labelled P), some wedge tiles (labelled W) and some cubic tiles. Note that triangular loops may arise naturally (many linear algebra algorithms use triangular loops) or through other program transformations (such as loop skewing [6]).

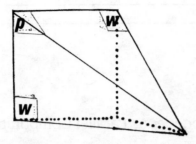

fig. 1. Right pyramid iteration space of 3-dimension triangular loop.

6. Summary. Block algorithms are designed to improve the memory hierarchy performance of a program. Many of the performance benefits of block algorithms can be realized through automatic program transformations, by properly applying strip mining and loop interchanging. Special rules apply when triangular loop indices appear, as they frequently do in linear algebra algorithms. Program restructurers, such as Parafrase [4], developed at the University of Illinois, PFC [3], developed at Rice University, and KAP, developed at Kuck and Associates, Inc., can be tuned to tile iteration spaces in such a way as to generate block algorithms automatically.

REFERENCES

[1]   W. A. Abu-Sufah, D. J. Kuck and D. H. Lawrie, "On the Performance Enhancement of Paging Systems Through Program Analysis and Transformations," *IEEE Trans. on Computers*, Vol. C-30, No. 5, May, 1981, pp. 341-356.

[2]   John R. Allen and Ken Kennedy, "Automatic Loop Interchange," *Proc. of the ACM SIGPLAN '84 Symposium on Compiler Construction*, Montreal, Canada, June 17-22, 1984, SIGPLAN Notices Vol. 19, No. 6, June, 1984, pp. 233-246.

[3]   Randy Allen and Ken Kennedy, "Automatic Translation of FORTRAN Programs to Vector Form," *ACM Trans. on Programming Languages and Systems*, Vol. 9, No. 4, pp 491-542, October 1987.

[4]   David J. Kuck, Ahmed H. Sameh, Ron Cytron, Alexander V. Veidenbaum, Constantine D. Polychonopoulos, Gyungho Lee, Tim McDaniel, Bruce R. Leasure, Carol Beckman, James R. B. Davies and Clyde P. Kruskal, "The Effects of Program Restructuring, Algorithm Change and Architecture Choice on Program Performance," *Proceedings of the 1984 Int'l Conf. on Parallel Processing*, Robert M. Keller (ed.), St. Charles, IL, Aug. 21-24, 1984, IEEE Computer Society Press, Washington, DC, 1984, pp. 129-138.

[5]   Michael Wolfe, "Advanced Loop Interchanging," *Proc. of the 1986 Int'l Conf. on Parallel Processing*, Kai Hwang, Steven M. Jacobs, Earl E. Swartzland(eds.), St. Charles, IL, IEEE Computer Society Press, Wash., DC, Aug. 19-22, 1986, pp. 536-543.

[6]   Michael Wolfe, "Loop Skewing: The Wavefront Method Revisited," *Int'l Conf. on Parallel Programming*, Vol. 14, No. 6, Dec. 1986.

[7]   Michael Wolfe, "Vector Optimization vs. Vectorization," to appear in *Proc. of the 1987 Int'l Conf. on Supercomputing*, Athens, Greece, June 1987.

CHAPTER 55

Synchronization of Nonhomogeneous Parallel Computations*

Dan C. Marinescu[†]
John R. Rice[†]

Abstract. The paper investigates the problem of synchronization in parallel computing. Several distributions of the execution time inside a fork-join pair are examined and closed form solutions for the processor utilization are derived in case of complete synchronization.

1. Overview. The potential for high performance is the strong motivation for parallel computation. It has been recognized early that excessive communication results in a substantial penalty to the performance of parallel computations. In addition to communication we analyze here another aspect of parallel computing which can affect adversely the performance.

We consider a set of computations which may be executed in parallel and which are related, but not homogeneous. We focus our attention on the effects of imposing synchronization upon such computations. In the past, attention has been given to the cost of initializing parallel computation, the "start-up" cost of a fork-join pair. We focus instead upon the effects of terminating parallel computation, when the execution time inside a fork-join pair is a random variable with a given distribution.

We have observed [2], [3] that synchronization is costly for a domain decomposition application on parallel machines with multiple domains of memory.

2. Synchronization Effects In Multi-Level Machines, The SHLPN Approach. Multi-level machines use a hierarchy of memories to achieve high parallelism without excessive communication. The authors have investigated such a machine [1] and a class of applications, namely the solutions of partial differential equations (PDEs) based upon the Schwartz splitting algorithms. We have used a modeling technique based upon Stochastic High Level Petri Nets (SHLPN) [5].

---

* This research was supported in part by the Strategic Defense Initiative Office through ARO grant DAAG-0386-K-0106.

† Department of Computer Science, Purdue University, West Lafayette, IN 47907

The hardware consists of clusters of shared memory multiprocessor systems. Each processor may access local, locally shared and global memory. The particular configuration investigated consists of 8 interconnected clusters, each with 9 processors. The application is partitioned in such a way that each processor needs to access both its local memory and different levels of shared memory, this is in order to pass on the results of its own computations and to use the results produced by other processors.

The objective of the analysis is to determine the effect of contention for shared resources upon the processor utilization. The modeling technique used is based upon the idea of a processor migrating among a set of execution domains. We say that a processor executes in a given domain when it accesses storage in that domain. The access time and consequently the execution speed, differs in different domains. It increases as we move from the local domain to the locally shared and to the global one. *Slow-down* factors for the locally shared and for the global domain reflect the increase of the access time.

Our application is the solution of PDEs using the Schwartz splitting for 2D and 3D problems. In the 2D case, a large domain is divided into pieces, which overlap to cover the domain and each piece is further subdivided into regions, which overlap to cover the piece and its boundary.

A numerical method further subdivides each region into N^2 elements or mesh points (assuming the regions are square). The work to solve the PDE depends on the method used and is modeled by $C(N^2)^k$, where C and k are constants depending upon the method used. Realistic values of k are from 1 (for FFT methods on "easy" problems) to 2 (for Gauss elimination type methods). The worst and the best cases for Schwartz splitting are $k = 1$ and 2 respectively, and we study these two cases which later are called *pessimistic* and *optimistic*. We consider values of N from 10 to 100, which are typical for common problems.

The amount of information exchanged between iterations is clearly proportional to the perimeters of the regions (on the local piece level) or to the perimeters of the pieces (on the higher, inter-piece level). Different numerical methods and types of boundary values require different amounts of information to be exchanged for one piece, however, a model of the form $4 \times N \times K$ is reasonable where K is a small integer, we consider the range 2 to 5 for K.

The problem data is organized so that locally shared memory contains all the boundary information for the regions in one piece and the global memory contains all the boundary information for regions which overlap piece boundaries.

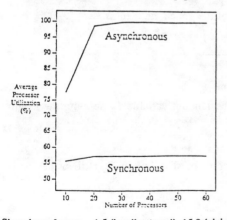
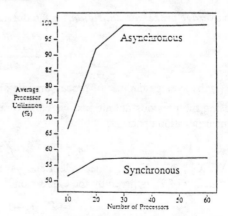

Slow-down factors = 1.5 (locally shared), 15.0 (global)   Slow-down factors = 4.0 (locally shared), 40.0 (global)

FIG. 1 Average processor utilization versus problem size. 1a. Slow-down factors 1.5 (locally shared), 15.0 (global). 1b. Slow down factors 4.0/40.0.

Two models have been investigated. In the first model [3], we assume that each processor's behavior is independent of the others; as soon as it has finished execution in a given domain, it migrates to the next. In the second model [2], we introduce synchronization. When a processor finishes execution in a certain domain, it migrates to the next one when all processors in its "group" have completed execution in the current domain. The "group" is determined by the semantics of the computation performed by the application.

Figure 1 presents the results of our analysis for the two models. The effect of synchronization is quite significant.

3. The Model. To better understand this phenomenon, we present and analyze another model of the computation. Consider a computation C which at a time t_0 is forked into n subcomputations $C_1, \ldots, C_n$, which are executed in parallel using n processors. The execution times of $C_1, C_2, \ldots, C_n$ are independent random variables $X_1, \ldots, X_n$. The computations have to be synchronized in the sense that all C_i have to complete before C can proceed at time t_1, as shown in Figure 1. Partial synchronization of order k occurs when the computation may proceed after k of the C_i complete.

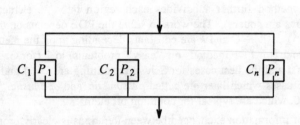

FIG. 2 Schematic of the model of a fork-join computation involving related processes running independently on separate processors.

We want to determine the average length of the synchronization epoch defined in terms of the *stopping time*

$$T_S = E[t_1 - t_0]$$

and the *average processor utilization* for the duration of the epoch, defined as

$$U = \frac{\int_0^{T_S} (n - W(t))\, dt}{n\, T_S}$$

with $W(t)$ the average number of processors idle are time t.

Using the previous definitions, we can express the random variable T_n defining the length of a synchronization epoch as:

$$T_n = \max\{X_1, X_2, \ldots, X_n\}$$

When $X_1, X_2, \ldots, X_n$ are independent and have a common distribution $F_X(t)$ then the distribution function of T_n is given by the expression

$$F_{T_n}(t) = [F_X(t)]^n$$

and the expected value of T_n is given by

$$T_S = E(T_n) = n \int_0^\infty t\, F_X^{(n-1)}(t)\, dF_X(t)$$

Then the average processor utilization can be expressed as

$$U = \frac{1}{T_S} \int_0^{T_s} (1 - F_X(t))\, dt$$

4. Analytical Results. There are virtually no experimental studies of the distribution of execution times for different parallel numerical algorithms and their implementation on multiprocessor systems. Consequently, we investigate several distributions. The uniform distribution and the standard normal distribution are intuitively very appealing. The exponential distribution is investigated since analytical models of program execution are often based upon Markov models. For each distribution, we give the closed form solution for the stopping time in case of complete and partial synchronization, then the average processor utilization. The details of the analysis are presented in [4].

5. Uniform Distribution. In this case, the distribution function of the random variable X is:

$$F_X(t) = Pr[X \le t] = \begin{cases} 0 & t < a \\ \dfrac{t - a}{b - a} & a \le t \le b \\ 1 & b < t \end{cases}$$

In case of complete synchronization, the average length of a synchronization epoch is:

$$T_S = E(T_n) = n\, \frac{1}{(b - a)^n} \int_a^b t\, (t - a)^{n-1}\, dt = b - \frac{b - a}{n + 1}$$

The average processor utilization can then be expressed as:

$$U = \frac{(a + b)(n + 1)}{2(a + nb)}$$

The average processor utilization versus the number of processor is plotted in Figure 2.

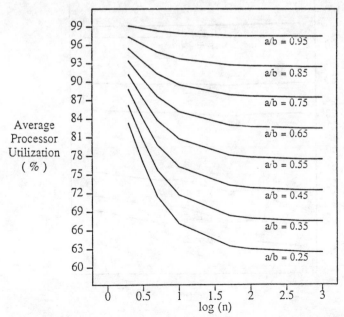

FIG. 3 Average processor utilization versus the number of processors for uniform distribution.

In the case of partial synchronization of order k, T_S has the following expression

$$T_S = b - \frac{(b-a)(n-k+1)}{n+1}$$

6. Normal Distribution. In case of normal distribution, the probability density function is

$$f_X(x) = \frac{1}{\sigma\sqrt{2\pi}} \, e^{\frac{-(x-\mu)^2}{2\sigma_2}}$$

The average length of a synchronization epoch in case of complete synchronization, is given by the expression:

$$T_S = (2 \log n)^{1/2} - 1/2 (2 \log n)^{-1/2} (\log \log n + \log 4\pi - 2C)$$
$$+ \, O(1/\log n)$$

C is Euler's constant, $b = 3.5$ is cut off of distribution

The average processor utilization is:

$$U = \frac{\mu + T_S \, g(T_S) + \sqrt{2/\pi} \, e^{-T_S^2/2} - 3.5}{\mu + T_S}$$

with

$$g(t) = \int_0^t e^{-u^2/2} \, du/\sqrt{2\pi}$$

Figure 4 presents the average processor utilization in case of a $(\mu, 1)$ normal distribution for different values of μ.

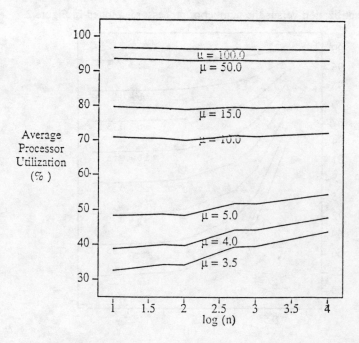

FIG. 4 Average processor utilization function of the number of processors for the normal distribution.

7. Exponential Distribution. The cumulative distribution function is

$$F_X(t) = 1 - e^{-t}$$

The stopping time and the average processor utilization are respectively

$$T_S = \frac{1}{\lambda} \left[\log n + C + 0(1/n) \right]$$

$$U = (1 - e^{-\lambda T_S})/\lambda T_S$$

In case of partial synchronization, we have

$$T_S = \frac{1}{\lambda} \sum_{j=k}^{m} \begin{bmatrix} m \\ k \end{bmatrix} \sum_{i=0}^{j} (-1)^{i+1} \begin{bmatrix} j \\ i \end{bmatrix} \frac{1}{(n-j+i)}$$

The results are plotted in Figure 5.

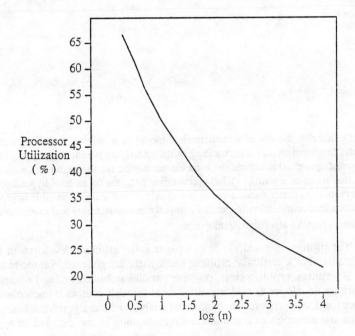

FIG. 5 Average processor utilization in case of an exponential distribution.

Literature

[1]   J.R. Rice, "Multi-Flex Machines: Preliminary Report", CSD-TR-612, Computer Science, Purdue University (1986).

[2]   J.R. Rice and D.C. Marinescu, "Analysis and Modeling of Schwartz Splitting Algorithms for Elliptic PDE's", In: Advances in Computer Methods for Partial Differential Equations, VI (Stepleman and Vischnevetsky, eds.), IMACS (1987), pp. 1–6.

[3]   D.C. Marinescu and J.R. Rice, "Domain Oriented Analysis of PDE Splitting Algorithms", Journal of Information Sciences, 43 (1987), pp. 3–24.

[4]   D.C. Marinescu and J.R. Rice, "Non-homogeneous Parallel Computations: I Synchronization Analysis of Parallel Algorithms", CSD-TR-683, Computer Science, Purdue University (1987).

[5]   C. Lin and D.C. Marinescu, "Stochastic High Level Petri Nets and Applications", IEEE Transactions on Computers (1988), to appear.

CHAPTER 56

Using Mathematical Modeling to Aid in Parallel Program Development

Elizabeth A. Eskow[*]
Robert B. Schnabel[*]

Abstract. We briefly describe the use of a mathematical model of a rather complex parallel computer program to assist in the development and understanding of the underlying parallel algorithm. The example we use is a parallel global optimization algorithm. First we summarize the formation of a model that accurately matched execution times of a parallel global optimization program on an Intel hypercube. Then we discuss the use of this model to detect weaknesses in the algorithm and analyze possible improvements to it. Our contention is that this combination of parallel computer implementation and mathematical modeling is a useful approach in parallel algorithm development.

1. Introduction. The problem we address in this paper is the efficient development of parallel algorithms and programs. For a particular problem class, there are generally far too many combinations of possible algorithms, problem sizes, possible parallel architectures, and numbers of processors for the algorithm developer to evaluate all of them through computer implementation and testing. On the other hand, conventional mathematical analysis of the algorithms, using counts of arithmetic operations and measures of communication costs, may be too detailed an approach for sizable problems.

We will briefly describe an approach that falls in between these two extremes of exhaustive testing and detailed complexity analysis. First we use the results of a relatively small number of parallel computer experiments to create a rather high level model of the parallel algorithm. Then we use this model to simulate the behavior of the parallel algorithm in various new situations, and use these simulations to better understand the method, suggest improvements to it, and evaluate possible improvements. This may lead to the need for additional computer tests. We believe that this process results in increased understanding of the parallel algorithm, and helps in producing good parallel algorithms efficiently.

The problem we use to illustrate this approach is the global optimization problem. It is to find the lowest minimizer of a function $f(x)$ of n variables that has multiple local minimizers, lowest points in some open subregion of the domain of interest. We have chosen to study this problem because we have considerable experience and interest in developing parallel methods for it ([1]), and also because its complexity seems well suited to test our modeling approach. This is because the parallel algorithm is non-trivial, having several distinct sections that must be

---

* Department of Computer Science, University of Colorado, Boulder CO 80309. Research supported by AFOSR grant AFOSR-85-0251, and NSF cooperative agreement DCR-8420944.

modeled separately, but not so complex that its complexity will overwhelm the case study. In addition, it has both deterministic and non-deterministic sections; the latter will be seen to present interesting challenges for our approach.

The type of approach we describe has recently been advocated by others, for example Reed [2]. To our knowledge, however, few case studies of this type of approach have been described in the literature.

Section 2 very briefly describes the parallel global optimization algorithm that we will model. Section 3 summarizes the formulation of the mathematical model of this algorithm, and describes how well it fit the experimental data. In Section 4 we describe some of the many applications we made of the model in order to understand the limitations of the algorithm and predict the effects of improvements to it. In Section 5 we briefly draw several conclusions. A more detailed and comprehensive description of this work will be available in a forthcoming paper.

2. The Parallel Global Optimization Method.   The parallel global optimization algorithm that we modeled, which was developed by Byrd, Dert, Rinnooy Kan, and Schnabel [1], is summarized in Algorithm 1. It is derived from the sequential, stochastic global optimization method of Rinnooy Kan and Timmer [3].

The basic idea of both the sequential and parallel methods is to evaluate the function at randomly chosen sample points in the domain of interest, select some of these sample points to be start points for local minimizations, and perform a standard local minimization algorithm from each of these start points, each terminating at some local minimizer. Then the whole process may be repeated one or more times; when it is terminated the lowest local minimizer that was found is reported as the global minimizer. The key to the efficiency of this approach is the selection of the start points for local minimizations; the goal is to select start points so that each local minimizer is found exactly once. The procedure that is used is to select a sample point as a start point if and only if it has the lowest function value among all sample points within some "critical distance" (calculated in the algorithm) of itself.

The parallel method retains the basic structure of the sequential method, and parallelizes each of its main steps. The parallelism in the sampling and start point selection phases is obtained by partitioning the feasible region among the processors. This leads to an important new phase : start points are first selected within a subregion, and then points near subregion borders are communicated to neighboring subregions to resolve their status. For the local minimization phase, all the start points are collected centrally and redistributed to the processors; the alternative of having each processor perform the minimizations for start points in each subregion would often lead to uneven distribution of work among the processors. The information we will need about the method for the purposes of this paper is contained in Algorithm 1.

3. Summary of Model Formation.   We formulated the mathematical model of Algorithm 1 based upon 48 test runs of this algorithm on an Intel hypercube. These 48 cases are for 3 standard test functions (Branin, Goldstein-Price, and Shekel 5), each run with all combinations of sample sizes 200 or 1000, 4,8, 16 or 32 nodes of the hypercube, and with inexpensive or expensive function evaluation. The reason for including variants with inexpensive and expensive function evaluation is that both situations are of considerable practical importance. In the inexpensive case, function evaluation took about 0.001-0.01 seconds, in the expensive case it required about 1 second but returned the same function value. All the problems required only one iteration of Algorithm 1.

The independent variables that we used in the model include the number of processors p, the number of variables n, the total sample size per iteration s, and the cost of an evaluation of $f(x)$, f. In addition, we require the number of function evaluations, $fcns$, and local search

Algorithm 1 -- A Parallel Global Optimization Algorithm

Given $f : R^n \to R$, hyper-rectangle D, p processors ; at each iteration

1. Generate sample points and function values

    Partition D into p subregions, one per processor

    Each processor randomly chooses s/p additional sample points in its subregion and evaluates $f(x)$ at each new sample point.

2. Select start points for local minimizations

    Each processor selects start points in its subregion (any sample point for which there is no lower sample point in its subregion within the critical distance).

    Resolve start points near borders between subregions (start point is removed if there is a lower sample point within critical distance in neighboring subregion).

3. Perform local minimizations from all start points

    Collect all start points and distribute one to each processor, which performs local minimization from that point.

    If there are more than p start points, distribute a remaining start point to each processor as soon as it completes its current local search. until all local searches have been completed.

4. Decide whether to stop

    If stochastic stopping rules are satisfied, report lowest minimizer found as global minimizer, otherwise begin next iteration.

---

iterations, *itns*, of the processor whose search phase was longest, because these characteristics cannot be predicted as functions of the other independent variables.

To formulate the model, we decided what terms to include to model each main step. Then we determined the unknown parameters in the model by fitting the model to the 48 runs times.

The cost of Step 1 of Algorithm 1 is dominated by the function evaluations at the s/p sample points. Hence, Step 1 is modeled by $(s/p)f$. Note that no unknown parameter is involved.

Step 2 is the most difficult portion of the algorithm to model because it is nondeterministic, with considerable variation possible in the amount of work required to select start points for local minimizations from a particular size sample. The worst case complexity of this step is $O(s^2/p)n$ in the case that each x_s must be compared to all lower points in the subregion. (We actually further subdivide the entire domain into $K \geq p$ parts which accounts for the $1/p$ as opposed to $1/p^2$ term.) On the other hand, the best case would be $O(s/p)n$ in the case that each x_s must be compared to only one point. We chose to model this step by $s^2/p\ n \cdot c_1$, with unknown parameter c_1, because this gives a better fit to this phase. As we would expect, however, the fit for this phase is not as good as the fits for other steps of the algorithm.

The other major cost of Step 2 is the comparison of candidate start points near subregions boundaries to sample points in neighboring subregions. This phase also is nondeterministic; we have modeled it by $s\ n \cdot c_2$, which results if we assume that we must compare $O(p)$ border points to $O(s/p)$ points each. This phase also includes some interprocessor communication which we discuss below.

Step 3 of Algorithm 1 is easier to model. Its main costs are the function evaluations, which we model by $fcns \cdot f \cdot c_3$, where we expect c_3 to be slightly greater than 1 because some additional work is associated with each function evaluation, and $itns \cdot n^2 \cdot c_4$ which models the predominant cost of the linear algebra calculations associated with each iteration.

The only cost that remains to be discussed is the interprocessor communication. The algorithm requires 6 messages between the master process and the node processes, and 5 $lg\,p$ internode messages, per iteration. Using the known message costs on the Intel hypercube, it was easily seen that this cost is small in comparison to any of the overall run times of the test problems, and can be regarded as a constant. Thus it was combined with the other overhead costs of the algorithm to obtain the final constant term in the model, c_5.

Thus we have modeled an iteration of Algorithm 1 by

$$\frac{s}{p}f + \frac{s^2}{p}\,n\cdot c_1 + s\cdot n\cdot c_2 + f\cdot fcns\cdot c_3 + itns\cdot n^2 c_4 + c_5 \tag{3.1}$$

where $c_1, \cdots, c_5$ are the unknown parameters. We determined these parameters by fitting the relative difference between the 48 measured computer times and the model prediction, using linear least squares. (Relative difference was used because run times varied widely.) The parameter values we obtained were

$$c_1 = 0.000016\,, \quad c_2 = 0.0034\,, \quad c_3 = 1.11\,, \quad c_4 = 0.0080\,, \quad c_5 = 0.54 \tag{3.2}$$

and resulted in a median relative error of 0.02, with a range of 0.002 to 0.19 on the 48 problems. We find this fit of the data to the model to be sufficient for the uses that we wish to make of the model.

4. Some Applications of the Model.   One main use we have made of our model is to help understand the limitations of our parallel algorithm, by tabulating the contributions of each of the six terms of the model for a representative sample of problems. For example, Table 1 shows the run times predicted by the model for the 8 problems that have all possible combinations of $f = 0.001$ or 1, $s = 200$ or 1000, $p = 8$ or 32, with $n = 3$, $itns = 20$, and $fcns = 100$.

Even though we had already worked with this parallel algorithm for a long time, we learned a considerable amount from tabulations like Table 1. When we began this work, the start point selection used p as opposed to the present $K \geq p$ subdivisions. Tabulations of the model for that algorithm made it very clear that the $O(s/p)^2$ cost of this approach was unacceptable for large s/p, so we changed the method to use K subdivisions. Now that we have made that change (and revised the model), our biggest surprise with Table 1 concerns the border resolution phase. We recognized that this phase, although parallel, does not speed up with p, but the cost of this phase was thought to be insignificant. The model tabulation results shows that this phase is significant for large values of p and s and small values of f. Consequently we have begun investigating better approaches to this phase.

The other use of the model that we discuss here is to predict the results of proposed improvements to the algorithm, prior to implementing and running them on a parallel computer. The main improvement that we have considered is parallelizing each local search by performing a speculative evaluation of the finite difference gradient in conjunction with each function evaluation. This means that the n function evaluations for the finite difference gradient are performed in parallel with the standard function evaluation, before it is known whether this information will be required. This modification enables each local search to utilize up to $n+1$ processors as opposed to one processor in Algorithm 1.

We modified the model to account for this change to the algorithm. The sample generation, start point selection, border resolution, linear algebra cost per search iteration, and the original overhead costs are unaffected by this change, so these terms in the model remain the same. The alterations are in the costs of function evaluations and communication. It is rather easily shown that the fourth term in equation (3.1) should become $f\cdot 1.25\cdot itns\cdot\lceil(n+1)w/p\rceil\cdot c_3$, and that the approximate additional communication cost is $1.25 \cdot itns \cdot \lceil lg(n+1) \rceil \cdot .004$. (The factor 1.25 comes from an estimated 20% failure rate in the line search.) An important point is that the model's constants can reasonably be assumed to have the same values as before, so that the revised model can be used to predict the performance of the modified algorithm without doing any computer experiments.

The final column of Table 1 shows the speedups that the new model predicts for the new speculative gradient algorithm in comparison to Algorithm 1, on the same 8 problems, assuming that $w = 4$. In general, the models show that the extra communications overhead involved in doing the speculative gradients would cause a slight degradation in performance for very inexpensive functions. Once function evaluations require even 0.01 - 0.1 second, however, the model shows that the additional communication would become insignificant, substantial speedups would result in the cases where p is greater than $n + 1$ times the number of local minimizations, and some speedup would occur as long as p is greater than or equal to $n + 1$.

We have also made other uses of the model. They include determining how expensive function evaluation must be to dominate the other costs of the algorithm, and evaluating the impact of changes in the communication cost of the parallel computer.

Table 1 -- Model Performance Predictions
($n = 3$, $itns = 20$, $fcns = 100$)

f	s	p	Sample Generation	Start Pt Selection	Border Resolution	Local Searches	Overhead	Additional Speedup from Parallel Gradients
.001	200	8	0.03	0.24	2.04	1.55	0.54	0.97
.001	200	32	0.01	0.06	2.04	1.55	0.54	0.97
.001	1000	8	0.13	6.11	10.20	1.55	0.54	0.99
.001	1000	32	0.03	1.53	10.20	1.55	0.54	0.99
1.0	200	8	25.0	0.24	2.04	112.29	0.54	1.65
1.0	200	32	6.0	0.06	2.04	112.29	0.54	3.18
1.0	1000	8	125.0	6.11	10.2	112.29	0.54	1.28
1.0	1000	32	31.0	1.53	10.2	112.29	0.54	2.14

5. Conclusions. Mathematical modeling of a parallel algorithm based on actual parallel computer measurements has proven to be a useful technique in the continued development of the algorithm. We were able to construct a fairly simple mathematical model that modeled a rather complex parallel algorithm fairly well. Using this model to simulate additional computer runs of the algorithm gave us considerable new insight into the algorithm's performance, and allowed us to assess the performance of possible changes to the algorithm before implementing and running them. We believe the overall effect was to make both the algorithm development process, and the resultant algorithm, more efficient.

REFERENCES

(1) R. H. BYRD, C. DERT, A. H. G. RINNOOY KAN, and R. B. SCHNABEL, *Concurrent stochastic methods for global optimization*, Tech. Rpt. CU-CS-338-86, Dept. Comp. Sci., Univ. Colorado, 1986.

(2) D. A. REED, *Parallel systems : performance modeling and numerical algorithms*, presented at Third SIAM Conf. on Parallel Processing for Scientific Computation, 1987.

(3) A. H. G. RINNOOY KAN and G. T. TIMMER, *Stochastic methods for global optimization*, Am. J. Math. Mgmt. Sci. 4, 1984, pp. 7 - 40.

SCHEDULE: An Environment for Developing Transportable Explicitly Parallel Codes in Fortran

J.J. Dongarra and D.C. Sorensen

SCHEDULE is a software package designed to aid in programming explicitly parallel algorithms for numerical calculations. The design goal of SCHEDULE is to aid a programmer with implementation of explicitly parallel algorithms in a style of Fortran programming that will lend itself to transporting the resulting programs across a wide variety of parallel machines. The approach relies upon the user adopting a particular style of expressing a parallel program. Once this has been done the subroutines and data structure provided by SCHEDULE will allow implementation of the parallel program without dependence on specific machine intrinsics. The basic philosophy taken here is that Fortran programs are naturally broken into subroutines which identify units of computation that are self–contained and which operate on shared data structures.

A graphics post processor has been developed to analyze the execution pattern and structure of a parallel program. A graphics pre–processor is under development that will aid in constructing a parallel program directly from the large grain data dependency graph.

CHAPTER 57

Multiprogramming and the Performance of Parallel Programs*

Muhammad S. Benten†
Harry F. Jordan†

Abstract Tight synchronization in parallel programs executed on multiprogrammed multiprocessors may result in catastrophic performance losses as a result of the absence of swapped out processes. Our work introduces a programming methodology that utilizes computational synchronization and avoids tight control flow synchronization in parallel programs. In this methodology, each phase of the computation is assigned a status that can be ready, blocked or completed, and tasks in each computational phase are selfscheduled to ensure computational progress by the available executing processes. Results obtained indicate that this methodology avoids the catastrophic performance losses resulting from the swapping of processes in multiprogrammed multiprocessors.

1. Introduction Recent advances in hardware technologies and the advances in parallel languages and algorithms are promising to make shared memory multiprocessors the future computation machines. Currently, there are many shared memory multiprocessors available, ranging from supercomputers to the minisuper and superminicomputers, and all have proven to be able to deliver very high throughputs that satisfy time sharing and batch programs needs. It has also been seen that these machines are capable of supplying speedup of single parallel programs when their execution is implemented by a set of cooperating processes. However, most of these machines run multiprogrammed operating systems and thus there is no certainty that processes of a parallel program will be executed concurrently on more than one hardware processor. In many instances, multiprogramming, which is a property of the operating system aimed towards the improvement of the overall system efficiency and throughput, interferes with parallel programs, and slowdowns in the execution time of these programs occur.

In this paper we will be discussing the effects of multiprogramming on the performance of parallel programs, where the performance of a parallel program is determined in terms of the wall clock time speed up of executing the parallel program over the sequential version. We will show this effect by presenting the results of executing an example parallel program on three multiprocessors; an Alliant FX/8, a 20 processor Encore Multimax and an 8 processor Sequent Balance 21000. These results will be given for situations where the parallel program processes were executed concurrently on separate processors and for situations where they were multiprogrammed on a smaller set of CPU's. In a later section, we will also present a new programming

* This work was supported in part by NASA Langley Research Center under NAG-1-640, and by the Office of Naval Research under N00014-86-K-0204.

† Electrical and Computer Engineering Department, University of Colorado at Boulder, Boulder, Colorado 80309-0425

methodology that will improve the performance of parallel programs and is suitable for multiprogrammed environments.

Throughout this paper we will assume the use of a parallel language, where parallelism is at the top of the program hierarchy, in particular we will be using the Force[1]. The Force is a parallel language that is currently being developed at the University of Colorado, Boulder, and is currently implemented on several machines. Among these are Flex's, Cray's, Alliant's, Sequent's and Encore's machines. In the Force and similar parallel programming paradigms, such as IBM EPEX Fortran[2], and Butterfly Uniform System library[3], the user writes a single program that is to be executed by an arbitrary number of processes, where the number is not specified until run time and stays fixed throughout the execution period. These processes are the parallel instruction streams that will participate in the execution of the parallel program. In the parallel programmer's view, these streams are supposed to be executed concurrently on separate CPU's during the execution of his parallel program. However, as a result of multiprogramming, coscheduling of these processes may not be possible and the performance of parallel programs may be severely affected.

2. Example Parallel Program Consider the LU factorization of a matrix with partial pivoting as implemented in the Linpack SGEFA subroutine[4]. The algorithm is composed of three phases that are repeated for all rows of the matrix to be factored. These phases must be completed in order for each row, before they can start for the next one. The search for the pivot row is done in the first phase and the swapping of the pivot row and the current row is done in the second phase. In the third phase, the actual Gaussian elimination computations are performed. Although, there are other papers that consider the possibility of overlapping computational phases[5], our aim in this paper is to show the effects of multiprogramming on tightly synchronized and strictly ordered phases.

Using the Force, this algorithm was written by Jordan[6], as a Force subroutine. This Force subroutine starts with a barrier[8], which is a Force construct that is used to synchronize the instruction streams. The barrier will let the force of processes executing a Force program wait until they all execute the barrier statement. One process of the Force can then execute a sequential code section, if any, and the Force is released again. The barrier provides a very clear and simple way to mark the completion of a computational phase and the availability of all the processes to start the next phase of the computation. The barrier in this algorithm initializes shared variables before the factorization process starts. The barrier is followed by the main loop over the rows of the matrix. In this loop the first phase is done using a prescheduled Do loop[9], followed by a critical section. The second phase consists of swapping the pivot row and the current row if swapping is necessary. Obviously, this can not be started before the pivot row is located, hence, a Force barrier is used to detect the completion of the first phase by detecting the arrival of all the processes that participated in finding the pivot. The third phase is that of performing the actual row reductions. Again the third phase is separated from the second phase by a barrier which will also calculate the pivot element required for the elimination. The Force subroutine which implements the SGEFA algorithm is shown in Table-I.

2.1 Performance of the LU example The results of executing the LU factorization algorithm on an Encore's Multimax, Alliant's FX/8 and Sequent's Balance 21000 is shown in Figures (1,2,3), respectively. In these figures, the light bars show the results for the case where coscheduling of processes on separate CPU's is guaranteed and multiprogramming is avoided by making sure that no other jobs are running on the machine when this program is run. Clearly, these results are excellent, and almost linear speedups are obtained, up to the number of processors in these machines. This algorithm, which performs and speeds up very well, is executed by all the processes in a lock step fashion, and the coscheduling of processes involved in the computation on separate processors is required to obtain this performance. When there are fewer processors

TABLE-I Jordan's LU Factorization Of a Matrix

```
Barrier
  Initialize;
End barrier;

for k=1,.....,n-1 do
  begin

    DOALL i=k,.....n-1
    find l such that ABS( a_{l,k} ) > max ( a_{i,k} );

    Critical
    If l > ipvt(k) then
      save l in ipvt(k);

Barrier
End barrier;

    If l <> k
    DOALL j=k,... n-1
    interchange a_{k,j} and a_{l,j};

Barrier
  piv = -1.0 / a_{k,k};
End barrier;

    DOALL i=k+1,....,n
      begin
        m_i = -a_{i,k} * piv
        for j=k+1,....n do
          a_{i,j} = a_{i,j} + m_i * a_{k,j};

      end i;

Barrier
  ReInitialize;
End barrier;
end k.
```

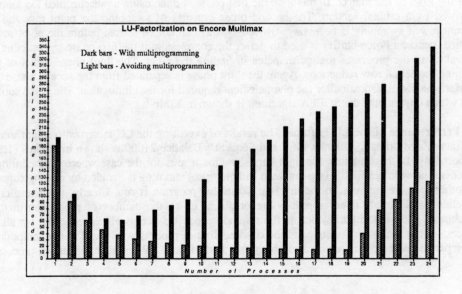

Figure(1) LU-Factorization on Encore's Multimax using Barriers

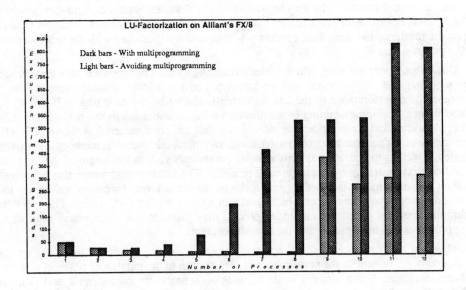

Figure(2) LU-Factorization on Alliant's FX/8 using Barriers

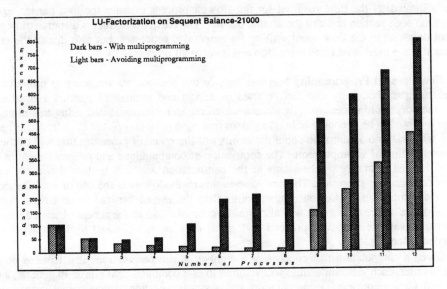

Figure(3) LU-Factorization on Sequent's Balance-21000 using Barriers

available than processes, these processes will be multiprogrammed on the available processors. The performance degradation due to multiprogramming can be seen from the previous figures when the number of processes exceeded the number of CPU's. Although, the user usually restricts the number of processes he uses to be less or equal to the number of hardware processors available, in an environment like Unix, he can't prevent other users or system processes from running on the machine, and thus, these processes will be multiprogrammed with the processes of his parallel program.

Due to multiprogramming, which is the scheduling of a set of processes on a smaller set of processors, some of the processes will be suspended in the middle of their execution. In this environment, the performance of the LU algorithm is shown by the dark bars in Figures(1,2,3). In these figures, multiprogramming is manifested by the deterioration in the performance of the parallel algorithm and the considerable slow down that has been incurred. Although this effect varies for the machines that have been used, these variations are related to memory and cacheing strategies as well as to operating system's tuning parameters such as the length of the time quantum and context switching overhead on each machine. The locking mechanism also plays a very important role and since it is based on spin locks on these machines, processes waiting for locks will be consuming their time slices which results in a lot of wasted CPU cycles. Thus, in tightly synchronized parallel programs, suspended process may preclude the advancement of other executing processes and limit progress in the overall program.

The light bars in figures (1,2,3), represent the execution time when the machines were not loaded, and improvement in the performance is obtained up to the number of physical processors on these machines. When running with more than 8 processes on Alliant's and Sequent's machines, the performance started to deteriorate and by 9 processes the performance is slower than running with one process. On the Encore, running with 24 processes is as slow as running with one process. The dark bars, show the performance of Jordan's LU Factorization algorithm on a loaded machine. In this case, the algorithm shows improvements up to 5 processes on the Multimax and up to 3 processes on the Sequent and and the Alliant, above these numbers the performance is deteriorating. Evidently, a main cause of this degradation is synchronization. Critical sections are one of the causes since a suspended process that has a lock will block others waiting for that lock. However, barriers are the major cause as the barrier would force the executing processes to wait for each other. In fact, the execution time of a parallel code section that is enclosed between two barriers, or equivalent stream synchronizing constructs in a parallel program, is limited by the time required for the slowest process to enter the first barrier, execute through the code section and exit the second barrier. The barrier construct is a source of degradation not only due to the time spent waiting for suspended processes, but also due to the critical sections that are used in its implementation and involve all processes[8].

3. A New Parallel Programming Methodology In this section, we are going to present a new programming methodology that can be used to design and implement parallel algorithms on shared memory multiprocessors. Algorithms which are to be implemented using this methodology are assumed to be decomposable into phases that have to be completed in order. This methodology will make no assumption about the number or the speed of processes that will participate in the completion of a computation. The completion of computations and progress made in them will be decided primarily by the status of the computation and will be less dependent on the number of executing processes. The key issue in this methodology is the use of work barriers to detect the completion of previous phases rather than the use of the traditional stream barriers. The avoidance of stream barriers will also limit synchronization to the access of shared data, and will let the available processes proceed as long as they are not blocked by locks acquired by suspended processes.

Given a computation that consists of a number of subcomputations that have to be completed in order, each subcomputation will referred to as a computational phase. In general, a computational phase represents a computation and a synchronization structure that surrounds the computation to insure its scheduling and completion regardless of the number of processes in a parallel program that will execute or pass over it. Currently, computations which can be handled by this construct are assumed to consist of a group of parallel tasks which are represented by indices that cover a range of integers, similar to those problems that can be implemented using DOALL loops[9]. An optional critical code section can also be placed following the computation and will be executed by every process that entered the computational phase and executed the DOALL loop. The last section of a computational phase is a strictly sequential section of code that is referred to as the completion section of the computational phase. It is executed by one process after the computation has been completed.

A computational phase has a status that can be blocked, ready, or completed. When the status of a computational phase is found to be blocked (not ready), processes that try to execute this computational phase will wait until the status is set to unblocked (ready) by some other process. When it is unblocked (ready), the construct will insure that no process can proceed with the parallel program section following the computational phase before all the tasks in this subcomputation have been scheduled. Processes can enter a computational phase if it is unblocked, and will execute in it if its computation and its completion section have not been finished, otherwise, the computational phase will automatically become passive as its status will be marked as completed. Processes that encounter a computational phase that has been completed, will skip it and continue with the statement or construct that follows the computational phase. The completion section of the current phase and the status of the following phase together comprise a work barrier if the completion section of the current phase contains a statement that would mark the next blocked phase as ready. This will insure the integrity of the computation without the need for waiting for all processes at the end of each phase.

3.1 LU factorization and the New Methodology Utilizing the concept of a computational phase, the LU factorization algorithm was rewritten as a subroutine in an extension version of the Force. The algorithm starts by initializing shared memory using a structure referred to as the *Init* construct. This construct is characterized by blocking processes until the initialization body has been executed by a single process which succeeded in obtaining a shared lock before any other process could. Processes arriving after the initialization code has been executed will skip the body of the *Init* construct and proceed with the code following the *Init*. The structure of this construct is as follows:

    Each process will:
        Atomically do:
            - check if init done,
                if done then skip the next two steps.
            - do initialization.
            - mark init done.

The use of this construct in initialization will allow the section of code to be executed before any process can proceed and will not wait for all the processes to arrive. This construct will be reinitialized in an *Init* construct at the end of this subroutine so that other calls to this subroutine will execute correctly. The *Init* construct we are using for Gaussian elimination will do the following initializations:

(1)   Initialize the shared row marker (pivot row) to 1, the first row.

(2)   Initialize the status of the pivot search computational phase to *ready*.

(3)   Reinitialize the *Init* construct at the end of this Force subroutine so that the first process to exit this subroutine will execute it.

(4)   Finally, the process executing this construct will wait for the matrix to be factored to become ready.

The *Init* construct, in this algorithm, is followed by an iterative loop which is to be executed in parallel (n-1) times, where n is the number of rows in the matrix to be factored.

Processes arriving at the beginning of the main algorithm loop, after passing by the *init* construct, can start the LU factorization process because the matrix has been set up, the working row has been initialized, and no process will get into the previous *Init* construct to change the initialized shared information. The main body loop consists of three computational phases, the pivot search phase, row swapping phase and the row reduction phase. This algorithm is written such that one phase will be ready at a time while the other phases are blocked. Initially all phases are blocked except for the pivot search phase. A block of strictly sequential code terminating each phase will block the current phase and mark the following one as ready. At first, processes can only execute the pivot search section which is described below.

Processes that succeed in getting into the pivot search computational phase will be self scheduled to search for the pivot element, each process will search for the row with the maximum pivot element among the rows it is assigned. Processes will then execute the critical section that follows the DOALL loop, so that the global maximum pivot row is identified, and will proceed to the completion section of the phase. The body of this section is to be executed by the first process that reaches it after the computation has been completed. Early as well as late arriving processes will skip this section and proceed to the next phase. The body of this section consists of blocking this phase (pivot search), setting up the next phase (row swapping), doing any required initializations and marking it as ready. When this is done, processes waiting for the row swapping computational phase will be let go, and every other phase would have been blocked.

The row swapping phase is organized in a way similar to the pivot search phase. Its computational section consists of swapping elements of the pivot row if swapping is needed. Its completion section is again similar to the completion section of the pivot search phase and the only difference is that the process which will execute it will setup the row reduction phase and make it the next ready phase.

The row reduction phase is organized similarly. Its computational section consists of DOALL loop that will perform the reduction of rows. Again, the completion section preserves the general structure seen in the previous two phases, with the exception that the process which executes this section will check to see if the LU factorization has been completed. If it is, all phases are marked as completed and an exit flag is set so that processes will exit the subroutine, otherwise, the row marker (pivot counter) is incremented and the three phases are repeated until the LU factorization is complete. The outline of this parallel algorithm is as shown in Table II. The performance of the implemented LU factorization algorithm, on Encore's Multimax, Alliant's FX/8 and Sequent's Balance 8000, is shown in Figures (4,5,6), respectively. In these figures, the performance of the new algorithm is the same as Jordan's when multiprogramming is not an issue. When it is, however, its effect is apparent, but the performance of the algorithm is

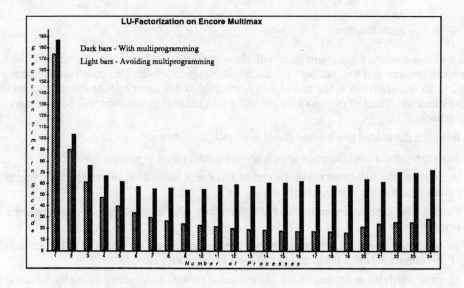

Figure(4) LU-Factorization on Encore's Multimax using the new method

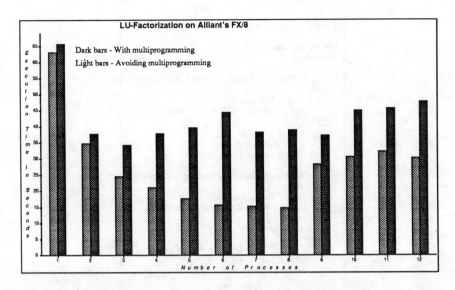

Figure(5) LU-Factorization on Alliant's FX/8 using the new method

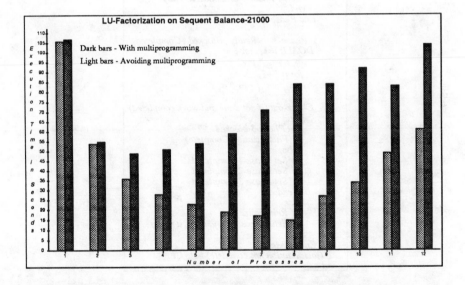

Figure(6) LU-Factorization on Sequent's Balance-21000 using the new method

far superior to the performance of Jordan's algorithm. The light bars show the execution time of the new LU algorithm with no other program running on the machine other than this job. In this case, on Alliant's and Sequent's machines, running with up to 12 processes is almost 80% slower than running with the minimum execution time with 8 processes but is still less than 2/3 of the single process time. Similarly, on a 20 processor Encore running with 24 is less than 1/3 of the single process time. The dark bars represent the execution time in an artificially loaded environment. In this case the minimum time was obtained when running with 7 processes on the Multimax and running with 3 processes on Alliant's and Sequent's machines. On all of these machines the execution time is less sensitive to the number of processes exceeding the number of available processors. Thus, there is a better balance between individual parallel program performance and overall system utilization.

```
TABLE-II The New LU Factorization Of a Matrix

Init- (if not done)
  Initialize;

L1:   p1: Phase-1  (Ready , Blocked , Completed)

        DOALL i=k,.....n-1
          find l such that ABS( a_{l,k} ) > max ( a_{i,k} );

        Critical
          If l > ipvt(k) then
            save l in ipvt(k);

        Check-out (if not done and work completed)
          begin
            Set Phase-1 blocked
            Setup Phase-2 and mark it ready
          end;
        End Phase-1

      p2: Phase-2   (Ready , Blocked , Completed)
          DOALL j=k,.....n-1
            interchange a_{k,j} and a_{l,j};

        Check-out (if not done and work completed)
          begin
            Set Phase-2 blocked
            piv = -1.0 / a_{k,k};
            Setup Phase-3 and mark it ready
          end;
        End Phase-2

      p3: Phase-3   (Ready , Blocked , Completed)
          DOALL i=k,.....n-1
            m_i = -a_{i,k} * piv
            for j=k+1,....n do
              a_{i,j} = a_{i,j} + m_i * a_{k,j};

        Check-out (if not done and work completed)
          begin
            Set Phase-3 blocked
            if (Factorization Complete)
              begin
                Set Phase-1 completed
                Set Phase-2 completed
                Set Phase-3 completed
              end
            else
                Reinitialize
                Setup Phase-1 and mark it ready
          end;
        End Phase-3
      If (factorization not complete) goto L1
    end .
```

4. Conclusions In this paper we presented a new programming methodology that can reduce inefficiencies that may result from running parallel programs on multiprogrammed shared memory multiprocessors. On these machines, the performance of parallel programs that are tightly synchronized may be severely affected and running with more than one process will result in a real execution time that is slower than running with a single process. Our methodology reduces this effect and depends on work completions as the basis for progress in parallel programs. This methodology will take into account the possibility of swapping processes by avoiding pre-scheduling of computations and avoiding the use of barriers that tightly synchronize instruction streams.

REFERENCES

(1)   H. F. Jordan, "The Force," *in The Characteristics of Parallel Algorithms*, L. H. Jamieson, D. B. Gannon and R. J. Douglass, Eds., Chap. 16, MIT Press (1987).

(2)   G. F. Pfister, W. C. Brantley, D. A. George, S. L. Harvey, W. J. Kleinfelder, K. P. McAuliffe, E. A. Melton, V. A. Norton and J. Weiss, "The IBM Research Parallel Processor Prototype (RP3): Introduction and Architecture," *Proceedings of the 1985 International Conference on Parallel Processing, August 1985*.

(3)   W. Cowther, J. Goodhue, E. Starr, R. Thomas, W. Williken and T. Blackadar, "Performance Measurements on a 128-Node Butterfly Parallel Processor," *Proceedings of the 1985 International Conference on Parallel Processing, August 1985*.

(4)   J. J. Dongarra, J. R. Bunch, C. B. Moler and G. W. Stewart, *LINPACK Users Guide*, SIAM Pub., Phil., PA (1979).

(5)   W. H. Jones, "Increasing Processor Utilization During Parallel Computation Rundown," *Proceedings of the 1986 International Conference on Parallel Processing, August 1986*.

(6)   H. F. Jordan, "Structuring parallel algorithms in an MIMD, shared memory environment," *Parallel Computing, Vol. 3, No. 2, pp. 93-110, May 1986*.

(7)   H. F. Jordan, M. S. Benten and N. S. Arenstrof, "Force User's Manual," *ECE Tech. Rept. CSDG 86-1-4, Computer System Design Group, Electrical and Computer Engineering Department, University of Colorado, Boulder, Oct 1986*.

(8)   N. S. Arenstrof and H. F. Jordan, "Comapring Barrier Algorithms," *ECE Tech. Rept. CSDG 87-1-2, Computer System Design Group, Electrical and Computer Engineering Department, University of Colorado, Boulder, June 1987*.

(9)   C. D. Polychronopoulos, D. J. Kuck and D. A. Padua, "Execution of Parallel Loops on Parallel Processor Systems," *Proceedings of the 1986 International Conference on Parallel Processing, August 1986*.

CHAPTER 58

The Cerberus Multiprocessor Simulator*

Eugene D. Brooks III[†]
Timothy S. Axelrod[†]
Gregory A. Darmohray[†]

Abstract. We describe a simulation facility for scalable shared memory multiprocessors. The processors are a RISC architecture with a minimum of additional instructions added to support synchronization. The functional units of the processors, including access to the shared memory, are fully pipelined. The multiprocessor is simulated at a very detailed level, with pipeline delays and conflicts in the packet switched shared memory server being accurately accounted for. Applications are written in C. The simulator can be used to examine algorithm performance as the number of processors in the multiprocessor is scaled, providing results for machine sizes not currently available in real hardware.

1. Introduction. Over the past few years we have seen the introduction of several shared memory multiprocessors into the marketplace. The availability of real shared memory multiprocessors, at prices that an individual working group can afford, is accelerating the development of both hardware and software for parallel computation. Unfortunately, the true needs of parallel computation (both for hardware and software) are not yet well understood. Most of the existing machines are to be regarded as engineering hacks; they were designed and built with virtually no underlying parallel code development experience upon which to base design decisions. Each of the available machines has its own custom synchronization operations, along with its own architectural limitations. Nearly all of the available general purpose machines contain memory subsystems that limit the number of processors one can have in the machine.

One can learn a great deal about the performance of shared memory parallel programs using the available commercial machines. In the course of using these machines, however, we have found that most of the programming effort is spent trying to make the best use of the available synchronization primitives and working around any architectural limitations. This programming effort, which can be incredibly painful, must be done essentially by hand as efficient automatic tools to handle the compilation of a parallel program do not yet exist. As parallel code is ported to other machines, the available synchronization primitives and other architectural constraints change. This causes a serious glitch in the learning curve if one has reached the processor limit for a particular machine, requiring movement of the code to another multiprocessor in order to investigate a larger range of N. Sometimes, one must completely rewrite a code in order to obtain acceptable performance on a new multiprocessor.

* Work performed under the auspices of the U. S. Department of Energy by the Lawrence Livermore National Laboratory under contract No. W-7405-ENG-48.

† Parallel Processing Project, Lawrence Livermore National Laboratory, Livermore, California 94550

We wish to thank Ridge Computers Inc., of Santa Clara, California, for generous assistance.

One goal of the Cerberus multiprocessor simulator is to provide code developers with a scalable shared memory multiprocessor architecture on which to benchmark their algorithms. The size of the machine that can be simulated is limited only by the users patience. The processor model used in the simulator is a RISC [1] with fully pipelined functional units, much like what is available in current super-computers. The shared memory subsystem is based on a packet switched network which shows promising adaptive behavior when being subjected to parallel vector addressing patterns [2,3,4]. The simulated machine is fully parameterized, allowing the user to examine the effects of changing functional unit pipe-line delays of the processors, the network topology and the nodal resource limits of the shared memory subsystem. Using the Cerberus multiprocessor simulator one can explore a wide range of architectural parameter space that is not available in real multiprocessor systems.

By simulating the target multiprocessor, we can provide many features which are difficult to obtain with real hardware, such as perturbation free execution monitoring. Although one may know that a paral-lel program is not performing well it can be very difficult to pin down the source of the trouble. One usu-ally has to resort to educated guesses, making changes to the algorithm to remove a suspected problem and then examining execution times to see if significant improvements occur. With a multiprocessor simulator, a wide variety of detailed execution statistics can be used to lead the algorithm developer directly to the source of a performance problem.

A second goal of the Cerberus project is to evaluate the need for and effectiveness of special syn-chronization primitives. If one finds that an application can not be made to run efficiently using the exist-ing instruction set a new instruction can be added, which would hopefully be of general utility and allow the application to run efficiently. Using the simulator, the need for and the effectiveness of special syn-chronization operations can be carefully evaluated. This provides very useful information to the manufac-turers of shared memory parallel computers.

2. The CPU model. The processor architecture for our simulator was chosen with two considerations in mind. We needed an instruction set that would be compatible with the indeterminate access delays for shared memory through a packet switched network. As a compiler and simulator for the architecture would have to be developed, the instruction set should be as simple as possible. The load/store architec-ture of a RISC fits these two criterion quite well. Provided that there is enough parallelism available in the instruction stream, the independent loads, computations and stores can be scheduled by an optimizer to minimize the effects of shared memory latency. The RISC architecture, by definition, presents a minimum of instructions which one has to develop compiler and simulator support for.

The processor instruction set for the Cerberus machine was bootstrapped from that of the Ridge 32, manufactured by Ridge Computers of Santa Clara, California. By starting with an existing instruction set we were able to take advantage of compiler support which Ridge Computers made available to us, thereby greatly reducing our development effort. To create the Cerberus instruction set we increased the number of registers to 256 and generalized the computation instructions to a three register $Rz \leftarrow Rx\,OP\,Ry$ for-mat. The large number of registers and extended three address format for the computation instructions greatly simplified the task of creating a high quality optimizer.

Three instructions were added to support processor synchronization. The first of these is the SWAP instruction, *swap register with memory location indivisibly*, used in the creation of a lock construct. The second is WAIT, *wait for pending shared memory accesses to complete*, used to assert that any modifications to shared store have completed before releasing another processor to access the data. The last instruction we have added (so far) is the BARRIER instruction. In the course of some preliminary application studies we found barrier synchronization to be quite heavily used. As we obtain experience with the simulator we may find more critical synchronization operations and support them directly in the Cerberus instruction set.

3. The shared memory server. The simulator is constructed in a modular way which allows one to sub-stitute models for the various subsystems of the multiprocessor easily. The topology of the shared memory subsystem and the model used for the memory modules themselves can be changed at will. We are currently using the indirect k-ary n-cube packet switched shared memory server, described in [3,4], as the shared memory subsystem for the multiprocessor. A 2-ary 4-cube system is shown in figure 1. The request network of the shared memory subsystem is shown, with the processors on the left and memory modules on the right. The memory addressing is handled in such a way that sequential accesses to 32 bit memory words run round-robin through the memory banks.

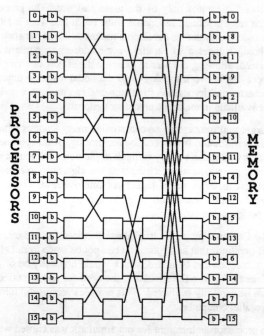

Fig. 1.   The indirect 2-ary 4-cube architecture.

When a processor issues a load or store instruction to the shared memory, a *request* is placed on the request network and it is routed to the appropriate memory module where the request is satisfied. A response, the data for a read or just a return receipt for a write, is routed back to the processor on a separate response network. The boxes labeled *b* are interface buffers used to establish a uniform interface between the packet switching network and the processors and memory modules. The minimum latency for a shared memory access will be one clock for each switch node and buffer and one clock for the memory module itself. For the system shown in figure 1 the minimum shared memory latency is 13 clocks. If the shared memory subsystem is heavily loaded, the latency of a shared memory access will in general be greater than the minimum, caused by conflicts at the switch nodes and memory modules. These conflicts are modeled in the Cerberus multiprocessor simulator exactly without any stochastic approximations.

In addition to the normal memory operations, load and store byte, half word, word and double words, the shared memory subsystem supports the indivisible swap of a word in memory with a 32 bit processor register. In this operation the request packet carries the registers value to the memory module where it replaces the target memory location. Before the actual write at the memory module, the previous memory value is read and placed in the response packet that is returned to the processor. Multiple swap with register requests may overlap in the shared memory subsystem as indivisibility is supported at the memory module end of the packet switching network. Combining [5] is not currently supported in the shared memory server network, although we may examine the utility of this feature in the future.

4. The programmer model. We use a simple but powerful parallel programming model to write programs for the simulator. The same programming model is available on shared memory multiprocessors manufactured by Sequent Computer Systems Inc. and Alliant Computer Systems Inc. and we have found this to be very useful when bug shooting a program to be run on the simulator. As the simulator runs too slowly to be used for interactive debugging, being able to run the same source code on a real shared memory multiprocessor is a great help. In the programming model all code is written in the C programming language with parallelism explicitly controlled through the use of macros. All static data is shared between the processors and automatic data is private to a processor.

The basic paradigm used in the model is that of a *team* of processors, the exact number of processors being unknown at compile time. A team of processors runs through the code executing it in the same manner as a single processor does in a serial program. The notion of *team splitting*, the splitting of a team into two independent subteams, is supported in this programming paradigm. The use of team splitting in combination with the more well known *forall* construct provides an efficient parallel programming paradigm that can be used to efficiently exploit concurrency at many levels.

Due to lack of space we can not provide a detailed exposition of the parallel programming model, but will give one example that serves to demonstrate some of its capabilities. Consider the parallel execution of the two vector expressions $A = M1 \cdot (B = M2 \cdot C)$ and $D = M1 \cdot (E = M2 \cdot F)$. The two expressions are independent and we would like to exploit this parallelism if we can. The exploitation of parallelism at a high level usually has a low synchronization and scheduling cost relative to the work performed. There is also parallelism within each vector expression we would like to exploit, the independent computation of dot products. Finally there is parallelism within each dot product which we want our compiler to exploit through vectorization. The code to accomplish this might be as follows.

```
float a[DIM], b[DIM], c[DIM];
float d[DIM], e[DIM], f[DIM];
float m1[DIM][DIM], m2[DIM][DIM];
work()
{
        int i, j;
        split {
                forall(i = 0; i < DIM; i += 1) {
                        b[i] = 0;
                        for(j = 0; j < DIM; j += 1) {
                                b[i] += m2[i][j] * c[j];
                        }
                }
                barrier;
                forall(i = 0; i < DIM; i += 1) {
                        a[i] = 0;
                        for(j = 0; j < DIM; j += 1) {
                                a[i] += m1[i][j] * b[j];
                        }
                }
        }
        and {
                forall(i = 0; i < DIM; i += 1) {
                        e[i] = 0;
                        for(j = 0; j < DIM; j += 1) {
                                e[i] += m2[i][j] * f[j];
                        }
                }
                barrier;
                forall(i = 0; i < DIM; i += 1) {
                        d[i] = 0;
                        for(j = 0; j < DIM; j += 1) {
                                d[i] += m1[i][j] * e[j];
                        }
                }
        }
        barrier;
}
```

A team of processors enters the at the entry point *work()*. The team of processors, with the number of team members totally hidden in the user code, is split into two subteams. The first subteam executes the code for $A = M1 \cdot (B = M2 \cdot C)$ in the *split* block. Exploitation of the parallelism in the independent dot products is explicitly indicated through the use of a *forall* loop. The need for synchronization is explicitly indicated by the use of *barrier*. On a production vector multiprocessor we would expect the dot product itself to be efficiently vectorized, in order to mask memory and functional unit latency, but in the Cerberus simulator we would have to explicitly unroll this innermost loop. The second subteam executes the similar code for $D = M1 \cdot (E = M2 \cdot F)$ in the *and* block. The two subteams join up at the last *barrier* and return to the calling program at the closing brace of the function *work()*. If the team entering *work()* has only one member, the *split* block and the *and* block are executed sequentially.

There are many other features built into the model and we lack the space to describe them here. This example is meant to show how concurrency can be exploited at many levels using the programming model available on the Cerberus simulator. See [6] for a detailed description of the features of this parallel programming paradigm along with its efficient implementation on several real shared memory multiprocessors.

5. An example. As a simple application of the Cerberus multiprocessor simulator, consider the performance of a parallel matrix vector product algorithm. We consider this rather trivial application because its memory reference patterns and load balancing issues are well understood and we will be able to quickly analyze the results in the remaining space. For this example we will consider how the execution speed for a matrix vector product of dimension 128 varies as we increase the number of processors in the multiprocessor. The code which was used to time the matrix vector multiply was as follows.

```
#include <pcp.h>
#define DIM 128
float x[DIM], y[DIM], m[DIM][DIM];
main()
{
        BEGIN;
        int i;
        BARRIER;
        MARK(0);
        FORALL(i, 0, DIM, 1) {
                y[i] = dot16(m[i], x, DIM);
        }
        BARRIER;
        MARK(1);
        MASTER {
                printf("Elapsed time %d clocks\n", markdiff(1,0));
        }
        BARRIER;
        exit(0);
}
```

Some explanations are in order. A team of processors, in this case all of the processors in the simulated machine, enter the main program. The statically allocated data x, y and m reside in shared storage. The automatic variable i is local to each processor. The macros BEGIN, BARRIER, FORALL, MARK and MASTER are defined in the include file <pcp.h>. The FORALL macro expands to the appropriate code for a statically scheduled *forall* loop, where the indices 0 through DIM-1 stepping by 1, are interleaved through the processors. The BARRIER is a barrier synchronization, either the software barrier described in [7] or the hardware barrier support which is built into the Cerberus simulator. The MARK() macro causes the executing processor to store the current clock value into an internal processor register. This takes no simulated time and is very useful in obtaining perturbation free execution monitoring. The BEGIN macro takes care of some team book keeping.

After the initial BARRIER to put all team members in sync, the timed FORALL loop and second BARRIER are executed. The *dot16()* routine is a dot product that has had the basic loop unrolled 16 ways with the operations carefully overlapped to mask functional unit and shared memory latency. For each chomp of 16 vector elements, the *dot16()* routine lines up the 32 fetches first, performs the 16 multiplies and then 16 partial sums. These operations are efficiently pipelined through the functional units of the processor and approach the asymptotic issue rate of one instruction per clock. The 16 partial sums are added to compute the returned value when the vector elements have been exhausted. The MASTER block admits one team member, called the master of the team, to print out execution statistics. All team members have in fact executed the MARK instructions, and printing out these values for all of the team members is sometimes quite useful. The *markdiff()* call is a simulator system call used to return the differences of times stored by MARK instructions.

We consider now the performance of this algorithm on the Cerberus machine with a 2-ary network as the shared memory server. We compare the execution speed of the multiprocessor to a run which was done using the single CPU simulator which is part of the Cerberus package. Using a single CPU, the code achieved an average floating point operation rate of .37 operations per clock, 74% of peak performance possible for dot products. Since each dot product element involves two fetches, a multiply and an add, the peak performance possible is .5 floating point operations per clock. In figure 2 we show the speedup of the algorithm, compared to the single cpu simulation, as the number of processors is scaled from 2 to 128. The solid line is the ideal speedup, the speedup the manufacturer of the Cerberus machine guarantees you will never exceed. The minimum shared memory latency varies from 7 clocks for the two processor system to 19 clocks for the 128 processor system. The actual memory latency can be quite higher, especially considering the memory reference pattern of this algorithm which we will discuss shortly. There are three sets of data shown in the figure, labeled *software*, *hardware* and *cooled*. The data labeled *software* and *hardware* are the measured performances for the software and hardware barrier support respectively. The difference between these data are solely due to the overhead of the software barrier. The latency of the hardware barrier is one clock (after all processors arrive). The 31% improvement in execution speed for 128 processors justifies the inclusion of barrier support in the instruction set.

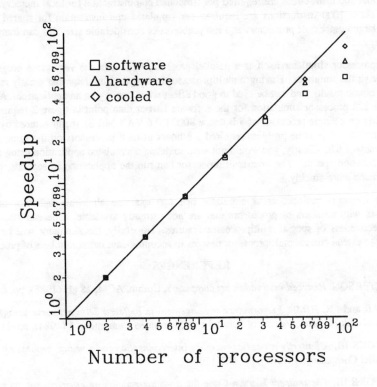

Fig. 2.   Parallel performance of the matrix vector product.

What is causing the roll off in speedup at 128 processors even though the hardware barrier support has only a one clock latency? The algorithm is perfectly load balanced, with every processor doing exactly the same amount of work. An examination of the memory reference pattern for this algorithm uncovers the trouble. For the 128 processor system the number of memory modules matches the length of the vectors we are computing dot products for. The first element of the vector x and the first element of each row of the matrix m start at the same memory bank. This causes each processor to start at this memory bank and proceed through the banks in an orderly fashion to compute their independent dot products. A hot spot travels through the memory banks producing severe vector startup costs, which we will call a *hot vector startup*. If the vectors are long enough to wrap around the memory banks several times, the relative cost of the hot vector startup can be made quite small. This is why the problem runs well on machines with smaller numbers of processors.

Given this analysis, it would appear that we could improve performance if we padded each row of the matrix with an extra word so that matrix rows started in different memory banks. Doing this, we found very little difference in execution time. The hot vector startup for the fetch of x was still ruining performance. To further test this hypothesis, the vector x was doubled in length and the code modified so that each processor would start at $x[i]$ to compute the dot product for $y[i]$. This would compute incorrect results, but allows us to examine the hot vector startup costs for x with a minimal perturbation to the code. This is what was done for the data labeled *cooled* in figure 2. Note that this change has very little impact for the smaller numbers of processors, where the hot vector vector startup cost is offset by long vector lengths and multiple vector operations for each processor. Removal of the hot vector startup for x and the rows of m improves the performance of the 128 processor machine by about 28%.

6. Discussion. We have presented a simulation system for a scalable shared memory multiprocessor. The package is composed of a single processor simulator and a multiprocessor simulator along with complete optimizing compiler and library support for a parallel extension of the C programming language. A detailed one pass simulation of the complete multiprocessor system is performed, with functional unit pipeline delays and shared memory conflicts being correctly accounted for. Such a detailed simulation has its costs. About 100 instructions are required per simulated cpu instruction for local memory memory references, and about 1000 instructions are required per simulated cpu instruction for shared memory references. For large numbers of processors and test problems of considerable size, this can translate into long execution times.

The multiprocessor simulator itself is a parallel program, written in the very same programming model used to drive the simulator. Having a multiprocessor to host the simulator can greatly reduce the time required to obtain results, and can be used to good effect when debugging an application. As a point of reference, the 128 processor simulation for the software barrier data point in figure 2 required about 12.2 hours running on a single processor of a Balance 8000 ($^-$0.6 VAX MIPS), manufactured by Sequent Computer Systems Inc. The same problem required 1.8 hours using 8 processors on the same machine, achieving a speedup of 6.8. Clearly, you would not want to debug a simulated application using a simulator with these execution speeds. The host multiprocessor can run the application in seconds, making a debug cycle go much more quickly.

Using the Cerberus multiprocessor simulator one can examine algorithm behavior on shared memory machines with numbers of processors that are not currently available. One can also use it to evaluate the effectiveness of special multiprocessor features. Hopefully, the simulator will be a useful tool in the effort to extend current multiprocessor designs to accommodate larger numbers of processors.

REFERENCES

[1] D.A. PATTERSON, *Reduced instruction set computers*, Comm. ACM, **28** (1) (1985), pp. 8-21.

[2] M. KUMAR and J.R. JUMP, *Performance enhancement in buffered delta networks using crossbar switches and multiple links*, Journal of Parallel and Distributed Computing, **1** (1984), pp. 81-103.

[3] E.D. BROOKS III, *A butterfly processor-memory interconnection for a vector processing environment*, Parallel Computing, **4** (1987), pp. 103-110.

[4] E.D. BROOKS III, *The indirect k-ary n-Cube for a vector processing environment*, to appear in: Parallel Computing, Lawrence Livermore Laboratory UCRL-94529, Oct. 1986.

[5] M. KUMAR and G.F. PFISTER, *The Onset of Hot Spot Contention*, Proc. 1986 International Conference on Parallel Processing, IEEE August 1986, pp. 35-41.

[6] E.D. BROOKS III and G.A. DARMOHRAY, *A parallel extension of C that is 99% fat free*, In preparation.

[7] E.D. BROOKS III, *The butterfly barrier*, International Journal of Parallel Programming, **15** (4) (1986), pp. 295-307.

PART VII

Architectures

CHAPTER 59

Neural Computing and Parallel Computation*

Michael A. Arbib[†]

Abstract. This paper addresses two complementary questions: what is the appropriate set of tools for the study of the networks of animal and human brains; and what are the strategies for building computers with "intelligence" ? We argue that there are overall architectural principles which unite both sides of this study, namely, that a computer be thought of as a network of more specialized devices, and that many of these devices be structured as highly parallel arrays of interacting neuron-like components. We illustrate this with a brief discussion of the following topics: the architecture of the frog's brain as revealed in studies of the mechanisms of visuomotor coordination; stability analysis of neural networks for depth perception; and the implications of associative memory techniques for the design of intelligent computers.

1. Neural Computing and the Sixth Generation. In Neural Computing the aim is to develop ideas from neuroscience into a form that is promising for the development of the next generation of computers and programs. In Computational Neurobiology models are developed which stand or fall in relation to the data of experimental study of the brain. To focus the discussion of neural computing, I would like to suggest that the study of the brain defines a Sixth Generation of computers[@], whose most salient characteristic is *cooperative computation* : the computer will be a problem-solving network, rather than a single serial machine. Inspiration

---

* Preparation of this paper was supported in part by NIH under grant 7 R01 NS24926 from NINCDS. My grateful thanks to the colleagues with whom much of the research reported herein was conducted. A fuller exposition of the material discussed here may be found in [3] and/or the forthcoming second edition of my book *The Metaphorical Brain* [5].

† Departments of Computer Science, Neurobiology, Physiology, Biomedical Engineering, Electrical Engineering, and Psychology, University of Southern California, Los Angeles, CA 90089-0782

@ When I visited China in November of 1984, colleagues there asked me to report on the status of Japanese work on the Fifth Generation Computer, which marries the "state of the art" in AI software of the early 1980s (e.g., expert systems) to appropriately engineered hardware (e.g., MIMD hypercubes). It was Tang Yi Qun of Wuhan University who asked, "And what then of the Sixth?" leading me to realize that the work my colleagues and I were doing in brain theory and distributed artificial intelligence was indeed defining a Sixth Generation (the computer systems for 1995-2005).

from the brain leads away from an emphasis on a single "universal" machine and suggests that the computer of the future will be a device composed of different structures, just as the brain may be divided into cerebellum, hippocampus, motor cortex, etc.   Thus we can expect to contribute to neural computing as we come to better chart the special power of each structure.

Like the brain, each sixth generation computer will thus be a networks of subsystems, including general-purpose engines and special-purpose machines some of which (such as the front ends for perceptual processors, and devices for matrix manipulation) will be highly parallel machines.   Some subsystems will use optical computing (lenses for Fourier transforms; holograms for high-density connectivity, etc.); some will be designed using the emerging principles of neural computing; while, on a longer time-scale, some will employ biomaterials.   Just as the interface to computers has progressed from bits to symbols to interactive graphics, the next generation will be more "action oriented" with computers including robotic actuators and multi-modal intelligent interfaces among their subsystems.   The design of these perceptual robotic systems will be heavily influenced by the analysis of neural mechanisms for perception and the control of movement.   Finally, sixth generation computers will be machines that learn, with principles of adaptive programming that incorporate lessons from brain theory and connectionism.

Since billions of human brains already exist, the goals of such computer design must be to develop computers that complement brain function, rather than simply emulate it.   Thus the field of neural computing must seek not to copy the brain, but to look for promising contributions from neural nets, physics, and "classic" computer science.   There is no ready-made body of wisdom in neuroscience waiting to be applied.   Thus we cannot simply follow the "neural blueprint" to build a better computer.   Rather, fundamental research in neuroscience must continue, but now structured to extract key principles of information processing that can enter into the design of the next generation of computers, building, for example, on the interaction between neuroscientists and AI workers in vision [6].

Our general approach involves the use of schemas [2,3,5] as an intermediary between overall task specification and neural networks.   We state the issues in terms of computationalneuroscience.   When we turn to neural computing, thre emphasis shifts from biological validity to technological efficacy:

1. An overall analysis of some behavior yields a functional decomposition in terms of interacting computing agents called *schemas.*   A given schema, defined functionally, may be distributed across more than one brain region; conversely, a given brain region may be involved in many schemas.   Top-down analysis advances specific hypotheses about the localization of (sub)schemas in the brain, and these may be tested by lesion experiments.

2. In some cases we will then proceed to model each schema by interacting layers of neuron-like elements, or by nets of "intermediate-level" units.   Even if the nets are little constrained by anatomy or physiology, such studies can be valuable in extending our understanding of "parallel distributed processing/connectionist" approaches to cognition and of the properties of neural networks, better preparing us to handle new data as they are available.

3. We model a brain region by seeing whether its known *neural circuitry* can indeed be shown to implement the posited schemas. At the most detailed level, then, we integrate neural network models with data of neuroanatomy and neurophysiology. Such models suggest properties of the circuitry that have not yet been tested, thus laying the ground for new experiments.

4. Once a number of models have been established, further modelling should be incremental, in that new models should refine, modify and build upon prior models, rather than being constructed ab initio.   Incremental modelling will not always work, however, and new data may lead to extensive reformulation. The less often this happens, the better the modeling strategy.

2. Visuomotor Coordination in Frog and Toad.   This section focuses on *Rana Computatrix,* a developing computational model of neural mechanisms of visually guided behaviour in frog and toad, to exemplify the methodology of "evolving" an integrated account of a single animal, meeting the challenges posed by combining different aspects of vision with mechanisms for the control of an expanding repertoire of behaviour.   Lettvin, Maturana, McCulloch and Pitts [17] initiated the behaviourally oriented study of the frog visual system with their classification of retinal ganglion cells into four classes each projecting to a retinotopic map at a different depth in the optic tectum, the four maps in register.   We view the analysis of such interactions between layers of neurons as a major approach to modelling "the style of the brain."

2.1. Schemas for prey recognition in *Rana Computatrix.* Group 2 retinal cells responded best to the movement of a small object within the receptive field and group 4 cells responded best to the passage of a large object across the receptive field [17].   It became common to speak of R2 cells as "bug detectors" and of R4 cells as "enemy detectors," although subsequent studies make it clear that a given frog or toad behaviour will depend on far more the than activity of a single class of retinal ganglion cells [11].

Given the mapping of retinal "feature detectors" to the tectum and the fact that tectal stimulation could elicit a snapping response, it became commonplace to view one task of the tectum to be directing the snapping of the animal at small moving objects — it being known that the frog would jump away from large moving objects and would not respond when there were only stationary objects.   This might suggest that the animal is controlled by, *inter alia* , two schemas, one for prey catching, which is triggered by the recognition of small moving objects, and one for predator avoidance, which is triggered by large moving objects. However, Ewert [11] has observed that animals with lesions of the pretectum will snap at large moving objects that a normal toad will avoid. This suggests a new analysis in terms of a prey-selection schema that can be activated by moving objects of any size; and a predator-recognition schema, that serves not only to activate avoidance behaviour but also to inhibit prey acquisition behaviour. Thus, even gross lesion studies can distinguish between alternative top-down analyses.

Of course, such an analysis can be refined by more detailed behavioural studies which let us determine what features of a moving object serve to elicit one form of behaviour or another. For example, Ewert placed a toad in a perspex cylinder from which it could see a stimulus object being rotated around it. He then observed how often the animal would respond with an orienting movement (this frequency being his measure of how "prey-like" the object was) for different stimulus objects. The worm-like stimulus (rectangle moved in the direction of its long axis) proved increasingly effective with increasing length; whereas for 8° or more extension on its long axis, the antiworm stimulus (rectangle moved in the direction orthogonal to its long axis) proved ineffective in releasing orienting behaviour. The square showed an intermediate behaviour, the response it elicits rising to a maximum at 8°, but being extinguished by 32°.

With such quantitative data to hand, Ewert and von Seelen [12] postulated that retinal output was passed in parallel to a tectal "worm filter," and a thalamic "antiworm filter," with the output of the latter serving to inhibit tectal (type II) activity excited by the former.   A worm stimulus would then tend to yield much excitation of the worm filter which would be little inhibited by the thalamic antiworm response, thus yielding a vigorous output; the antiworm would yield weak tectal type I activity, strong thalamic activity, and resultant weak tectal output. The square would yield intermediate behaviour. Ewert and von Seelen were able to adjust the parameters in this model to fit the data over a linear subrange of the results. However, the model is "lumped" in both space and time. That is, while the average rate of response of the output correlates well with the average turning rate of the toad, the model can neither explain the spatial locus at which the toad snaps nor the time at which it snaps.

2.2. Neural circuitry in anuran tectum. We discuss mechanisms for facilitation of prey-catching behavior found at a single locus of tectum. We will then sketch how nteractions among a number of such loci ("tectal columns") "unlump" the Ewert-von Seelen model. Presenting a worm to a frog for 0.3 sec may yield no response, whereas orientation is highly likely to result from a 0.6 sec presentation. Ingle [15] observed a facilitation effect: if a worm is presented initially for 0.3 sec, then removed, and then restored for only 0.3 sec, the second presentation suffices to elicit a response as long as the intervening delay is at most a few seconds. Ingle observed tectal cells whose time course of firing accorded well with this facilitation effect, leading us [16] to a model in which the "short-term memory" is encoded as reverberatory neural activity.

Each column comprises one pyramidal cell (PY) as sole output cell, one large pear-shaped cell (LP), one small pear-shaped cell (SP), and one stellate interneuron (SN). All cells are modelled as excitatory, save for the stellates. The retinal input to the model is a lumped "foodness" measure, and activates the column through glomeruli with the dendrites of the LP cell. LP axons return to the glomerulus, providing a positive feedback loop. A branch of LP axons also goes to the SN cell. There is thus competition between "runaway positive feedback" and the stellate inhibition. Glomerular activity also excites SP which likewise sends its axon back to the glomerulus. SP also excites LP to recruit the activity of the column. PY is excited by both SP and LP. The cells in our model interact only via their "firing rates," while the firing rate of a cell depends only on its own membrane potential. Thus in this model we do not need separate variables to hold "old" and "new" values, but instead cycle through two steps:

Step 1: Combine current values of firing rates to compute new values of membrane potentials.

Step 2: Using new values of membrane potentials, form the new values of the corresponding firing rates.

Step 1. Updating the Membrane Potentials: The membrane potential of each cell is described by a differential equation of the form

$$\tau_m \, dm(t)/dt = -m(t) + S_m(t)$$

which we will replace by the difference equation

$$\tau_m[(m(t+\Delta t) - m(t))/\Delta t] = -m(t) + S_m(t)$$

which yields

(1) $m(t+\Delta t) = (1 - \Delta t/\tau_m)m(t) + (\Delta t/\tau_m)S_m(t).$

The term $S_m(t)$ incorporates the input from all the cells to which the given cell is connected.

Step 2. Updating the Firing Rates: The "firing rate" of a cell is obtained by passing the membrane potential through a "thresholding function" of one of the types below:

f(x) = if x ≥ 0 then 1 else 0

h(x) = if x ≥ 0 then x else 0

s(x,A,B) = if x ≤ A then 0 else (if x ≥ B then 1 else (x-A)/(A-B)).

The updating of the firing rate is then accomplished simply by applying a thresholding function to the corresponding membrane potential:

Glomerulus Here we simply take GL = gl, but we still need two distinct values because of our updating convention.

Large Pear-Shaped Cell	LP = f(lp - 1.0)
Small Pear-Shaped Cell	SP = f(sp - 2.0)

Stellate Neuron	SN = h(sn - 0.2)
Pyramidal Neuron	PY = s(py, 2.3, 5.0)
Pretectal Cell	TP = h(tp - 3.8).

The overall dynamics will depend upon the actual choice of excitatory and inhibitory weights in each $S_m(t)$ and of membrane time constants to ensure that excitation of the input does not lead to runaway reverberation between the LP and its glomerulus as well as to ensure that this activity is "chopped" by stellate inhibition to yield a period of alternating LP and SN activity. SP has a longer time constant and is recruited only if this alternating activity continues long enough. In one simulation experiment, we graphed the activity of cells as a function of how long a single stimulus is applied. There is, as in the experimental data, a critical presentation length below which there is no pyramidal response. Input activity activates LP, which re-excites the glomerulus but also excites the SN, which reduces LP activity. But if input continues, it builds on a larger base of glomerular activity, and so over time there is a build-up of LP-SN alternating firing. If the input is removed too soon, the reverberation will die out without activating SP enough for its activity to combine with LP activity and trigger the pyramidal output. If input is maintained long enough, the reverberation may continue, though not at a high enough level to trigger output. However, re-introducing input shortly after this "subthreshold" input can indeed "ride upon" the residual activity to build up to pyramidal output after a presentation time too short to yield output with an initial presentation.

It required considerable computer experimentation to find weights that yield the neural patterns discussed below. More recently, Cervantes-Perez [7] has given a mathematical analysis of how weighting patterns affect overall behaviour. We hope our hypotheses concerning the ranges of the parameters involved in the model will stimulate more detailed anatomical and physiological studies of tectal activity.

To see the "evolutionary" development of *Rana Computatrix*, we refer the reader to [8] (see [3] for an exposition) which shows how an array of tectal columns may be connected to model how the tectum may interact with pretectum in subserving "prey schemas."   Rather than using a single column to represent the tectum, we now use an 8x8 array of columns, and we provide the column not only with the R2 input of our initial study of facilitation, but also with R3 and R4 input. In addition, we represent the pretectum by an array of TH3 cells receiving R3 and R4 input. The retinal input is based on overall ganglion cell response curves, in which only the average rate of firing of a cell is given for each stimulus, rather than the temporal pattern of that response. The connections of R3 and R4 to TH3 are tuned to yield the observed TH3 responses. (Since the response of tectum must depend on the spatiotemporal pattern of retinal ganglion cell firing, current work in my laboratory is aimed at more detailed modeling of the response of frog retina to varied activity.)

The input to the model comprises three retinal arrays R2, R3 and R4 . The tectum-pretectum model comprises six layers, each of which is characterized by two arrays: the first, labelled with a lower case letter, represents membrane potentials, while the second, bearing the corresponding upper case label, represents the firing rates. There are five layers in the tectum — glomeruli (gl, GL), large pear-shaped cells (lp, LP), small pear-shaped cells (sp, SP), stellate neurons (sn, SN), and pyramidal cells (py, PY) — while there is just one layer (tp, TP) of cells in the current model of the pretectum. Recalling the equation

$$m(t+ \Delta t) = (1- \Delta t/ \tau_m)m(t) + (\Delta t/ \tau_m)S_m(t)$$

for updating the membrane potential of each cell, we must now specify how each term $S_m(t)$ incorporates the input from all the other cells to which the given cell is connected. The influence of one layer, a, on another, b, will be represented by a matrix $W_{a.b}$. In the present model, each matrix is 3x3, and is constant save for its central element. In what follows we shall use the abbreviation $W\{x,y\}$ for the matrix

$$\begin{bmatrix} y & y & y \\ y & x & y \\ y & y & y \end{bmatrix}$$

Glomerulus: Here $\tau_{gl} = 2.3$ while

$$S_{gl} = W_{r2.gl}{}^*R2 + W_{lp.gl}{}^*LP + W_{sp.gl}{}^*SP$$

where $W_{r2.gl} = \{6.7, 0\}$, $W_{lp.gl} = W\{8, 5.3\}$ and $W_{sp.gl} = W\{0.7, 0.7\}$.

Large Pear-Shaped Cell: Here $\tau_{lp} = 0.3$ while

$$S_{lp} = W_{gl.lp}{}^* gl + W_{r2.lp}{}^* R2 + W_{sp.lp}{}^*SP - W_{tp.lp}{}^*TP - W_{sn.lp}{}^*SN$$

with $W_{gl.lp} = W_{r2.lp} = W\{1, 0\}$, $W_{sp.lp} = W\{0.8, 0.6\}$, $W_{tp.lp} = W\{0.1, 0\}$, and $W_{sn.lp} = W\{8.0, 8.2\}$

Small Pear-Shaped Cell: Here $\tau_{sp} = 0.9$ while

$$S_{sp} = W_{r2.sp}{}^* R2 + W_{gl.sp}{}^*GL - W_{tp.sp}{}^*TP - W_{sn.sp}{}^*SP$$

with $W_{r2.sp} = W\{1, 0\}$, $W_{gl.sp} = W\{1.0, 0.5\}$, $W_{tp.sp} = W\{0.1, 0\}$, and $W_{sn.sp} = W\{20.0, 0\}$.

Stellate Neuron: Here $\tau_{sn} = 1.6$ while

$$S_{sn} = W_{lp.sn}{}^*LP$$

with $W_{lp.sn} = W\{5.2, 5.2\}$.

Pyramidal Neuron: Here $\tau_{py} = 0.12$ while

$$S_{py} = W_{r2.py}{}^*R2 + W_{r3.py}{}^*R3 + W_{r4.py}{}^*R4 + W_{sp.py}{}^*SP + W_{lp.py}{}^*LP - W_{tp.py}{}^*TP$$

where $W_{r2.py} = W\{3.5, 0\}$, $W_{r3.py} = W\{0.3, 0\}$, $W_{r4.py} = W\{7.0, 0\}$, $W_{sp.py} = W\{2.0, 0\}$, $W_{lp.py} = W\{0.7, 0.56\}$, and $W_{tp.py} = W\{0.9, 0\}$.

Pretectal Cell: Here $\tau_{tp} = 0.02$ while

$$S_{tp} = W_{r3.tp}{}^*R3 + W_{r4.tp}{}^*R4$$

where $W_{r3.tp} = W\{0.3, 0\}$ and $W_{r4.tp} = W\{5.0, 0\}$.

With the above setting of parameters, the model does indeed exhibit in computer simulation responses to moving stimuli of different types that match well the neural data. However, the model is only approximate at a quantitative level so that — if our goal is prediction of detailed neural firing rather than just a general understanding of pattern recognition networks — further work must be done on tuning the model parameters.

3. Layered Models of Depth Perception: A Stability Analysis. The problem for many models of binocular perception is to suppress ghost targets. The essence of one solution, the Dev scheme [10], is to have those neurons which represent similar features at nearby visual directions and approximately equal depths excite each other, whereas those neurons which correspond to the same visual direction but different depths are (via interneurons) mutually inhibitory. In this way, neurons which could represent elements of a surface in space will cooperate, whereas those which would represent paradoxical surfaces at the same depth will compete. The result is that, in many cases, the system will converge to an adequate depth segmentation of the image.   However, such a system may need extra cues. For example, in looking at a paling fence, if several fenceposts are matched with their neighbors on the other eye in a systematic fashion, then the cooperative effect can swamp out the correct pairing and lead to the perception of the fence at an incorrect depth. In animals with frontal facing eyes such ambiguity can be reduced by the use of vergence information to drive the system with an initial depth estimate. Another method is to use accommodation information to provide the initial bias for a depth perception system; this is more appropriate to the amphibian, with its lateral-facing eyes. It is the latter possibility we pursue here, as part of our continuing concern with *Rana Computatrix*.

The Dev model of stereopsis (as analyzed in [1]) receives input from two one-dimensional retinas, each with coordinate axis x. It comprises a two-dimensional "excitatory field" where the membrane potential $u_d(x,t)$ of the cell at position (x,d) represents the confidence level at time t that there is an object in direction x on the left retina whose disparity on the right retina is d; and a one-dimensional inhibitory field with $v(x,t)$ the activity of the cell at position x at time t. We think of d as taking a set of discrete values. The output of these cells are given as $f(u_d(x,t))$ and $g(v(x,t))$ respectively, where $f(u) = 1$ if $u > 0$ else 0, while $g(v) = v$ if $v > 0$ else 0. The model can then be expressed mathematically as:

$$\tau_u \partial u/\partial t = -u + w_1 {}^*f(u) - w_2{}^*g(v) - h_u + s$$

$$\tau_u \partial v/\partial t = -v + \Sigma_d f(u_d) - h_v.$$

Here h_u and h_v are threshold levels (constants) and s_d is the input array. We set

$$R_L(x) = 1 \text{ if some object projects to point x on the Left retina else 0}$$

and similarly for R_R. We then define the stereo input to the model to be

$$s_d(x) = R_L(x) \, R_R(x+d)$$

which is 1 only if there is an object at position x on the left retina as well as at x plus disparity d on the right retina. A more refined model would make $s_d(x)$ a measure of the correlation of the features of the pattern near position x on the left retina and those centered at x+d on the right retina.

In many cases, a stereo system such as that provided by Dev will converge to an adequate depth segmentation of the image.   However, such a system may need extra cues. For example, in looking at a paling fence, if several fenceposts are matched with their neighbors on the other eye in a systematic fashion, then the cooperative effect can swamp out the correct pairing and lead to the perception of the fence at an incorrect depth. In animals with frontal facing eyes such ambiguity can be reduced by the use of vergence information to drive the system with an initial depth estimate. Another method is to use accommodation information to provide the initial bias for a depth perception system; this is more appropriate to the amphibian, with its lateral-facing eyes.

House's Cue Interaction Model [14] uses two systems, each based on Dev's stereopsis model, to build a depth map. One is driven by disparity cues, the other by accomodation cues, but corresponding points in the two maps have excitatory cross-coupling. The model is so tuned that binocular depth cues predominate where available, but monocular accomodative cues remain sufficient to determine depth in the absence of binocular cues. The model produces a complete depth map of the visual field, and so is appropriate for building a representation of barriers for use in navigation. An accommodation-driven field, M, receives information about accommodation and — left to its own devices — would sharpen up that information to yield relatively refined depth estimates. Another system, S, of the type posited by Dev uses disparity information as input and seeks to suppress ghost targets. Further, the systems are so coupled that a point in the accommodation field M will excite the corresponding point in the disparity field S, and vice versa. Thus a high confidence in a particular (direction, depth) coordinate in one layer will bias activity in the other layer accordingly. The result is that the system will converge to a state affected by both types of information — although the monocular system can, by itself, yield depth estimates.

S is thus given by

$$\tau_s \partial S/\partial t = -S_d + w_{s1}{}^*f(S) - s_{S2}{}^*g(V) - h_s + K_{ms}f_c(M) + s$$

$$\tau_v \partial V/\partial t = -V + \Sigma_d f(S_d) - h_v$$

which just differs from the Dev model in the addition of the coupling term $K_{ms}f_c(M)$ from M, where the nonlinearity f_c is given by $f_c(S) = 0$ if $M < 0$ else M/h_{sat} if $0 \leq M \leq h_{sat}$ else 1.

M is given by

$$\tau_m \partial M/\partial t = -M + w_{m1}{}^*f(M) - w_{m2}{}^*g(U) - h_m + K_{sm}f_c(S) + a$$

$$\tau_u \partial U/\partial t = -U + \Sigma_d f(M_d) - h_u$$

which differs from S in that the input is now provided by an accomodation measure:

$$a_L(x,d) = \exp[\{-(d-d_a(x))/\beta\}^2]$$

where $d_a(x)$ is the actual distance of an object in direction x from the left eye, and ß is a suitable constant.   We set object in direction x from the left eye if there is no object in direction x from the left eye. Thus $a_L(x,d) \leq 1$, and equals 1 only for $d = d_a(x)$. It represents the sharpness of the image in direction x when the focal length of the left eye is set at d. Similarly, we may define $a_R(x,d)$ for the right eye.

With this background, we turn to a brief statement of results from the mathematical analysis of Chipalkatti and Arbib [9].   They introduce a number of plausible assumptions which we shall not reproduce here. With these assumptions they prove the following:

Theorem: When discordant stimuli are supplied to the two fields, say $a_i > 0$ and $s_j > 0$, the selected module depends on the strength of the input.

• For small inputs and weak cross-coupling, no module is active at equilibrium.

• For strong inputs and strong cross-coupling, we have at equilibrium that:

  Module i is excited in both fields if $a_i > s_j$ and $K_{ms} > s_j$.

  No activity can be sustained if $a_i = s_j$.

  Module j is excited in both fields if $a_i < s_j$ and $K_{sm} > a_i$.

Active points compete with each other along the d-dimension by trying to suppress the activity of other points and by inhibiting unexcited points from becoming active. However, due to excitatory interactions, nearby points along the x-dimension tend to excite each other. Adapting techniques set forth in [1] for the Dev model, we are able to prove:

Theorem: Let stimulation be restricted to the x-dimension for a fixed d. If a stimulus is applied to M such that at equilibrium the excitation lies in the interval $[x_{m1}, x_{m2}]$ with $x_{m1} - x_{m2} = a_m$, then the induced excitation in the field S must lie within the interval $[x_{m1}, x_{m2}]$ and at its endpoints, the magnitude of the cross-coupling input must be less than h_s and the field M there must lie in the interval $(0, h_{sat})$. (Recall that h_{sat} is the lowest value of

M for which the cross-coupling $f_c(M)$ reaches its saturation level of 1.)

With the assumptions in place for the separate analysis of the x-dimension and d-dimension, we can indeed prove that the Cue Interaction model satisfactorily processes the "two-worm" situation:

Theorem: For two targets placed at (x_1, d_1) and (x_2, d_2) in the visual field, the peaks corresponding to the "ghost targets" in the field S are suppressed at equilibrium, assuming that all points in the two fields are initially in the same state.

The model succesfully prevents the emergence of "ghost images" for multiple targets, e.g., a fence, as well.   The two fields influence each other's activity in both dimensions.   Along the competition-dimension they cooperatively select the surviving module, whereas along the cooperation-dimension they directly control each other's excitation lengths.   These properties of the model lead to the suppression of activity due to cues representing "ghost targets."

4. Adaptation and Programming.   Finally, having briefly discussed networks for vision and visuomotor coordination, I want to now place the study of adaptive networks in the perspective of our introductory discussion of sixth generation computers.   Such studies of adaptive networks (see Chapter 5 of [4] for an exposition) as the Boltzmann machine [13] or backpropagation networks [18] are examples of "neurally-inspired" modelling, not modelling of actual brain structures.   It is well to summarize here what these learning methods achieve — it is really rather simple.   We are given a network N which, in response to the presentation of any x from some set X of input patterns, will eventually settle down to produce a corresponding y from the set Y of the network's output patterns.   A training set is then a sequence of pairs (x_k, y_k) from X x Y, $1 \leq k \leq n$.   Results for networks of the above kind say that in "many cases" (and the bounds are not yet well defined), if we train the net with repeated presentations of the various (x_k, y_k), it will converge to a set of connections which cause N to compute a function f: X → Y with the property that, as k runs from 1 to n, the $f(x_k)$ "correlate fairly well" with the y_k.   Of course, there are many other functions g: X → Y such that the $g(x_k)$ "correlate fairly well" with the y_k, and they may differ wildly on those x in X that do not equal an x_k in the training set.

Some workers in neural computing, over-impressed by the success of these new methods of training hidden units in layered neural networks, speak as if the days of programming will soon be over.   Simply present a trainable net with a few examples of solved problems, and it will adjust its connections to be able to solve all problems of a given class! However, this glosses over three main issues:

(a) *complexity* : is the network complex enough to encode a solution method?;

(b) *practicality* : can the net achieve such a solution within a feasible period of time? and

(c) *efficacy* : how do we guarantee that the generalization achieved by the machine matches our conception of a useful solution?

All this suggests that the art of programming will survive into the age of neural computing, but greatly modified. Given a complex problem, the programmer will still need to decompose it into subproblems, specify an initial structure for a separate network for each subproblem; place suitable constraints on (the learning process for) each network; and, finally, apply debugging techniques to the resultant system. Given all this (and much of it will be automated), we may expect that the initial design and constraint processes may in some case suffice to program a complete solution to the problem without any use of learning at all. Thus, the "domain-specific" analysis of networks exemplified in Sections 2 and 3 will remain a vital part of neural computing, with schemas providing the intermediate programming language for the specification of "neural programs."

REFERENCES

(1) Amari, S. and Arbib, M.A., 1977, Competition and cooperation in neural nets. In Systems Neuroscience (Metzler, J., ed.), pp. 119-165, Academic Press, New York.

(2) Arbib, M.A., 1981, Perceptual structures and distributed motor control. In Handbook of Physiology — The Nervous System II. Motor Control (ed. V.B. Brooks), Bethesda, MD: Amer. Physiological Society. pp. 1449-1480

(3) Arbib, M.A., 1987, Levels of modeling of mechanisms of visually guided behavior, The Behavioral and Brain Sciences, in press.

(4) Arbib, M.A., 1987, Brains. Machines. and Mathematics, 2nd Edition, Springer-Verlag.

(5) Arbib, M. A., 1988, The Metaphorical Brain: An Introduction to Schemas and Brain Theory, Second Edition, Wiley Interscience.

(6) Arbib, M.A., and Hanson, A.R., Eds., 1987, Vision. Brain. and Cooperative Computation. A Bradford Book/MIT Press.

(7) Cervantes-Perez, F., 1985, Modelling and Analysis of Neural Networks in the Visuomotor System of Anuran Amphibia, Ph.D. Thesis and Technical Report 85-27, Computer and Information Science Department, University of Massachusetts at Amherst.

(8) Cervantes-Perez, F., Lara, R. and Arbib, M.A., 1985, A neural model of interactions subserving prey-predator discrimination and size preference in anuran amphibia J. Theoretical Biology 113, 117-152

(9) Chipalkatti, R., and Arbib, M.A., 1987, The Cue Interaction Model of Depth Perception: A Stability Analysis, J. Math. Biol., in press.

(10) Dev, P., 1975, Perception of depth surfaces in random-dot stereograms: A neural model Int. J. Man-Machine Studies 7: 511-528.

(11) Ewert, J.-P., 1976, The visual system of the toad: behavioural and physiological studies on a pattern recognition system. In The Amphibian Visual System (Fite, K., ed.), pp. 142-202. Academic Press, New York

(12) Ewert, J.-P. and von Seelen, W.,1974, Neurobiologie and System-Theorie eines visuellen Muster-Erkennungsmechanismus bei Kroten. Kybernetik 14: 167-183.

(13) Hinton, G.E., Sejnowski, T.J., and Ackley, D.H., 1984, Boltzmann Machines: Constraint Satisfaction Networks that Learn, Cognitive Science, 9:147-169.

(14) House, D., 1984, Neural Models of Depth Perception in Frog and Toad, Ph.D. Dissertation, Department of Computer and Information Science, University of Massachusetts at Amherst.

(15) Ingle, D. , 1975, Focal attention in the frog: behavioural and physiological correlates Science 188: 1033-1035

(16) Lara, R., Arbib, M.A. and Cromarty, A.S., 1982, The role of the tectal column in facilitation of amphibian prey-catching behaviour: a neural model J. Neuroscience. 2: 521-530.

(17) Lettvin, J. Y., Maturana, H., McCulloch, W. S. and Pitts, W. H., 1959, What the frog's eye tells the frog brain. Proc. IRE, 1940-1951.

(18) Rumelhart , D.E., Hinton, G.E., and Williams, R.J., 1986, Learning Internal Representations by Error Propagation, in Parallel Distributed Processing: Explorations in the Microstructure of Cognition (Rumelhart, D., and McClelland, J., Eds.), The MIT Press/ Bradford Books.

CHAPTER 60

Optoelectronics in Computing

W. Thomas Cathey[*]

Abstract. This paper discusses the promise and problems in using optoelectronics in computing systems. The advantages of optical interconnects are presented, but emphasis is given to research on three different optical computing systems. The first is a bit-serial, stored-program, optical computer. The second is the use of symbolic substitution, and the third is an optical symbolic logic system to do proofs by resolution.

Introduction. Optoelectronics is the combination of optics and electronics to employ the advantages of each. Optical signals have low dispersion, low electromagnetic interference, and can pass through each other without interaction. On the other hand, an electron is needed somewhere in the system to provide signal interaction. In so-called all optical computers, there is an electronic interaction at some level. To maximize speed, the electronic interaction should be minimized. For example, the electron may only cause the optical transmission of the device to change.

[*]Center for Optoelectronic Computing Systems, and Department of Electrical and Computer Engineering, University of Colorado, Boulder, Colorado 80309-0425.

The optical symbolic logic research was supported by AFOSR grant no. AFOSR-86-0189, and the optical bit serial research by NSF/CDR grant no. 8622236 and the Colorado Advanced Technology Institute (CATI).

The use of optoelectronics in computing systems has the potential of increasing computing power several orders of magnitude. In some cases, the reasons for this are low signal dispersion and higher speeds, and in others, the advantage comes from massively parallel operations. In the near term, optical interconnects between purely electronic processors will increase the data transfer rate between the processors and memory. In the longer term, the use of optical switches and highly parallel operations will affect computer design and architecture.

There are several groups doing research on optical general-purpose computing, special purpose computing, and signal processing. Most of the research is being done in signal processing and pattern recognition. In this paper, I will not discuss those areas, but will concentrate only on computing and, in the area of computing I will discuss only optical interconnects and a sampling of different architectural approaches. Three different architectures are used to illustrate different aspects of the impact of optics on architectures.

This paper is to make the reader aware of the applications of optics to computing and of the potential impact that optics may have on the design, architecture, and operating systems of computing systems. It is not a comprehensive review of the field.

First, the potential advantages of optics in general are given. Next, the progress of research in optical interconnects is reviewed and comparisons are made of electronic and optical interconnects. A bit-serial architecture for an optoelectronic computer is described, and some of the theoretical questions are delineated. A different approach of symbolic substitution is reviewed. Finally, one approach to the use of the parallel capabilities of optics in solving artificial intelligence problems is described.

Advantages of Optoelectronics. One of the biggest attractions of optoelectronics is that, in contrast to electronics, which is a mature field and is approaching the theoretical limits, optoelectronics is a newly developing field that has fundamental limits several orders of magnitude away from what is being done today. These fundamental limits are also several orders of magnitude greater than those of purely electronic systems. There are several specific advantages to optics. In particular, optical signals suffer much lower dispersion than do electronic signals in low-cost circuits. In devices, the speed can be higher because the electrons are not used to carry signals by moving, but simply change the absorption or index of refraction of a material. With the use of imaging systems, millions of parallel paths are available for data transfer and parallel data processing. These advantages are best illustrated by examples.

From an evolutionary point of view, optical data transmission is becoming the dominant means of transmitting data long distances. For large data bases, optical data storage is taking over larger and larger portions of the task. This means that when data are received or brought from memory, a photon-to-electron conversion must be made to allow electronic processing, and an electron-to-photon conversion must be made to store or transmit data. These conversions slow the over-all system speed. Elimination or reduction of the need to perform the conversions will greatly speed computer operation.

Optical Interconnects. In an optoelectronic computer, the interconnects would, of course, be optical. The computer need not be optical, however, to benefit from optical interconnects [8]. Parallel or time multiplexed optical data busses offer low EMC/EMI problems and reduce the impedance matching problems. There also is the potential of easily reconfigurable interconnects and a greatly reduced clock skew.

Parallel optical interconnects have been investigated for several years. Only a sample of the research is given here. Parallel data buses [1, 20] and optical distribution of clocks [4, 6] were among the first applications to be suggested. Holographic optical interconnects between chips have been explored by several investigators [2, 5, 14]. The use of LiNbO$_3$ switches [15, 23] and liquid crystals [12, 19] for optical crossbar switching has been investigated. Parallel interconnects have obvious advantages, but as a first step, many industries are focussing on time-division multiplexing of electrical signals on a high-speed optical bus [9, 13, 22]. Optical busses require more power at present, but it has been shown that for n to 1 multiplexing, the total power consumed is less than for an n-wide electrical interconnect when n > 4. [9]

Optical Bit-Serial Computer. Most applications of optics in computing have been special-purpose processors or data storage mechanisms. One project has the goal of building a complete, stored-program optical digital computer [10]. This research is oriented toward studying the architectural problems that arise when the gate delay is no longer much greater than the delay encountered during signal propagation. One example of the effect when gate delay and propagation delay are comparable is that a parallel addition is not really parallel. When the carry propagation time is greater than or equal to the addition time, the operation is, in essence, a serial operation. In this case, the serial addition is as fast, as parallel, and can handle arbitrarily long inputs.

The state of the art for optical devices is approximately where electronic device development was with discrete transistors; there are no high-speed optical IC's. As a test or proof-of-principle system, a computer has been designed that uses 48 discrete, state-of-the-art switches [10]. Data storage is accomplished by switching optical signals into optical fiber storage loops as is shown in Fig. 1. The clock is provided by a mode-locked GaAs laser and, in the first implementation, the gates will be $LiNbO_3$ switches. Recirculating registers are built by using shorter fiber loops.

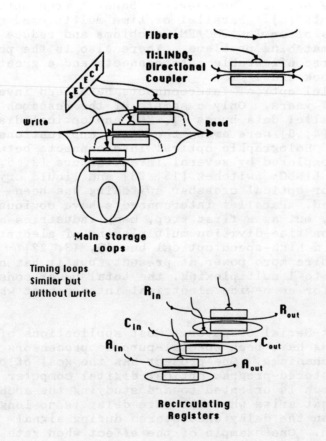

Fig. 1.   The main storage loops of a demonstration bit serial optical computer are optical fiber loops.

Optical Symbolic Substitution. An architecture that takes
advantage of the inherent parallelism of optics is that of
symbolic substitution. In this technique, one two-
dimensional set of symbols is replaces by another set of
symbols. By repeating this process, any computation can be
performed [11]. For example, addition can be performed by
the substitution rules given in Fig. 2. These rules can be
performed in parallel on two-dimensional patterns, thereby
using one of the strengths of optics.

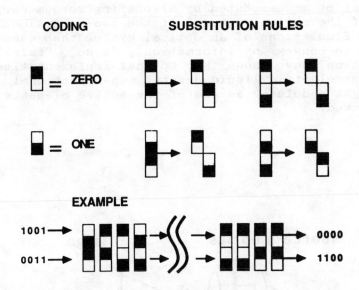

Fig. 2. Binary addition can be performed by
repetitive application of substitution rules.

Optical Symbolic Logic. Optical artificial intelligence is
a tantalizing application for optics because high accuracy
frequently is not required, but many operations must be
performed. One difficulty is that, in artificial
intelligence problems, the program is data dependent.
That is, the operation on the data depends upon the outcome
of previous operations. This normally is not easily done
with optical systems. Figure 3 illustrates the problem.
In optical systems, the "program" reflects what the optical
system does to the optical distribution. This can be a
Fourier transformation, a correlation, symbolic
substitution, or other operations, but they are not easily
changed. The data are in the form of temporal variations

or spatial variations of the optical signal coming into the system, or both. To use optics in artificial intelligence, the problems must be reformulated to reduce the interaction between the control and the data.

One program seeking to solve artificial intelligence problems is concentrating on optical symbolic logic [16, 17, 18]. In that program, the first problem being attacked is that of proof by resolution. One of the characteristics of proof by resolution is that the data base tends to increase exponentially as the facts are compared to derive other facts in an attempt to find a contradiction of the original assumption. This is a difficult problem in any computing implementation, but in an optical system using parallel data storage and manipulation, the expanding data base can not be accommodated by allocating more memory. The size of the optical "image" of the two-dimensional data is fixed. Simulations of an optical system that uses dual-rail logic to convey "no information", "true", "false", and "contradiction" have shown that optical implementation using a ferroelectric liquid crystal two-dimensional spatial light modulator as one of the active elements can be constructed.

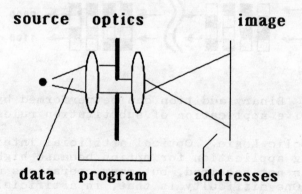

Fig. 3. Optical working environment showing the fixed "program" normally associated with optical systems.

Optical Devices. The architectures discussed require optoelectronic devices. There are several that are being pursued. In addition to the $LiNbO_3$ switches that exist, newer devices are being developed. Multiple quantum well spatial light modulators have the promise of providing high speed spatial light modulators in the future [24]. For the immediate future, ferroelectric liquid spatial light modulators have low loss and high contrast with switching speeds in the microseconds [3, 7]. This speed, when used in a 1000 X 1000 array, permits very high data processing speeds. Other devices use resonators to increase the optical effects [21].

Conclusions. The use of optoelectronics in computing offers great promise not only in the area of interconnects, but also in computing systems. Optoelectronic computers offer greater speed and much easier parallelism than purely electronic computers. Optoelectronics, however, is not developed to the extent of electronics, and is about where electronics was with the development of the transistor. With the investment in devices for optical communications and optical data storage, much of the work needed for optoelectronic computing will be done.

REFERENCES

1. W. T. Cathey and B. J. Smith, *High concurrency data bus using arrays of optical emitters and detectors*, Appl. Optics, 18 (1979), pp. 1687-1691.

2. H. J. Caulfield, *Parallel N^4 weighted optical interconnections*, Appl. Optics, 26 (1987), pp. 4039-4040.

3. N. A. Clark and S. T. Lagerwall, *Submicrosecond bistable electro-optic switching in liquid crystals*, Appl. Phys. Letters, 36 (1980), pp. 899-901.

4. B. D. Clymer and J. W. Goodman, *Optical clock distribution to silicon chips*, Opt. Engr., 25 (1986), pp. 1103-1108.

5. M. R. Feldman and C. C. Guest, *Computer generated holographic optical elements for optical interconnection of very large scale integrated circuits*, Appl. Optics, 26 (1987), pp. 4377-4384.

6. J. W. Goodman, F. J. Leonberger, S. Y. Kung, and R. A. Athale, *Optical interconnections for VLSI systems*, Proc. IEEE, 72 (1984), pp. 850-866.

7.   M. A. Handschy and N. A. Clark, *Structures and responses of ferroelectric liquid crystals in the surface-stabilized geometry*, Ferroelectrics, 59 (1984), pp. 69-116.

8.   D. H. Hartman, *Digital high speed interconnects: a study of the optical alternatives*, Opt. Engr., 25 (1986), pp. 1086-1102.

9.   P. R. Haugen, et al., *Optical interconnects for high speed computing*, Opt. Engr., 25 (1986), pp.1076-1085.

10.   V. P. Heuring, H. F. Jordan, and J. P. Pratt, *A bit serial architecture for optical computing*, Submitted to IEEE Trans on Computing.

11.   A. Huang, *Parallel algorithms for optical digital computing*, IEEE 10th Int. Optical Computing Conf., 13 (1983).

12.   H-I Jeon and A. A. Sawchuk, *Optical crossbar interconnections using variable grating mode devices*, Appl. Optics, 26 (1987), pp. 261-269.

13.   S. K. Korotky, et al., *Fully connectorized high-speed Ti:LiNbO₃ switch/modulator for time-division multiplexing and data encoding*, IEEE J. Lightwave Tech., LT-3 (1985), pp. 1-5.

14.   E. Marom and N. Konforti, *Dynamic optical interconnections*, Optics Letters, 12 (1987), pp. 539-541.

15.   L. McCaughan and G. A. Bogert, *4X4 Ti:LiNbO₃ integrated-optical crossbar switch array*, Appl. Phys. Lett., 47 (1985), pp 348-350.

16.   R. A. Schmidt and W. T. Cathey, *Optical representations for artificial intelligence problems*, SPIE 625 (1986), pp. 226-233.

17.   R. A. Schmidt and W. T. Cathey, *Optical implementations of mathematical resolution*, Applied Optics, 26 (1987), pp. 1852-1858

18.   R. A. Schmidt and W. T. Cathey, *Optical representations for symbolic logic*, Opt. Engr., to be published, Optical Engr.

19.   K. M. Johnson, M. Surette, and J. Shamir, *Optical Interconnection network using polarization-based ferroelectric liquid crystal gates*, to be published, Applied Optics.

20.   B. J. Smith and W. T. Cathey, *Some thoughts on time-multiplexed optical buses*, Internat. Optical Computing Conf., 23-25 April 1975, Washington, D.C.

21.   S. D. Smith, et al., *The demonstration of restoring digital optical logic*, Nature, 325 (1987), pp. 27-31.

22.   T. Tamura, M. Nakamura, S. Ohshima, T. Ito, and T. Ozeki, *Optical cascade star network — a new configuration for a passive distribution system with optical collision detection capability*, IEEE J. Lightwave Tech., LT-2 (1984), pp. 61-66.

23.   R. A. Thompson, *Traffic capabilities of two rearrangeably nonblocking photonic switching modules*, AT&T Tech. Jour., 64 (1985), pp. 2331-2371.

24.   K. Tai, et al, *1.55-μm optical logic etalon with picojoule switching energy made of InGaAs/InP multiple quantum wells*, Appl. Phys. Lett., 50 (1987), pp. 795-797.

Graph Theory plus Group Theory equals Answers to Communication Problems
Vance Faber

A primary obstacle in obtaining significant speedup of highly parallel computer programs on massively parallel distributed processors is global communication. For this reason, we have been investigating theories which enable us to exhibit provably optimal global communication schemes on a wide class of interconnection networks.

One such class for which a theoretical treatment is successful are those networks which can be represented as the directed graph of a group. The elements of the group represent processors (the nodes of the graph) while the generators correspond to communication lines (the arcs of the graph). The author has shown that if the group processes a certain sequencing property P then various global communications tasks (e.g., universal broadcast, accumulation or exchange) can be performed in minimal time.

The hypercube is a prime example of a network which can be represented as the graph of a group (an elementary abelian 2–group) and the author has been able to show that this group does in fact have property P. At this time no groups are known to the author that do not have property P.

Plans are to extend these notions to networks which cannot be represented as the graph of the group, but may still possess optimal global communications schemes. Ultimately, one should expect to be able to measure the communication performance of an interconnection network in these terms, in order to identify those networks that are in some sense "optimal" in terms of global communication.

Parallel Processing Research at IBM
Tilak Agerwala

Two highly parallel processor prototyping efforts are currently underway at the IBM T.J. Watson Research Center: GF11, a Single Instruction
Multiple Data Stream (SIMD) array of processors which can achieve close
to 10 billion floating point operations per second on certain computations,
and the Research Parallel Processor Prototype (RP3), which is a Multiple Instruction Stream Multiple Data Stream (MIMD) system consisting
of conventional microprocessors. A high bandwidth, low latency switch
together with a flexible memory structure makes this machine potentially
suitable for a variety of applications. The talk will briefly describe these
two projects and report on their current status. Motivations, goals, and
technical plans will be discussed to provide an overall perspective for these
efforts.

CHAPTER 61

A Parallel Architecture for Optical Computing

Ahmed Louri*
Kai Hwang*

Abstract. In this paper, we present a new architecture for supporting massively parallel computations. The architecture exploits optics for its ultra-high speed, massive parallelism, and dense connectivity. The system manipulates 2-D arrays as fundamental computational entities. The processing is based on symbolic substitution. New *optical bit-slice substitution* rules are introduced for 2-D processing. 2-D symbolic substitution algorithms are developed for primitive arithmetic/logic operations. The performance of the system is considered.

1. Introduction Electronic technology seems to be reaching its fundamental limits in switching speeds and packaging requirements[1]. This fact coupled with the limitations of electronic computers in processing large amounts of data such as data-intensive applications and multi-dimensional structured data at high speed, have stimulated interset in optics. Optics can provide a quantum leap over electronicss in high-speed processing and massive parallelism.

In this paper, we present a new optical architecture and an extended 2-D *symbolic substitution* technique[2] for implementing massively data-parallel computations. These applications exhibit a high degree of *data-parallelism*, where the parallelism comes from the large amounts of data that can be handled concurrently. Optical systems can simultaneously perform the same operation on all the entries of an image, hence are attractive for implementing data-parallel computations. The proposed system processes binary images (images of 0's and 1's) called *bit planes* as fundamental computational entities.

2. The Optical Bit-Slice Array Architecture. Figure 1 depicts a block diagram of the basic components of the system. Up to three binary images can be processed simultaneously. The heart of the architecture is the processor array unit. Locally, this unit can be viewed as a bit-serial or bit-slice processor, since it performs one logical operation on one, two or three single-bit operands. Globally, it can be viewed as plane-parallel processor, since large sets of operands stored as bit planes are fed through and executed in parallel. Data is stored as bit planes. Each bit plane, i, corresponds to a

*The authors are both with the Computer Research Institute, University of Southern California, Los Angeles CA 90089

weight factor 2^i in the binary representation as shown in Fig.2.

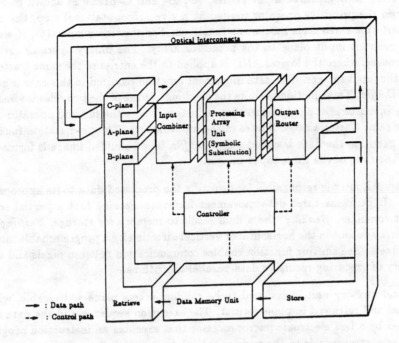

Figure 1: The architecture of an optical bit-slice processor array

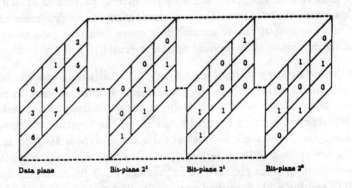

Figure 2: Data representation as a stack of bit planes

2.1 The Processor Array Unit: This unit operates in a SIMD mode (single instruction stream multiple data stream), where the same operation is applied to all the data entries. The processing is based on the optical symbolic substitution technique[2]. Information is coded as spatial symbols in a 2-D plane. A symbol corresponds to a spatial arrangement of the binary bits 0 and 1. Computation proceeds in transforming symbols into other symbols according to a set of substitution rules specifying how to replace every symbol or pattern. The processor array is equipped with three fundamental operators, a unary operator, logical NOT, a binary operator, logical AND, and a ternary operator, full add.

2.2 Input/Output Data Routing: The data represented as binary images is fed to the processing array through three input planes, **A-**, **B-** and **C-plane** as shown in Fig.1. Depending on the primitive operator needed at a given computational step, the input combiner performs three data movement functions: For the unary operator, it simply latches the relevant input plane to the processor array. The binary operator involves two input images, where the logical AND is applied to the entries of the same Cartesian position of the input images. The data movement function required in this case is called *2-D perfect shuffle*. This function permits the shuffling of rows of the two relevant images. The ternary operator is applied to three input images, where the full add operator adds 3 bits of the same Cartesian coordinates in the input images. The permutation function required to partition the data is called *2-D 3-shuffle*. The 3-shuffled image is formed by alternating the rows of the three input images.

The Output Router is responsible for directing the processed data to its appropriate destination. It performs three data movement fuctions. *Feeding back* a partial result to the input combiner. *Sending* the a final result to memory for storage. *Shifting* the processing array output in the horizontal or vertical direction by a programmable integer number of pixels. The shifting function enables communication between pixels and adds the capability of executing recursive data-parallel algorithms.

The optical memory unit is assumed to be random-access plane-addressable, where data is stored and retrieved in plane format. The execution sequence and the data flow are controlled by a fast electronic microprocessor that executes an instruction program and generates control signals to the various optical devices.

3. Optical Implementation Considerations.

In order to process information optically, we use the positional coding [3] of the logical values 0 and 1, where 0 is represented by a dark next to a bright pixel, and 1 by the inverse pattern as shown in Fig.3.(a). In what follows, we concentrate on the implementation of the processor array. The input combiner and the output router assume only space-invariant data movement functions, which can be implemented by several optical means[4].

3.1 Symbolic Substitution Array Unit : Symbolic substitution consists of two processing steps, a *recognition phase*, where the parallelism of optics is used to recognize all the occurrences of a pattern within a 2-D input image, followed by a *substitution phase*, where a replacement pattern is substituted in all the locations of the search pattern. In this paper, we introduce a new set of rules called *optical bit-slice substitution* rules. These rules are applied to large sets of operands simultaneously. Fig.3(b-d) depicts the bit-slice substitution rules required to implement the 3 fundamental operators (logical AND, NOT, and full add) described previously. These rules are derived from the truth-table specifications of the three fundamental operators. The left-hand-side patterns (or search patterns) of these rules represent input combinations and the right-hand-side (replacement patterns) represent the table entries, i.e. rule 1 (r_1) of the full add operator is obtained as follows: given 3 bits 0, 0, 0 then their sum generates a sum bit of 0 and a carry bit of 0. Implementation of the two processing steps involved in symbolic substitution (recognition and substitution phases) can be done in several ways[3,4,5]. The whole unit can be implemented as shown in Fig.4, where the rules are divided into modules. The full add, logical AND, and NOT modules. An incoming plane of data is directed to the appropriate module depending on the operation required. Only one module is active at a time. Within each module, all the rules are fired in parallel.

Therefore, every combination of the input operands is searched and replaced by its corresponding replacement pattern in parallel.

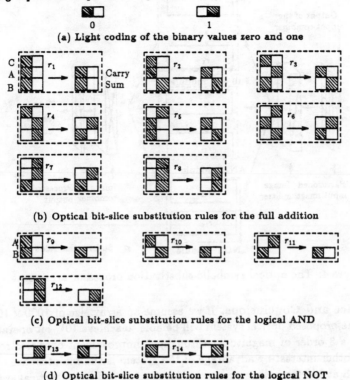

(a) Light coding of the binary values zero and one

(b) Optical bit-slice substitution rules for the full addition

(c) Optical bit-slice substitution rules for the logical AND

(d) Optical bit-slice substitution rules for the logical NOT

Figure 3: Optical bit-slice substitution rules for the fundamental operators

4. 2-D Arithmetic/Logic Symbolic Substitution Algorithms.

The proposed architecture exploits spatial parallelism. This property can be exploited to devise 2-D arithmetic operations that can be applied across a large number of operands represented as 2-D planes. We show here an example of such operations, more details can be found in [4].

2-D Addition: This operation refers to the component-wise addition of two data planes. It is similar to the conventional integer matrix addition. Given two data planes $X = \{x_{ij}\}$ and $Y = \{y_{ij}\}$, for $i, j = 1, \ldots, n$, their addition results in a data plane $Z = \{z_{ij}\}$ where $z_{ij} = x_{ij} + y_{ij}$. Let X be a q-bit planes $X_{q-1}, X_{q-2}, \ldots, X_0$, where X_{q-1} and X_0 are the least significant and most significant bit planes respectively. Similar considerations take place for Y. The procedure starts by initializing the C-plane to zero, and loading bit planes X_0, Y_0 into A-plane and B-plane respectively. The input combiner shuffles the three input images. The processing array applies the full add substitution rules simultaneously to the resulting 3-shuffled image. Thus the bit-slice addition is performed on all the operand pairs in parallel. The sum bit plane is stored in memory location S_0, and the carry bit plane is fed back to the C-plane for the next iteration, while the memory unit loads the bit planes X_1 and Y_1 in the A-plane and B-plane respectively. The whole process continues until X_{q-1} and Y_{q-1} are added and the sum $S_0, S_1, \ldots, S_q$ stored as stacks of bit planes in the memory. The addition of two data planes, of q word length is done in q iterations, regardless of the number of operands to be added.

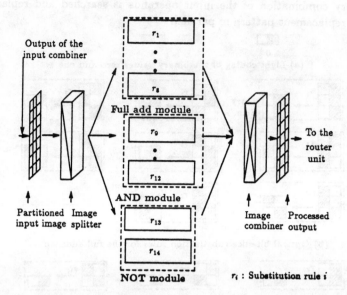

Output of the
input combiner

r_1

r_8

Full add module

r_9

r_{12}

AND module

r_{13}

r_{14}

NOT module

To the
router
unit

Partitioned   Image
input image splitter

Image   Processed
combiner   output

r_i : Substitution rule i

Figure 4: The optical symbolic substitution processor array unit

5. Performance and Conclusions. If we assume an array size of 1000×1000 and a 10 Mhz rate, the proposed optical system will be able to achieve 10^{13} bit operations/sec. This results in a 3 order of magnitudes throughput improvement over electronic array processors. Another interesting advantage of the system is the maintenance of the 2-D parallelism both at I/O as well as processing levels. For and $n \times n$ optical system, this advantage results in a I/O speedup of n over electronic array processors such as the MPP, CLIP, and ICL DAP, since these systems load the data into the array one column (or row) at a time. This speedup advantage can be potentially enormous due to the possible large value of n with the optical system.

Acknowledgments. This research was supported by an ONR Contract No. N14-86-k-559 and in part by a NOSC Contract No. 85-D-203

References

[1] K. C. Saraswat and F. Mohammadi, "Effect of scaling of interconnections on the time delay of VLSI circuits," *IEEE Transaction on Electron Devices*, vol. ED-29, no. 4, pp. 645–650, 1982.

[2] A. Huang, "Parallel algorithms for optical digital computers," in *Proceedings IEEE Tenth Int'l Optical Computing Conf.*, pp. 13–17, 1983.

[3] K. H. Brenner, A. Huang, and N. Streibl, "Digital optical computing with symbolic substitution," *Applied Optics*, vol. 25, 15 Sept 1986.

[4] A. Louri and K. Hwang, "A bit-slice architecture for optical implementation of data-parallel algorithms," *Submitted to Applied Optics for publication*, Jan. 1988.

[5] K. H. Brenner, "New implementation of symbolic substitution logic," *Applied Optics*, vol. 25, 15 September 1986.

CHAPTER 62

Architecture and Operation of a Systolic Sparse Matrix Engine

Steven W. Hammond[*][†]
Robert J. Dunki-Jacobs[*]
Robert M. Hardy[*]
Terry M. Topka[*]

Abstract. We present the architecture and operation of the FEM. It is a high-performance linear systolic array designed to execute the operations of the conjugate gradient method for solving large, sparse, linear systems of equations. Matrix computations are typically very simple and regular and thus ideally suited for systolic computation. The FEM is designed to efficiently maintain a regular flow of data and instruction regardless of the sparsity structure of the matrix that is being operated on. It is not restricted to banded or other "structured" matrices. The computational requirements are explained as well as the data flow through the architecture.

1. Introduction. A systolic array is a network of processors which rhythmically compute and pass data through the network. Every processor regularly receives some data and instruction, performs some simple computations, and then passes the instruction and data to a neighboring processor(s). A regular flow of data and instruction is maintained in the network. These systolic arrays enjoy simple and regular communication paths, and almost all processors used in the array are identical. Additionally, one seeks to increase computational throughput without an associated increase in external memory bandwidth. The virtues and applications of systolic arrays as a high-performance yet cost effective parallel computing structure have been discussed principally by H. T. Kung [7,6], but by others as well [1,2,5,8,10,11]. We want to exploit this powerful computing structure in operating on sparse matrices.

In order to make the solution of very large, sparse, matrix problems (resulting from the discretization of PDE's) computationally feasible with respect to execution time and memory requirements, algorithms for such problems should store and operate only on the nonzero elements of the matrix and should try to minimize the creation of new nonzero elements as computations proceed. The conjugate gradient method (CG) is a robust solution technique that has these characteristics [3,4,5] and it will be the focus here. CG is an

---

[*] GE Corporate Research and Development Center, Schenectady, New York 12301
[†] On leave of absence and enrolled in the PhD program of the Computer Science Department at Rensselaer Polytechnic Institute, Troy, NY

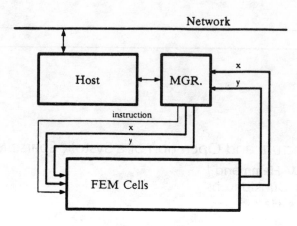

FIG. 1. *System Architecture*

iterative technique for solving systems of equations such as $\mathbf{Ax} = \mathbf{b}$ for the vector of unknowns, $\mathbf{x}$. It searches the space of vectors $\mathbf{x}$ in such a way as to minimize the residual error. The system matrix remains static through the computation and thus we want to store all the nonzero matrix elements into the memories of the systolic array just once, at the beginning of the iterative process. Also, the calculation of a sparse matrix-vector product is the most time consuming part of the CG method. It is essential that this operation be handled efficiently and this concern was the primary factor in the system design.

2. Architecture. The architecture of the FEM is shown in figure 1. It is similar to the Warp developed at CMU [1]. The individual calls in the arrays are different due to the intended applications. The Warp was designed for image and signal processing which typically involve relatively small, dense matrices. The FEM is targeted for operations involving very large, sparse matrices (on the order of tens to hundreds of thousands of rows/columns). The FEM is intended to sit on a network and serve as a shared, high-performance computing resource. It is composed of three elements. The host is a workstation and acts as a network server managing remote compute requests. The Array Manager (MGR) controls the FEM cells; generates the instructions and handles I/O between the array and the host. The FEM cells are identical printed circuit boards connected in a linear array that operate homogeneously.

The FEM cell itself is shown in figure 2. Each cell contains 4 memory partitions, 2 arithmetic logic units (ALU) and 1 multiplier (MPY), a crossbar to connect them, and simple address generation capabilities. Thus, each cell contains only three chip types: memory, arithmetic, and interconnect. The interconnect or glue chip is a single, socket-selectable, 10K gate, gate array used many times. It is used to implement the cross bar, counters, fifo's, and buffers. All data paths are 64 bits wide to support double-precision data types. The second ALU provides the user with the ability to compute partial results distributed across the cells and then accumulate them concurrent with another part of the calculation.

Address generation has been optimized for matrix computations. It typically consists of setting a counter and then incrementing it for the total rows (or columns) of the matrix. Therefore, counters and fifo's serve as addressing units to memories Mem1 - Mem3 as shown

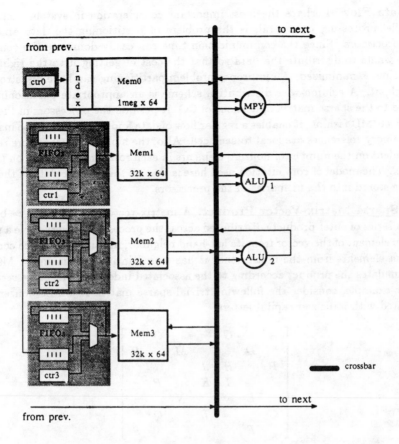

FIG. 2. *Cell Architecture*

in the shaded region in figure 2. There are two fifo's in the address generation unit used to hold addresses for a read-modify-write pipelined arithmetic operation. The fifos match the pipeline delays through the crossbar and arithmetic units. The first one can be used to hold a *read* address long enough to match the pipeline delay through the MPY before reading a value from one of the *vector* memories. The second fifo grabs a *read* address and holds it long enough to match the pipeline delay through the ALU. The contents of the second fifo (previous read address) is then used as a *write* address.

There are four memory partitions in each cell as shown in figure 2. Mem1 - Mem3 are *vector* memories, each one dual ported. They can store up to 32K 64-bit words each and are used to store the various intermediate vectors created during the CG algorithm. Mem0 is used to hold the nonzero data of the stiffness matrix. It is capable of storing up to 1 million 64-bit words of data. To further simplify the addressing on each cell, a 16-bit index is stored with each nonzero datum in Mem0. The index is used as a memory address into the *vector* memories to implement indexed addressing. Thus, when an element of the matrix is fetched, some $a_{i,j}$ to be multiplied by x_j, the associated vector element, b_i (into which the product is accumulated) can be automatically read at the same time using the index. The fifo provides a key function here. It enables the address generation unit to hold a *read* (*write*) address and then use it as a *write* (*read*) address some number of clock cycles later without having to regenerate it.

3. Data Flow. Perhaps the most important consideration in systolic computation, and parallel processing in general, is the problem of partitioning the data amongst the multiple processors. Since the communication time can easily dominate the computation time, one wants to distribute the data so that the cost of getting it to the right place at the right time is minimized. An improper problem partitioning can have disastrous effects on throughput. A column-wise distribution scheme is an appropriate partitioning that is well suited to the sparse matrix operations of CG. It permits the processors in the array to operate in a SIMD fashion, it enables a regular flow of data and instruction to be maintained, and all memory references are local to each cell. Also, the number of processors required is not dependent on the number of nonzero elements in each row (or column) or on the size of the matrix. The model of computation used here is that all the nonzero data of the *stiffness* matrix are stored into the memories of the processors.

3.1. Sparse Matrix-Vector Product. A matrix-vector product, $\mathbf{Ax} = \mathbf{b}$, is computed as a series of outer products distributed across the processors – every cycle a processor receives an element of the vector from its left-hand neighbor, multiplies it by the corresponding column elements from the matrix that it has stored in its local memory, Mem0, and then accumulates the product according to the associated index into one of its *vector* memories. For example, consider the following trivial sparse matrix $\mathbf{A}$ whose nonzero entries are indicated with italicized capital letters

$$\begin{bmatrix} A & & G & & & O & \\ & D & & & M & & R \\ B & & H & J & & & \\ & & I & K & & P & \\ & E & & & N & & \\ C & & & L & & Q & \\ F & & & & & & S \end{bmatrix}$$

Suppose that the FEM array has 3 cells. We begin by dealing out the nonzero elements of the matrix much like one would deal a deck of cards. Start down the first column and give the first nonzero element to the first processor, the second nonzero element to the second processor, etc. This is detailed in [5]. The matrix multiply is carried out by feeding the elements of the vector one at a time to the array. The first cell gets $\mathbf{x}_1$ on time step 1 and multiplies it by element A. At step 2 in cell 1, the product feeds into the ALU to be accumulated into $z(1)$ while $\mathbf{x}_2$ is multiplied by element D. Within a particular cell the product computed at time i is denoted by $\star_i$. At step 2 in cell 2, $\mathbf{x}_1$ is multiplied by B, etc. The first five time steps are detailed in table 3.1: After all elements of $\mathbf{x}$ are fed through the array, each processor has a partial result stored in vector z in mem1. The result $\mathbf{b}$ is computed as the vector sum of z from each of the processors. The time to compute a sparse matrix-vector product is a function of:

- n_i, the number of nonzero elements in column i,
- M, the number of processors in the array, and
- N, the number of columns (or rows) in the matrix.

It takes $\left(\sum_{i=1}^{N} \frac{n_i + M - 1}{M} \right) + M + N$ steps*. The final M is the pipeline latency and the N is the pipelined sum of the z elements to compute $\mathbf{b}$.

---

* The division is integer division

	cell1	*cell2*	*cell3*
step 1	$\star_1 \leftarrow A \cdot \mathbf{x}_1$	$\cdot$	$\cdot$
	$\cdot$		
step 2	$\star_2 \leftarrow D \cdot \mathbf{x}_2$	$\star_2 \leftarrow B \cdot \mathbf{x}_1$	$\cdot$
	$z(1) \leftarrow z(1) + \star_1$	$\cdot$	$\cdot$
step 3	$\star_3 \leftarrow G \cdot \mathbf{x}_3$	$\star_3 \leftarrow E \cdot \mathbf{x}_2$	$\star_3 \leftarrow C \cdot \mathbf{x}_1$
	$z(2) \leftarrow z(2) + \star_2$	$z(3) \leftarrow z(3) + \star_2$	
step 4	$\star_4 \leftarrow J \cdot \mathbf{x}_4$	$\star_4 \leftarrow H \cdot \mathbf{x}_3$	$\star_4 \leftarrow F \cdot \mathbf{x}_2$
	$z(1) \leftarrow z(1) + \star_3$	$z(5) \leftarrow z(5) + \star_3$	$z(6) \leftarrow z(6) + \star_3$
step 5	$\star_5 \leftarrow M \cdot \mathbf{x}_5$	$\star_5 \leftarrow K \cdot \mathbf{x}_4$	$\star_5 \leftarrow I \cdot \mathbf{x}_3$
	$z(3) \leftarrow z(3) + \star_4$	$z(3) \leftarrow z(3) + \star_4$	$z(7) \leftarrow z(7) + \star_4$

TABLE 1
First steps in multiplication

4. Conclusions. We have presented the FEM, a systolic array tuned to operations involving very large, sparse matrices. The individual cells in the array are identical printed circuit boards arranged in a linear array whose length can be chosen to meet some desired performance and/or cost limitations. We have also shown a sparse matrix manipulation scheme tuned to the Conjugate Gradient method for solving linear equations. Our approach is independent of the bandwidth and the sparsity structure of the matrix.

REFERENCES

[1] M. ANNARATONE, E. ARNOULD, R. COHN, T. GROSS, H. KUNG, O. LAM, M. MENZILCIOGLU, K. SAROCKY, J. SENKO, AND J. WEBB, *Warp architecture: from prototype to production*, in Proceedings of the National Computer Conference, June 1987.

[2] E. ARNOULD, H. KUNG, O. MENZILCIOGLU, AND K. SAROCKY, *A systolic array computer*, in Proceedings of 1985 IEEE International Conference on Acoustics, Speech and Signal Processing, March 1985, pp. 232–235.

[3] G. BEDROSIAN, *FORTRAN Subroutine Package for Solving Large, Sparse, Symmetric Linear Systems*, Tech. Rep. 84CRD284, General Electric Co., Corporate Research and Development Center, 1984.

[4] G. H. GOLUB AND C. F. VANLOAN, *Matrix Computations*, Johns Hopkins University Press, Baltimore, Maryland, 1983. Second Printing.

[5] S. W. HAMMOND AND G. BEDROSIAN, *Solution of Large, Sparse, Linear Systems of Equations on Systolic Arrays*, Tech. Rep. 86CRD166, GE Company, Corporate Research and Development Center, October 1986.

[6] H. KUNG, *Why systolic architectures?*, IEEE Computer, (1982), pp. 37–46.

[7] H. KUNG AND C. E. LEISERSON, *Systolic arrays (for VLSI)*, in Sparse Matrix Proceedings 1978, SIAM, 1979, pp. 256–282.

[8] K. H. LAW, *Systolic arrays for finite element analysis*, Comp. Struct., 20 (1985), pp. 55–65.

[9] T. A. MANTEUFFEL, *The Shifted Incomplete Cholesky Factorization*, Tech. Rep. SAND78-8226, Sandia Laboratories, May 1978.

[10] R. G. MELHEM, *On the design of a pipelined/systolic finite element system*, Comput. Struct., 20 (1985), pp. 67–75.

[11] R. SCHREIBER, *Solving eigenvalue and singular value problems on and undersized systolic array*, Siam J. Sci. Stat. Computing, 7 (1986), pp. 441–450.

CHAPTER 63

An Augmented Tree Multiprocessor for Parallel Execution of Multigrid Algorithms

H. C. Wang*
Kai Hwang*

Abstract We deal with the problem of implementing multigrid algorithms on a new multiprocessor system. The architecture is a binary tree augmented with additional links to connect nodes at the same level of the tree. It also incorporates an orthogonal memory organization to afford conflict-free memory accesses. We describe the the architecture and present methods for mapping multigrid algorithms onto the target architecture. It is demonstrated how pipelining and parallelism can be used to boost the performance by a slight modification of the usual relaxation scheme for two-dimensional PDE problems.

1. Introduction With the advent of multiprocessor systems, a lot of work has gone into parallelizing multigrid algorithms so as to make efficient use of the available resources, thereby improving the performance. Implementation of multigrid algorithms on different types of multiprocessors has been studied quite extensively. The major thrust of effort has been concentrated on adapting multigrid algorithms to existing architectures. This is best exemplified by the use of binary reflected gray codes to assign grid points to processor nodes in a hypercube multiprocessor system. For example, see [1]. On the other hand, it is also desirable to design a multiprocessor with an architecture that directly embeds the computation and communication structures of multigrid algorithms. Indeed the two approaches are complementary to each other.

In this article, a multiprocessor architecture is proposed and shown to be efficient for implementing multigrid algorithms. In section 2, we first give a brief overview of multigrid algorithms, justifying the rationale for the architecture under consideration. The architecture is then presented in section 3. We describe in some detail the overall configuration and interconnection among processors. Realization of multigrid algorithms on the architecture is considered in section 4. Two forms of concurrency, parallelism and pipelining, can be readily exploited. We will elaborate on the decomposition of problem domains and other performance-related issues. Section 5 summarizes the results and concludes the paper.

---

* The authors are affiliated with Computer Research Institute, University of Southern California, Los Angeles, CA 90089.

2. Multigrid algorithms   In this section, a short account of multigrid algorithms is given. Our purpose is to emphasize the computation structure of the algorithms and its impact on architecture design. For a more thorough treatment of the algorithms per se, see [5].

Multigrid algorithms use a layer of grids with different spacings to damp error components of different wavelengths eficiently. The ensuing discussion is based on the standard multigrid algorithms wherein every other grid point on a fine grid is left out in each dimension to form the next coarser grid. Also a V-cycle iteration is assumed. Numerous variants of the algorithms exist. For instance, Frederickson and McBryan [3] proposed an algorithm that operates on the same number of grid points throughout the solution procedure by relaxing on several coarse grids at the same time.

The basic operations, relaxations on each grid, require information exchanges among neighboring grid points. If the grid points are distributed to different processors, they will give rise to inter-processor communications. Thus it is important to keep adjacent grid points on nearby processors in order to reduce the overhead. Another type of communication arises in inter-grid transitions, including restrictions and interpolations. This is particularly true if pipeline operation is intended.

3. The augmented tree multiprocessor architecture   When designing architectures implementing multigrid algorithms, it is essential to keep in mind the computation and communication requirement of the algorithms. The most remarkable features of multigrid algorithms are (1) hierarchical grid structure and (2) localized communication. Hence an ideal multiprocessor should also possess a hierarchical structure and adequate local connectivity. Several types of architectures, notably tree and pyramid, do have the hierarchical property. The proposed architecture, as shown in Figure 1(a), is based on binary tree structure with supplementary linear connections among nodes at the same level of the tree. A similar structure was studied by Despain and Patterson [2] related to congestion in a binary tree.

Besides, processors in the leaf nodes also share orthogonally configured memory modules. Under proper control, such a memory organization allows simultaneous row or column accesses without any conflict. A detailed description of this type of orthogonal multiprocessors is given in [4]. The memory modules are used to store the initial data, intermediate results after each multigrid iteration, and the final solution. As will be seen later, the orthogonal memory access pattern is valuable in the efficient implementation of multigrid algorithms.

As shown in Figure 1(b), each node of the tree is composed of twin processors, which share a two ported memory module. Such an arrangement allows the processors to access the memory simultaneously as long as the memory addresses are different. The processor to the right of the memory will be referred to as processor R and the one to the left as processor L. All the R processors at the same level of the tree are connected by nearest-neighbor links, as are the L processors. In addition, processors R and L are connected to processors R and L in their corresponding parent nodes. The memory will be called "local memory". Of course, each processor has complete communication capabilities.

4. Implementation of multigrid algorithms   We observe that for a one-dimensional problem, the number of grid points decrease by a constant factor of 2 when moving from a fine grid to an immediately coarser grid. An important consequence of this phenomenon is that if we designate different grids to individual tree levels, the processing time will be the same at each level of the tree hierarchy. The uniformity of processing time and the

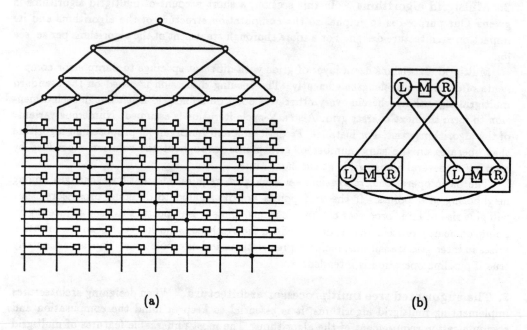

(a)   (b)

Figure 1. (a) The overall architecture; (b) configuration
of each node and interconnections among nodes

processors left behind at each grid transition immediately suggest pipelined operations as
an appropriate way to utilize the idle processors.

To illustrate how pipelined operations are carried out consider the following Poisson
equation defined over a rectangular domain:

$$\frac{\partial^2 u}{\partial x^2} + \frac{\partial^2 u}{\partial y^2} = f \quad \text{in } \Omega$$

$$u = g \quad \text{on } \partial\Omega.$$

Conventionally, the problem domain is discretized using 5-point or other schemes and
relaxations are performed at each point to update its value until convergence is achieved.
Thus, at point (i, j),

$$u_{i,j} = 0.25(u_{i-1,j} + u_{i+1,j} + u_{i,j-1} + u_{i,j+1} - f_{i,j}).$$

We decompose the above relaxation scheme into two half steps:

$$u_{i,j} = 0.5(u_{i,j-1} + u_{i,j+1} - \rho f_{i,j})$$

and

$$u_{i,j} = 0.5(u_{i-1,j} + u_{i+1,j} - (1 - \rho)f_{i,j}).$$

In other words, we first relax along x dimension and then along y dimension. Here
$\rho \in [0, 1]$ is a parameter specifying the relative weighting of f in each of the two half steps.
An obvious choice is 0.5, but other values can also be used. The decomposition is similar

in flavor to Gauss Seidel relaxation, since when relaxation in the y dimension is under way, the values involved have already been updated along the x dimension.

The decomposition permits us to treat two-dimensional problems as one-dimensional problems and hence the uniform speed property is retained in pipelined mode of operations. Since a complete V-cycle consists of a sequence of relaxation-restriction followed by a sequence of interpolation-relaxation operations, two data paths emerge. The first path consists of the R processors in each node. The L processors in each node together constitute the second path.

The operations performed by a typical R processor include:

1. Receive data from R processors of its son nodes
2. Relax to update the solution and write it to the local memory
3. Compute residual and restrict it to the R processor of its father node.

Similarly, the following operations are performed by an L processor:

1. Receive data from L processor of its father node
2. Read data from local memory and perform interpolation operation
3. Carry out extra relaxations and pass the data down to the next level.

In each relaxation, only one grid point need be exchanged between neighboring nodes, which is readily satisfied by the linear links among nodes at the same tree level. Moreover, the locality of communication is maintained on all grids. Inter-level connections cater to the data transfers necessitated at grid transitions. The operations at the leaf level and the root are a little different. For R processors on the leaf nodes, step (1) above is changed to reading from the orthogonal memory modules. L processors of the leaf nodes write the solution back to the orthogonal memory modules. At the root, no restriction or interpolation is necessary. Since the computation and communication demands are essentially the same for all processors, the pipeline works smoothly. Also the capacity demand for local memories is low.

In pipelined operations, it is highly desirable to keep "vectors" large in length so that processors will be constantly busy. To this end, we divide the grid points into two halves along each direction, with one grid point overlap along the internal interfaces. See Figure 2(a). The advantage of this scheme is obvious. By the time relaxation along the x direction has been completed for points lying in the left plane, some rows of grid points from the right plane are already in the pipeline. Before grid points from the right plane run out of supply, grid points on the left plane are ready for relaxation in y direction. Likewise, when y relaxation is being performed on points in the upper plane, those in the lower plane are ready for relaxation in x direction of the next iteration. In this way, high processor utilization efficiency is sustained.

At the beginning, the grid points are stored rowwise across the orthogonal memory modules. Simultaneous row accesses to the memory modules distribute an entire row of data uniformly among the leaf processors. After a complete multigrid iteration, the data are written back to the orthogonal memory modules in a columnwise fashion. Doing so ensures that after relaxation in x direction has been completed, the grid points are properly stored in the memory modules and can be accessed simultaneously for relaxation in y direction. Such row and column access patterns help eliminate the time-consuming matrix transpose operations. The storage scheme is depicted in Figure 2(b).

5. Concluding remarks   We have described a multiprocessor architecture and shown that it is suitable for implementing multigrid algorithms. In particular, both parallel processing and pipelined operations can be exploited efficiently on the architecture. All these translate into a multiprocessor system capable of realizing multigrid algorithms at a very

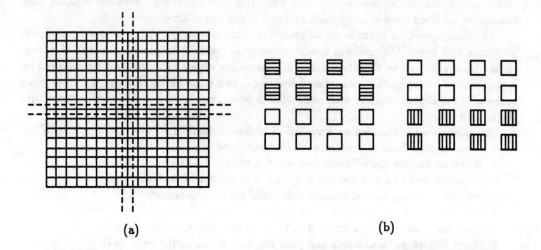

(a)   (b)

Figure 2. (a) Domain decomposition to increase vector length; (b) storage schemes in orthogonal memory modules for relaxations in x and y directions.

high speed. One noteworthy feature of the architecture is the full utilization of computing resources after an initial startup period.

Our treatment of two dimensional problems is different from the work done by other researchers. One focus of much research activity has been the relative communication overhead incurred by box and strip decompositions of the problem domain. In our approach, since only one dimension is involved in each relaxation, the problem is largely avoided, making decomposition and mapping a trivial undertaking. The orthogonal memory modules in the system further facilitate transitions between the two phases of relaxations.

Acknowledgement   This research was supported by an AFOSR grant under contract No. 86-0008.

References

[1] Tony F. Chan and Youcef Saad. Multigrid Algorithms on the Hypercube Multiprocessor. *IEEE Transactions on Computers*, 969–977, November, 1986.

[2] Alvin M. Despain and David A. Patterson. X-Tree: A Tree Structured Multi-processor Computer Architecture. In *5th Annual Symposium on Computer Architectures*, pages 144–151, 1978.

[3] Paul O. Frederickson and Oliver A. McBryan. *Parallel Superconvergent Multigrid*. Technical Report CTC87TR12, Cornell Theory Center, 1987.

[4] K. Hwang, D. Kim, and P.S. Tseng. *Parallel Processing on Orthogonal Multiprocessor Systems*. Technical Report CRI-87-5, USC Computer Research Institute, March, 1987.

[5] K. Stuben and U. Trottenberg. Multigrid Methods: Fundamental Algorithms, model Problem Analysis and Applications. In W. Hackbusch and U. Trottenberg, editors, *Multigrid Methods*, pages 1–176, Berlin: Springer-Verlag, 1982.